French Defence Policy into the Twenty-First Century

Also by Shaun Gregory

THE HIDDEN COST OF DETERRENCE

NUCLEAR COMMAND AND CONTROL IN NATO

French Defence Policy into the Twenty-First Century

Shaun Gregory
Senior Lecturer
Department of Peace Studies
University of Bradford

First published in Great Britain 2000 by
MACMILLAN PRESS LTD
Houndmills, Basingstoke, Hampshire RG21 6XS and London
Companies and representatives throughout the world

A catalogue record for this book is available from the British Library.

ISBN 0–333–69832–0

 First published in the United States of America 2000 by
ST. MARTIN'S PRESS, LLC,
Scholarly and Reference Division,
175 Fifth Avenue, New York, N.Y. 10010

ISBN 0–312–23468–6

Library of Congress Cataloging-in-Publication Data
Gregory, Shaun.
French defence policy into the twenty-first century / Shaun Gregory.
p.   cm.
Includes bibliographical references and index.
ISBN 0–312–23468–6 (cloth)
1. France—Military policy. 2. France—Military policy—Forecasting. I. Title.

UA700 .G72   2000
355'.033044'0905—dc21
                                                            00–027115

This book is printed on paper suitable for recycling and made from fully managed and sustained forest sources.

10   9   8   7   6   5   4   3   2   1
09   08   07   06   05   04   03   02   01   00

Printed and bound in Great Britain by
Antony Rowe Ltd, Chippenham, Wiltshire

For Elizabeth, Eve and Anna

There be none of Beauty's daughters
with a magic like thee

Lord Byron, *Stanzas for Music*

Contents

List of Figures and Table

Figures

Table

Acknowledgements

I owe a debt to Pascal Boniface at IRIS and Frédéric Bozo at IFRI for their support and advice during the conduct of my research for this book, to their respective institutes for hosting me during my sabbatical periods in 1997 and 1998, and to the Nuffield Foundation, Joseph Rowntree Trust, and W. Alton Jones Foundation for supporting parts of the research.

I would also like to thank the following who in innumerable ways have helped me with the preparation of this book: Bruno Barrillot, Laurent Barthélemy, Jean Betermier, Yves Boyer, Christophe Carle, Patricia Chilton, Emmanuel Cocher, Malcolm Dando, Dominique David, Michel Duclos, Marcel Duval, Jacques Fontanel, Réné Galy-Dejean, Anne Gilbert, Camille Grand, Robert Grant, Jean-François Guilhaudis, Trevor Harris, François Heisbourg, Jolyon Howorth, John Keiger, Paul Rogers, Bernard Sitt, Bruno Tertrais, Maurice Vaisse and Tom Woodhouse. I am indebted also to the library staff at the *Institut des Hautes Etudes de Défense Nationale* and *Institut Français de Relations Internationales* for unstinting help. Responsibility for any errors is entirely my own.

Finally, my love and thanks to Abigail for support throughout the production of this study.

SHAUN GREGORY
Bradford

List of Abbreviations

ABM	Anti-Ballistic Missile Treaty
ACEMF	Allied Command Europe [Mobile Forces]
AFCENT	Allied Force Central Command [NATO]
AN	*Arme Nucléaire*
APTGP	*Arme de Précision à Très Grande Portée*
ARRC	Allied Rapid Reaction Corps
ASAT	Anti-Satellite
ASE	*Agence Spatiale Européenne*
ASMP	*Air-Sol Moyenne Portée* [missile]
ASTARTE	*Avion Station Relais de Transmissions Exceptionelles*
ASW	Anti-Submarine Warfare
BMD	Ballistic Missile Defence
BWC	Biological Weapons Convention
CDE	Conference on Disarmament in Europe
CE	*Communauté Européenne* [EC]
CEA	*Commissariat à l'Energie Atomique*
CED	*Communauté Européenne de Défense* [EDC]
CEE	*Communauté Economique Européenne* [EEC]
CEI	*Communauté des Etats Indépendants* [CIS]
CEMA	*Chef d'Etat-Major des Armées*
CEMAA	*Chef d'Etat-Major de l'Armée de l'Air*
CEMAT	*Chef d'Etat-Major de l'Armée de Terre*
CEMM	*Chef d'Etat-Major de la Marine*
CEP	*Centre d'Expérimentations du Pacifique*
CFE	Conventional Forces in Europe Treaty
CFSP	Common Foreign and Security Policy
CIG	*Conférence Intergouvernementale* [IGC]
CINCENT	Commander in Chief Central Command [NATO]
CIS	Community of Independent States [non-Baltic ex-Soviet states]
CJTF	Combined Joint Task Force
COCONA	*Conseil de Coopération Nord Atlantique* [NACC]
CODA	*Commandement de la Défense Aérienne*
COFAS	*Commandement des Forces Aériennes Stratégiques*
COTAM	*Commandement du Transport Aérien Militaire*
CSCE	Conference on Security and Co-operation in Europe
CSG	*Centre Spatiale Guyanaise*
CTBT	Comprehensive Test Ban Treaty
CWC	Chemical Weapons Convention

DAM	*Direction des Applications Militaires*
DAS	*Délégation aux Affaires Stratégiques*
DCN	*Direction des Constructions Navales*
DGA	*Délégation Générale pour l'Armement*
DGSE	*Direction Générale de la Sécurité Extérieure*
DICOD	*Délégation à l'Information et à la Communication de la Défense*
DOM-TOMs	*Départements et Territoires d'Outre-Mer*
DPC	Defence Planning Committee [NATO]
DRM	*Direction du Renseignement Militaire*
EAG	European Air Group
EAPC	European-Atlantic Partnership Council
EC	European Community
ECSC	European Coal and Steel Community
ECU	European Currency Unit
EDC	European Defence Community
EEC	European Economic Community
EFA	European Fighter Aircraft
EMS	European Monetary System
EPC	European Political Community
ESDI	European Security and Defence Identity
ESI	European Security Identity
EU	European Union
FAN	*Force d'Action Navale*
FAP	*Force Aérienne de Projection*
FAR	*Force d'Action Rapide*
FATAC	*Forces Aériennes Tactiques*
FAWEU	Forces Answerable to the WEU
FIO	*Force d'Interposition de l'Otan* [IFOR]
FLA	Future Long-Range Aircraft
FLNC	*Front de Libération Nationale Corse*
FN	*Front Nationale*
FORPRONU	*La Force de Protection des Nations Unies*
FOST	*Force Océanique Stratégique*
FRR	*Force de Réaction Rapide* [RRF]
GAEFB	*Groupe Aérien Européen Franco-Britannique*
GASM	*Group d'Action Sous-Marine*
GFIM	*Groupe de Forces Interarmées Multinationales* [CJTF]
GPS	Global Positioning System
IEDS	*Identité Européenne de Sécurité et de Défense* [ESDI]
IFF	Identification Friend or Foe
IFOR	Implementation Force [NATO in Bosnia]
IGC	Inter-Governmental Conference
IHEDN	*Institut des Hautes Etudes de Défense Nationale*

IMS	Integrated Military Structure [NATO]
INF	Intermediate Nuclear Forces Treaty
IRBM	Intermediate Range Ballistic Missile
JCNPD	Joint Commission on Nuclear Policy and Doctrine
JNA	Yugoslav National Army
KFOR	Kosovo Force [NATO]
LMJ	*Laser Mega-Joule*
MEADS	Medium-Extended Air Defense System
MINREM	*Moyens Interarmées de Renseignements Electromagnétiques*
MIRV	Multiple Reentry Vehicle
MSC	Major Subordinate Command [NATO]
MTCR	Missile Technology Control Regime
MVK	*Mission de Vérification au Kosovo*
NAC	North Atlantic Council
NACC	North Atlantic Co-operation Council
NATO	North Atlantic Treaty Organisation
NPG	Nuclear Planning Group [NATO]
NPT	Nuclear Non-Proliferation Treaty
NSA	Negative Security Assurance
NWFZ	Nuclear Weapons Free Zone
OCCAR	*Organisme Conjoint de Coopération en matière d'Armement*
OSCE	Organisation for Security and Co-operation in Europe
OTAN	*Organisation du Traité de l'Atlantique Nord*
PALEN	*Programme d'Adaptation à la Limitation des Essais Nucléaires*
PAN	*Porte-Avions Nucléaire*
PCF	*Parti Communiste Français*
PCG	Policy Co-ordination Group
PCIAT	*Poste de Commandement Interarmées de Théâtre*
PESC	*Politique Etrangère et de Sécurité Commune* [CFSP]
PfP	Partnership for Peace
PJC	Permanent Joint Council
PS	*Parti Socialiste*
PSA	Positive Security Assurance
RAM	*Révolution dans les Affaires Militaires* [RMA]
RMA	Revolution in Military Affairs
RPR	*Rassemblement Pour la République*
RRF	Rapid Reaction Force
SACEUR	Supreme Commander Europe [NATO]
SALT	Strategic Arms Limitation Talks
SDI	Strategic Defence Initiative
SFOR	Stabilization Force [NATO in Bosnia]
SGDN	*Secrétariat Général de la Défense Nationale*
SHAPE	Supreme Headquarters Allied Powers Europe
SIRPA	*Service d'Information et de Relations Publiques des Armées*

SMA	*Service Militaire Adapté*
SNLE	*Sous-marin Nucléaire Lanceur d'Engins* [SSBN]
SPOT	*Satellite Pour l'Observation de la Terre*
SSBN	Nuclear Ballistic Missile Submarine
START	Strategic Arms Reduction Talks
SYRACUSE	*Système de Radio Communication Utilisant un Satellite*
TAAF	*Terres Australes et Antarctiques Françaises*
TASM	Tactical Air-Surface Missile
TN	*Tête Nucléaire*
TNP	*Traité de Non-Prolifération* [NPT]
UDF	*Union pour la Démocratie Française*
UE	*Union Européenne*
UEO	*Union de l'Europe Occidentale* [WEU]
UNPROFOR	United Nations Protection Force
VBCI	*Véhicule Blindé de Combat Infanterie*
VBL	*Véhicule Blindé Léger*
WEAG	Western European Armaments Group
WEU	Western European Union
WMD	Weapons of Mass Destruction
ZG	*Zone de la Guyane*
ZNC	*Zone de la Nouvelle-Calédonie*
ZPF	*Zone de la Polynésie Française*
ZSOI	*Zone Sud d'Océan Indien*

Introduction

At the beginning of the twenty-first century, after a decade of change since the end of the Cold War, France seems newly assertive and relevant in the military sphere. It has almost completed the adaptation of its armed forces to the new international context, it leads the construction of a European defence and has narrowed its differences with NATO, and its armed forces exercise influence across the globe in national, bilateral and multinational operations. French nuclear ballistic missile submarines patrol in the Atlantic, French peacekeepers monitor the streets of Bosnia and Kosovo, *Légionnaires* protect the European space-launch facility at Kourou, the Helios-1 military spy satellite scans Europe's unstable periphery, and French armed forces defend national territory and interests from the Mediterranean to the South Pacific and from the Caribbean to the Indian Ocean.

At the same time there are considerable tensions within, and powerful constraints acting upon, French defence policy. In adapting its armed forces France has struggled to reconcile the requirements of the new context with its traditional framework of thinking about defence and with shrinking defence budgets. In Europe France has found it difficult to adjust to the Europeanisation of national interests, to the renewed dominance of NATO and to the inability or unwillingness of European partners to keep pace with French ambitions for European defence. Outside Europe France has been buffeted by the multiplication of demands for the exercise of French military power, by the emergence of a multilateral imperative which is recasting the French military role on the international stage and by the evident contradictions between France's European and extra-European defence requirements. None of these issues has been resolved and each seems likely to shape French defence policy for decades to come.

This book is first and foremost an attempt to understand and explore these issues at the end of the twentieth century and to assess the trajectory of French defence policy into the twenty-first century. Running through the book are a number of core questions which broadly reflect the key dilemmas outlined above: (1) how successfully has France adapted its armed forces and defence policy to the new context and do Gaullist prescriptions remain relevant to the defence agenda France now faces?; (2) can France exploit the opportunities in multilateralism, both within and outside Europe, in pursuit of its defence objectives or will those objectives be eroded or lost in the adjustment and compromise of multilateralism?; and (3) can France reconcile the tensions evident in the erosion of the traditional distinctions between the '*trois cercles*' of France's national, European and global interests?

Seeking answers to these questions is greatly facilitated by the robustness and scope of the debates within France on defence matters and by the range of official, semi-official, academic and other material which is available. At the official level a wide range of speeches, statements, reports, webpages, and so forth, are available through the public relations outlets of the Presidential palace [*Elysée*], the Prime Minister's office [*Hôtel Matignon*], the Defence Ministry [*rue Saint-Dominique*], the Foreign Ministry [*Quai d'Orsay*], the National Assembly [*Assemblée Nationale*], the Senate [*Sénat*] and the Armed Forces [through DICOD, the *Délégation à l'Information et à la Communication de la Défense*].

At the semi-official level, large numbers of publications emanate from think-tanks either within or closely linked to government, amongst the most important of which are the *Fondation des Relations Stratégiques* [FRS, formerly FED/CREST], the *Institut Français des Relations Internationales* [IFRI], and the *Institut des Hautes Etudes de Défense Nationale* [IHEDN].

The academic sector, often part of and indistinguishable from the semi-official, contributes a further dimension to the debates. Important centres within Paris such as the *Centre d'Etudes et Recherche sur les Stratégies et les Conflits* [CERSC], the *Centre Interdisciplinaire de Recherche sur la Paix et d'Etudes de la Sécurité* [CIRPES], and the *Institut de Relations Internationales et Stratégiques* [IRIS], and some outside Paris such as the *Centre d'Etudes de Défense et de Sécurité Internationales* [CEDSI] in Grenoble and the *Centre d'Etudes et de Recherches sur l'Armée* [CERSA] in Toulouse, are arguably able to exercise greater scope in their analysis than those tied more overtly to government and thus contribute an important strand of ideas through close interaction with government and linked institutes.

The small French defence NGO community, foremost amongst which is the *Centre de Documentation et de Recherche sur la Paix et les Conflits* [CDRPC] in Lyon, also contributes some important critical material, though the elite discourse in France tends to be less permeable to these ideas than is the case, for example, in the United States or the United Kingdom.

Finally a great deal of material is available from the French media which both contributes to the debates directly and has an important role in embedding the expert discourse in the wider public context .

Amongst this abundance of material a number of sources stand out in defining or describing the contemporary French defence landscape. The government position is to be found in the 1994 *Livre Blanc sur la Défense*[1] and the 1996 *Une Défense Nouvelle*[2], the view from the left is expertly articulated by Pascal Boniface[3], that of the right by Pierre Lellouche[4], while the voice of the military is evident in the publications of Charles-George Fricaud-Chagnaud[5] and Pierre M. Gallois[6]. The adaptation of defence policy and the armed forces has been expertly analysed by François Heisbourg[7], Franco-NATO relations dissected by Frédéric Bozo[8], the role of France in European defence assessed by Nicole Gnesotto[9], and Thierry Garcin has perhaps most fully described France's wider place and security role in the post-Cold War world[10].

Outside France the English-language discourse about French defence and security, which has primarily a European or transatlantic focus, is dominated by a handful of writers, and it is barely an exaggeration to say that in the 1990s their work has become a virtual window through which much of the English-speaking security community looks in on the French situation. Principal amongst this group are Philip Gordon[11], Jolyon Howorth[12], Robbin Laird[13], Diego Ruiz Palmer[14] and David Yost[15]. English-language sources about contemporary French defence and security issues outside Europe are considerably less plentiful, but one can still point to important work by John Chipman[16], and Robert Aldrich and John Connell[17].

This book is a contribution to this literature. It is hoped that by reflecting on a wide range of issues, by bringing together an extensive source material, and by offering an assessment of the future of French defence policy, the book can meet the need for a single volume which explores French defence issues within and outside Europe in the contemporary context. It is thus intended to serve as a primer for those approaching French defence issues with little background, as an *entrée* for those seeking ideas about French thinking on particular defence issues, and as an overview of interest to the specialist.

The book is also intended to function as a research tool. It seeks actively to promote and encourage research interest in French defence and security issues by addressing a deliberately wide range of issues, by incorporating guidance on source material, and by providing, in an appendix, a short guide to conducting defence research in France.

To meet all the foregoing objectives the book is structured in the following way. To understand the parameters and elements of the debates it is first of all necessary to know something of the background and framework of French thinking about defence in the contemporary context. The first chapter lays this foundation by discussing French strategic culture[18], the decisive and persistent influence of the defence ideas of Charles de Gaulle, and the defence policy process. The chapter is also intended to provide a functional grasp of key ideas in French defence – such as the consensus, exceptionalism, and Gaullism – while recognising that many of these ideas remain subject to considerable debate.

The second chapter focuses on the turbulent years from 1989 to 1992 when French defence policy was transformed by events. It examines the effect of the end of the Cold War and the transformation of Europe on French thinking about the geopolitical landscape and its place within it, and the effects of the Gulf War of 1990/91 and the break-up of Yugoslavia on French thinking about NATO, European defence and the role of armed forces in the post-Cold War era.

Chapter 3 explains the reform of defence after 1992. It assesses the defence debate within France which led to the publication of the second defence white paper in 1994 [*Livre Blanc sur la Défense*], considers the effect on this debate of the transition from the Mitterrand to the Chirac presidency, discusses Chirac's impact on defence policy and his far-reaching reform of the armed forces set out in February 1996 in *Une Défense Nouvelle* which presages change through to 2015, and concludes with an assessment of the impact on defence of the Jospin administration and France's third period of political 'cohabitation' of the Fifth Republic.

The fourth chapter is concerned with the trajectory of French defence policy in the new Europe. It first examines Franco-NATO relations and assesses the prospects for, and implications of, France's return to full NATO membership. The chapter then looks at the French role in the construction of a European security and defence identity and at French ambitions for the Europeanisation of nuclear weapons, assessing respectively the prospects and implications of France realising an autonomous European defence capability and a European nuclear deterrent.

The fifth and final chapter considers the future of France's global military role by looking at the French military presence in Africa, in France's sovereign territory around the world [the DOM-TOMs], and in UN military operations. In each case the French presence and role is explained, the evolution of French policy in the 1990s is examined, and future trends are assessed. The chapter concludes by discussing trends in the projection of French conventional and nuclear military power outside Europe and the linkages between the two.

Four limitations of this study need to be made clear at this point. Firstly a broad definition of defence policy has been taken to encompass the issues circumscribed by military security and the question of 'security for whom?' is answered in terms of the French state. Secondly in opting for a wide-ranging discussion of issues the study cannot avoid some sacrifice of analytic depth. This problem has been mitigated in three ways: by limiting discussion to military rather than broader concepts of security; by confining analysis to those elements of the defence debate which are embedded within foreign policy rather than attempting to include domestic issues [such as defence reforms and the economy or unemployment], internal security issues [the purview of the gendarmerie] or domestic terrorism [such as the FLNC in Corsica]; and, by guiding the reader to the relevant literature where issues can be explored more fully than is possible in a single volume. Thirdly, throughout the study, 'Europe' is intended to mean the group of European states with membership of the EU, NATO or the WEU, and European interests are explored from a Western European perspective. Finally, this study has avoided the use of personal pronouns for the French state throughout [except in original quotes] and has used the masculine pronoun [for example, the President and *his* Cabinet] only to avoid the narrative inconvenience of the alternatives.

1
French Defence in Context

Introduction

To witness in the late 1990s the French celebration of Bastille Day on 14 July is to be offered a rich metaphor about the place and role of the armed forces in France at the end of the twentieth century. That an event in 1789 which came to symbolise the civilian overthrow of the existing political order should be celebrated more than 200 years later by a military parade may seem deeply incongruous, the more so because few European states today feel comfortable with a parade of military power at all. Yet the parade and its accompanying fly-past of military aircraft speaks volumes about the relationship between the French people and the French armed forces, or at least about the way that relationship is seen by political elites. It speaks volumes also about the importance of French military power as an instrument of statecraft and the way in which the French armed forces function as a symbol, perhaps *the* symbol, of French independence and rank.

Nor need one stretch the imagination too far to observe the significance of *tricolore* jet-trails in the Parisian sky above the *Avenue Charles de Gaulle* and the *Avenue de la Grande Armée* which link the *Arc de Triomphe*, symbol of victory in 1918 and nation restored in 1944, to the modernity of the *La Défense* business quarter. The elements of this spectacle – including the importance of history, the resonance of two costly wars, the coexistent ideas of continuity and progress, and the place of de Gaulle between old and new – can be read as a shorthand for those issues which first need to be understood before one can begin to make sense of French defence policy on the cusp of the new millennium.

The purpose of this opening chapter is to explore these issues in order to establish the background to contemporary defence debates in France and to understand something of French thinking about defence, what may here be termed the French strategic paradigm, and of the defence policy-making process in France. To do this the chapter is organised into three sections. The first considers French strategic culture, by which is meant the interplay of geographical, historical, socio-cultural and other factors which together determine how a nation and its elites conceive of national defence and the place and role of the armed forces. The second discusses the decisive and persistent influence on French defence of the ideas of Charles de Gaulle who created from the fabric of French strategic culture a framework of thinking about defence and a set of prescriptive policy tools from which the national consensus on defence was fashioned and which continues to inform defence debates to the present day. The third examines the defence policy-making process in France, largely the creation of de Gaulle, by identifying the key players and assessing their interaction, the checks and balances within the system, and the constraints on policy-making.

French strategic culture

Even a cursory glance at the position of France on the map is deeply instructive. *La France Métropolitaine* is a large country on the European land-mass and has long borders with three strong neighbouring states, Germany, Italy and Spain, delineated by the Rhine, Alps and Pyrenées. To its north, across *La Manche*, lies a fourth powerful neighbour, the United Kingdom. To the West the long French coast looks outwards into the Atlantic Ocean, while to the South its shoreline is lapped by the Mediterranean Sea. Geography condemns France to engagement; lacking the isolation of the island state or the remoteness of the peripheral state, France has been forced to look outwards to the affairs of its neighbours and beyond.

As Metternich understood, geography strongly influences a nation's strategic interests. France bridges the divide between the Northern and Southern European states epitomised by the division between the grain of the beer-drinking North and the grape of the wine-drinking South. In outlook France has naturally combined the preoccupations of its Northern neighbours – the British, Dutch, Germans and Benelux peoples – with those of its Southern neighbours – the Spanish, Italians and Greeks; while, between East and West, France has tempered an overarching interest in continental stability in Central and Eastern Europe with a maritime concern which has paralleled that of Britain, Portugal and Spain. Pulled in these directions and standing arguably at the intersection of these concerns France might be considered the quintessential European state.

Intimately linked to geography, and in many ways shaped by it, French history and the socio-cultural elements within it, form the second strand of the background of contemporary defence policy. Although one could reach back to Vercingétorix and the coming together of Gallic tribes to make arguments about the relationship between the French people and their armed forces, or to Charlemagne to make mischief about Franco-German ambitions for Europe, in all important respects the history which has the most bearing on the contemporary context is that since 1789.

Of the historical elements which still resonate in contemporary defence and security debates, five themes stand out: relations with Germany, the nation and the armed forces, the French empire, perfidious allies, and the decisiveness of military technology. Each has been seared into the French national consciousness by the fire of experience and each has served to reinforce the others in the construction of a framework of thinking about defence that has been stubbornly realist, state-centred, self-reliant and threat-focused. As one commentator noted: 'no other former great power has been so faithful to its attitudes of the past, no other so successful in transporting seemingly out-dated images of power into the political discourse of the late twentieth century'[1].

Arguably, since 1789 the historical constant which has underpinned this framework has been that of invasion. Five times between 1789 and 1944 – in 1814, 1815, 1870, 1914 and 1940 – France has been invaded and this has served in the words of one analyst to 'perpetuate and universalise the reality of the threat'[2]. That the instrument of the most recent, destructive and humiliating of these invasions has been Germany [in 1870, Prussia] has served further to place Germany at the heart of French thinking about its security, to such an extent that, in the historical context at least, Germany and the threat it poses could be considered a national *idée fixe*. André Fontaine articulated this view most clearly when in 1952 he noted that 'France has a Germany policy: she has no other'[3]. Despite years of rapprochement and the deepening relationship of co-operation which has developed between France and Germany since the Second World War caution about Germany has remained deeply influential in underlying thinking, and has surfaced in French rhetoric at times of tension between the two. Thus, for example, reacting in 1973 to an increasingly assertive Western Germany President Pompidou argued that 'the Germans must act tactfully, for one does not have to scratch too far for the French to once more uncover an old aversion'[4]. More recently, and wholly pertinent to the contemporary discourse, following the fall of the Berlin Wall in 1989 and in response to the manner of German reunification, President Mitterrand mined old seams to declare: 'facing the German problem, the French must be aware that they possess a great history, that they have always triumphed in the end over the obstacles in this history and over sometimes dangerous neighbours'[5].

After the failure to provide a long-term resolution of the German question in history through collective security, neutralisation, or disarmament, France in the second half of the twentieth century has embraced Germany through a variety of bilateral and multilateral arrangements intended to constrain Germany and to provide a means by which France can influence developments in Germany. These measures began with the 1952 Treaty of Paris, which created the European Coal and Steel Community, and deepened via the Western European Union [1954], West German accession to NATO [1955], the bilateral Elysée Treaty [1963], the CSCE [1975], the Single European Act

[1986], the Maastricht Treaty on European Union [1992] and Economic and Monetary Union [1999]. In the contemporary context one of the key themes in defence policy is the closeness of Franco-German relations in the European context and the tensions arising from French suspicions about Germany and the differences in interests and outlooks of the two[6].

A second element to flow from the consequences of repeated invasion, and the second of the historical strands informing contemporary defence, has been the forging of a highly idiosyncratic view of the place of the armed forces in the French state and of the relationship between the French armed forces and the French people[7]. Invasion and the 1791 *'levée en masse'* when French people took up arms in defence of the new *République* has inculcated the idea of collective responsibility for defence and, expressed through conscription, this idea has both profoundly informed national debates about defence issues and given the French armed forces a role beyond that of national defence and instrument of statecraft. The armed forces have functioned as an expression of national cohesion, and thus, by extension, of French identity. In this respect there is little to separate the observation that since its eighteenth-century origins 'conscription... [has been] the foundation stone of the Republic, the cultural glue of the nation state'[8] from the view stated in the 1994 *Livre Blanc sur la Défense* that military service remains 'an integrating melting-pot, a school for good citizenship, a paradigm of the Frenchman's allegiance to France'[9]. Along the way de Gaulle in his *La France et son Armée* alluded to the same thing when he noted that the French were 'a people whose genius, whether in eclipse or in glory, has always found a faithful reflection in the mirror of its army'[10]. The French armed forces have thus functioned as the embodiment of the *grandeur* and potency of the French state and at times of disaster as the embodiment also of her national humiliation or shame[11].

The third historical element of continued relevance is that of the French empire and the manner of its loss. At its height the French colonial empire comprised territories from the South Pacific, South America, the Caribbean, across much of North and West Africa, through the Middle East, Madagascar and into Indo-China, covering an area of 12 million square kilometres and incorporating more than 100 million people[12]. Although rationalised [and widely supported in France] on the grounds of France's *mission civilisatrice* and the not irrelevant principle that what France did not seize other European states would, the Empire itself was largely obtained and maintained by the same blend of occupation, subjugation, dispossession, and annihilation, which accompanied the expansion of other European empires, albeit finessed by the ideology of assimilation, the idea that ultimately every inhabitant of the Empire would become a French citizen, represented in the French parliament[13]. As late as 1944, on the point of the unravelling of its Empire, the basic charter of French colonial policy drafted that year at the Brazzaville Conference determined for colonial peoples that 'the purpose of the civilising work accomplished by France in the colonies excludes any idea of autonomy, any possibility of evolution outside of the French empire. The establishment, even in the remote future, of "self-government" in the colonies must not be considered'[14]. Few, in government at least, had problems squaring this with *La Déclaration des Droits de l'Homme et du Citoyen.*

As it was, France was powerless to prevent the transition to independence of all but the weakest and most dependent territories or its exclusion from certain regions. By 1946 France had been forced by the British and Americans out of Syria and Lebanon; by 1954 France was defeated in Indo-China after a disastrous seven-year war in Vietnam; by 1956 Morocco and Tunisia were independent, and by 1960 de Gaulle had abandoned formal colonialism, ceding independence to all African territories, save Algeria whose independence occurred two years later after eight years of bloody conflict. In its place a 'family' of Francophone states emerged comprising French-speaking states and territories from the DOM-TOMs in a sovereign relationship, through those bound to France by formal arrangements often with a strongly military element, to those bound to France by little more than a shared tongue, residual affection and an entangling network of personal relationships.

The relevance of this in the contemporary context is that *Francophonie*, now comprising perhaps 150 million people, provides France with claims to a global role as a consequence of its continued presence across the remnants of empire including, until recently, a virtually exclusive sphere of influence in much of North and West Africa. For France the idea of global interests is intimately linked to those of French *rang* [rank] and *grandeur* [greatness] and to the notion, with deep historical roots, that global interests enhance French power and influence in Europe.

The fourth historical strand informing contemporary defence and security issues is that of perfidious allies. History, particularly the history of the twentieth century, has for France confirmed the realist principle that, as the former Minister of Defence François Léotard put it as late as 1994, 'at the moment of truth a nation has no friends'[15]. In other words history has shown that France is unable to rely on its allies in times of crisis and need and this in turn has influenced it to sustain an emphasis on self-reliance and national interest. The object of much of this analysis has been the 'Anglo-Saxons', Britain and the United States, and the path which transports this view into the late twentieth century can be charted, at least in French eyes, from the reluctance of the British and Americans to engage Germany after 1913, via the American failure to commit itself to the League of Nations, and via the British withdrawal at Dunkirk in 1939 and the sinking of French ships at Mers el-Kébir, through the late arrival of the United States in the Second World War in 1941 to Diên Biên Phu and Suez.

History has one other insight to offer, and that is the role that military technology plays in the power and influence of states. The lessons of the failure to obtain and exploit the defining technology of the age were learned in 1914 when French mass infantry faced machine guns and relearned in 1940 when the impressive but static and anachronistic Maginot defences proved irrelevant to the security of France in the face of *Blitzkrieg*. Since then technological edge has been seen as underwriting French security, an idea with its clearest expression in French nuclear weapons and one at the end of the millennium finding its continued relevance in the French uptake of technologies associated with the so-called Revolution in Military Affairs [RMA][16].

Defence and de Gaulle

In terms of defence and security General Charles de Gaulle stands like a colossus at the mid-point of the French twentieth century and casts a long shadow over almost all that came after him. He was arguably unique in defining a set of ideas, a framework of thinking, and an institutional arrangement which together have not only persisted but have continued to shape a discourse separated from its origins by more than 40 years and by an upheaval as profound as the end of the Cold War. In the democratic tradition it is hard to think of another leader whose personality and presence has so shaped the subsequent political process or so consistently over-ridden even the most vociferous political critics in office. Certainly there exists no parallel in the United States or United Kingdom, nor amongst other states in Western Europe with a democratic history. Indeed the idea that the thinking of Truman, Eisenhower or Kennedy might still dominate debates in the United States or those of Churchill, Eden or Macmillan exercise similar influence in the United Kingdom seems more than faintly absurd.

De Gaulle's ideas on defence and security can best be understood in the wider context of his ideas about the place of France in the international system and the nature of relations between states. At the core of de Gaulle's ideas is the notion of *le fait national* [the reality of the nation], the belief that the nation-state functions as the ultimate object of human loyalty and that no other form of political organisation, nor any economic, ideological or social force was more important, nor had a better instrument of government emerged.

This emphasis on the nation-state is intimately linked to the idea of French exceptionalism, the belief that amongst the nations of the world France had been singled out, by implication divinely so, for an exceptional and uniquely important place in the world. The 'exceptional destiny', to which de Gaulle made many references, was premised on a particular reading of French history and culture which passed through the French Revolution, the birth of the Republic, the *Declaration des Droits de l'Homme et du Citoyens*, Napoleon Bonaparte, and the work of French philosophers, political theorists, writers, artists and others, to argue that France was historically the 'central source of European intellectual inspiration' and that consequently it functioned as a civilising and humanising force for the rest of humanity. It is a view summarised by Victor Hugo in 1877: 'Glory consists in the creation of daylight. France creates daylight. Thence her immense human popularity. To her civilisation owes its dawn. The human mind in order to see clearly turns in the direction of France'[17]. This kind of rhetoric informed the historic thread of France as the *nation phare* [literally a lighthouse or beacon nation] and shaped the colonial justification of *mission civilizatrice*. From it de Gaulle fashioned his *certain idée de la France*, a highly personalised concept of France, stated unambiguously at the beginning of his war memoirs, a concept which more than any other conditioned French policy-making under de Gaulle, and arguably since. In perhaps his most widely known text de Gaulle wrote:

> All my life I have thought of France in a certain way. This is inspired by sentiment as much as reason. The emotional side of me tends to imagine France, like a princess in the fairy stories or the Madonna in the Frescoes, as dedicated to an exalted and exceptional destiny. Instinctively I have the feeling that Providence has created her either for complete successes or for exemplary misfortunes. If, in spite of this, mediocrity shows in her acts and deeds, it strikes me as an absurd anomaly, to be imputed to the faults of Frenchmen, not to the genius of the land. But the positive side of my mind also assures me that France is not really herself unless in the front rank; that only vast enterprises are capable of counterbalancing the ferments of dispersal which are inherent in her people; that our country, as it is, surrounded by others, as they are, must aim high and hold itself straight, on pain of mortal danger. In short, to my mind, France cannot be France without greatness[18].

Of the many analysts who have given attention to this paragraph the most insightful for present purposes is perhaps de Porte who commented that this statement 'is more than an expression of the traditional love affair that the French of all times have had with their nation . . . what the General calls the positive side of his mind adds something new and important to that, namely, *a policy-generating insight into the relationship between the nation's inner life, its essence as France, and its place in the politics of nations. This gave him [de Gaulle] a guide to a lifetime of political action, at home and abroad* . . . [emphasis added]'[19]. In other words de Gaulle's *certain idée de la France*, premised on French exceptionalism, became his guiding approach to policy-making shaping what he did, most importantly, in foreign and defence policy.

That 'France cannot be France without greatness' set the agenda for de Gaulle, focusing his attention on the need for international status and influence. It both conditioned his policy objectives for the French nation-state, namely that in the post-Second World War context France needed to rebuild its own national power base, and it conditioned his policy towards other nation-states, namely that to recover greatness France needed independence and autonomy of action to ensure its own security and crucially to avoid subordination to other, objectively more powerful, states in the international system.

With respect to the former de Gaulle saw the need for a strong and stable domestic political system as the necessary foundation for the projection of French power and influence overseas. In his view this meant a powerful presidency exercising executive power on behalf of the nation, largely unfettered by the checks and balances which had, at least in his eyes, weakened France in the past. Fashioning a powerful state also meant the need to rebuild the French army and to equip it with the best weapons possible, and the need to exploit to the fullest extent the other instruments of French state [and colonial] policy.

De Gaulle understood nevertheless that the power of a state might be enhanced by close relations with others provided that the relationship implied neither loss of national sovereignty nor subordination. De Gaulle was consequently determined not to fall into the orbit of either superpower but to maximise French room for manoeuvre in the diplomatic no-man's-land between the two. With respect to smaller states, and again informed by national independence and autonomy, he articulated the imperative of 'Grand Ensembles', the idea that states should co-operate to aid one another in terms of economic development and security and work together for shared objectives beyond the reach of any single state alone.

The search for greatness also informed de Gaulle's thinking about empire. It was axiomatic to de Gaulle that France was a world power and the basis of that global role was France's colonial empire. De Gaulle was however shrewd enough to see the 'winds of change' blowing across the European colonies and practical enough to see that France had neither the resources nor ultimately the will to resist the pressure for change. He consequently determined to redefine France's relations with its colonial states but to do so in a way which retained influence and reach[20].

Once again in de Gaulle's memoirs one can find the distillation of these ideas. Writing about the period immediately after the Second World War when he was for a short time in office [1944–46] de Gaulle wrote:

I intended to assure French primacy in Western Europe by preventing the rise of a new Reich [Germany] that might again threaten her safety; to co-operate with the East and West and, if need be, contract the necessary alliances on one side or another without ever accepting any kind of dependence; to transform the French Union [colonial empire] into a free association . . . ; to persuade the states along the Rhine, the Alps and the Pyrénées to form a political, economic and strategic bloc; to establish this organisation as one of the three world powers and, should it become necessary, as the arbiter between the Soviet Union and Anglo-Saxon camps[21].

Within this well-defined foreign policy framework the elements of French defence policy were largely pre-configured. The empowerment of the instruments of state to restore France to her rightful place meant the rebuilding of the French armed forces, and total defence spending between 1946 and 1958 consequently rose by an order of magnitude from around 200 billion francs to 2000 billion francs [pre-devaluation][22].

Restoring French power also meant the acquisition of nuclear weapons, the defining technology of the age. De Gaulle was out of office between 1946 and 1958, the period when the French nuclear weapons programme materialised, but his role was decisive nevertheless. He initiated the development of French nuclear weapons while in office between 1944 and 1946, signing on 18 October 1945 the document which created the *Commissariat à l'Energie Atomique* [CEA] and tasked it with 'applying atomic energy in the areas of science, industry and national defence', a document which incidentally set the CEA in the largely autonomous position with respect to political and financial oversight which it continues to enjoy to the present time[23]. He returned to office in 1958 and became the first President of the Fifth Republic in January 1959 in time to oversee the detonation of France's first nuclear device in the Algerian Sahara at Reggane on 13 February 1960[24]. Thereafter he directed the weaponisation of French nuclear technology which produced a nuclear triad based on aircraft-delivered bombs operational in 1964, intermediate-range land-based ballistic missiles operational in 1971, and ballistic missile submarines operational in 1972[25], though he did not live to see the triad fully in place.

De Gaulle also led, together with those like Pierre Gallois and Lucien Poirier, the development of French nuclear doctrine and strategy through which a crude, but nevertheless impressive, technical achievement was fashioned into a powerful and complex policy tool[26]. French nuclear weapons became the ultimate symbol and instrument of national independence and rank and thus functioned to sanctuarise France from threats to its vital interests and underwrite France's place at the top table of international fora. From the outset French nuclear weapons were weapons of political influence designed to prevent not to fight war. The relatively small size of the French arsenal in relation to that of the Soviet Union, the clearest threat, gave birth to the proportional strategy, summarised as *dissuasion du faible au fort*, that France need not match the arsenal of its adversaries and could dissuade any attack by ensuring that French nuclear weapons could exact a price greater than any benefit an adversary might gain by attacking France. Limited means also underpinned the notion of *suffisance*, the idea that the French nuclear arsenal was the minimum deemed sufficient to meet France's dissuasive needs.

De Gaulle's emphasis on diplomatic distance from both superpowers created rhetorical ambiguities around the potential targets of French nuclear weapons which informed Ailleret's widely misunderstood idea of *'tous azimuts'* which, while never formally adopted as policy, proclaimed that French nuclear weapons could be targeted anywhere in the world and not just at the Warsaw Pact[27]. This did not mean that French nuclear weapons were pointed at the United States, but it did mean that de Gaulle and his advisers understood nuclear weapons to have a role in the protection of French vital interests irrespective of the quarter from which threats to those interests arose.

It was implicit in his ideas that de Gaulle would review France's existing defence obligations to other states when he returned to office in 1958. In this respect his response to France's role in NATO and the WEU, negotiated and entered into in his absence, is instructive about the thinking underlying his policy ideas. Although alike in many respects – both are multinational alliances with the goal of collective defence – NATO and the WEU differ in a number of important ways. NATO is a hegemonic alliance dominated by the United States while the WEU is a diffuse alliance with no clear hegemon. France could thus remain in the WEU without the risk of subordination, but for de Gaulle involvement in NATO's integrated command meant *de facto* subordination to the United States. NATO also in Article 5 of the North Atlantic Treaty has a relatively tight *casus foederis* in a commitment to collective defence under clearly described circumstances, circumstances in the nuclear age with inevitably integrational implications. The WEU has only a loose *casus foederis,* in that states agree to consult in the event of a threat to any party, and thus does not carry the same integration dynamic.

From the perspective of the 1960s NATO could not but look to de Gaulle as an alliance whose trajectory was bound to limit French autonomy and subordinate France to the United States. He consequently withdrew France from NATO's integrated military command in March 1966, but made sure through 1967 Ailleret-Lemnitzer accords that France would be able to fight alongside NATO if necessary[28]. The WEU on the other hand more closely fitted de Gaulle's *Grand Ensemble* ideas [membership implied the erosion of sovereignty, which de Gaulle was willing to accept, but not the loss of sovereignty nor subordination], and had not the perfidious British been members it might have become the vehicle of de Gaulle's European defence ambitions. These de Gaulle pursued instead through the 1961 Fouchet initiative which sought to encourage defence co-operation between the European Economic Community members and which was influenced by the failure of the 1950–54 French-led European Defence Community [EDC] proposal. The elements of these proposals and the reasons for their failure remain highly pertinent to contemporary French European security debates.

The final strand of de Gaulle's ideas about defence relates to notions of *grandeur* and *rang*. Despite the objective decline in France's global presence between 1945 and 1962 de Gaulle sought to translate what remained of France's international reach into power and influence in Europe. De Gaulle's tripartite initiative is insightful in this respect. On 23 September 1958 de Gaulle addressed a memorandum to NATO's Secretary-General Paul-Henri Spaak, US President Eisenhower and British Prime Minister Macmillan. The memorandum proposed the establishment of a tripartite directorate – comprised of the US, UK and France – to in effect co-ordinate and direct the entire global policy of the West in an attempt to position France as an equal of the United States and, less implausibly, the UK and to define a 'great power' role for France within and outside NATO[29].

That de Gaulle made the tripartite directorate proposal at all provides useful insight into an element of Gaullist defence policy which might usefully be called 'style' and which remains relevant to the way in which France directs defence at the beginning of the twenty-first century. Objectively in 1958 France and de Gaulle were in a position of considerable weakness: the nation was riven by the Algerian crisis which returned de Gaulle to power; it was little more than four years since the defeat at Diên Biên Phu and less than two years since the debacle of Suez; France was facing the end of empire as a result of decolonialisation; the French military contribution to NATO during the 'missing pillar' period was small and virtually irrelevant to front-line defence; and France was still two years away from developing a nuclear device and six years away from having an operational nuclear weapon. That de Gaulle thought the United States in particular would accept French co-direction of global affairs at such a point is deeply instructive about Gaullist *hauteur*.

Underpinning the proposal seems to be a separation of rhetoric and reality of the kind which led de Gaulle to proclaim at the liberation of Paris in 1944 'Paris liberated, liberated by itself, liberated by its people with the help of the armies of France, with the support and help of the whole of France, of France that is fighting, of France alone'[30], a declaration of course wholly at odds with the role the United States, Britain and others had played in the liberation of Paris and wholly at odds too with the reality of divisions inside France between the resistors, the collaborators and those in between. One can rightly explain this in part as rhetoric aimed at the uniting of France and the restoration of French pride and one can take from it too the point, often made, that de Gaulle well knew the weakness of his hand in 1944 and again in 1958 and that it was precisely this which made him press for France to be dealt with not as it objectively was but as de Gaulle held it to be. De Gaulle himself once explained this: 'limited and alone as I was, and precisely because I was so, I had to climb to the heights and never then to come down'[31].

The point here is less the explanatory value of tactics or of weakness recognised in de Gaulle's statements and policy-making than the implications for policy-making which followed from the existence of a gap between France as it was and de Gaulle's *certain idée de la France*. This matters the more so because de Gaulle's *certain idée* was puffed up by notions of exceptionalism, *grandeur*, and, by his own admission, by personal sentiment and thus was even further distanced from the objective France. It is little wonder the perfidious Anglo-Saxons found de Gaulle arrogant [one commentator memorably wrote of de Gaulle: 'an improbable creature, like a human giraffe, sniffing down his nostrils at mortals beneath his gaze'[32]] when he spoke and acted as though master of a great deal more power and influence than actually held. This element of policy-making style is important to the extent that it has persisted long after de Gaulle as one of the elements informing, and arguably clouding, French relations with her allies.

A second, and closely linked facet of Gaullist style is that identified by Harrison in his deeply insightful study *The Reluctant Ally*[33]. Harrison makes the case that from his position of relative weakness de Gaulle sought to enhance French power and influence 'through tactics of refusal, obstruction, intransigency, and the manipulation of *fait accomplis*'[34], the underlying idea being that while genuinely great powers can afford 'largesse and compromise with others because it costs them little . . . a weak middle power can give away nothing because each compromise brings greater dependence in its wake'[35]. One can of course take issue with the logic of this – not least because compromise can bring co-operation and agreement which in turn can enhance power and influence – but it is instructive as an explanatory factor in de Gaulle's policy-making. By refusing to go along with things, by insisting on being consulted and factored in, de Gaulle sought to exploit the value of intransigence as a means of making France count, of making the French voice heard.

A third and final element of style is more contentious and can hardly be said to be either an exclusively Gaullist or French issue. It is the capacity to critique the policy of others while simultaneously shaping one's own policy on the same basis, an approach which might be summarised as sitting somewhere between dichotomy and hypocrisy. One can find an historic strand of this in the tension between the self image of France as *nation phare*, the humanising and civilizing state *par excellence* and the objective history of France as an aggrandising and colonial state, a dichotomy which at times has threatened the French nation-state itself, as for example during the Algerian war[36]. One can find another version in the policy-making of de Gaulle who was fond of attacking the policies of allies [inevitably most usually the Anglo-Saxons] while conducting or planning policy of virtually identical character himself, as though the critiques he himself had made had not been uttered or as though France was somehow uniquely exempt from them. The 1958 tripartite directorate initiative illustrates this clearly. De Gaulle framed much of his policy-making on the rejection of subordination, the refusal to grant others special status, the rejection of condominium, and the insistence on equality and inclusiveness. These were more than mere words and conditioned de Gaulle's policy on issues as profoundly important as NATO membership and non-participation in international arms control[37]. Yet in the tripartite initiative he manifestly sought for France all that he rejected as unacceptable for others, not least the United States. The directorate he proposed meant the subordination of others to the 'big three', the granting of special status, the emergence of a *de facto* condominium, and the rejection of equality and inclusiveness. Evidently major power collusion was unacceptable if it constrained France, but acceptable if France constrained others. Similarly, hegemonic influence was unacceptable if exercised on France but acceptable if exercised by France.

The contention is thus that by insisting that others deal with the France of de Gaulle's mind rather than the France of objective fact, by exploiting the uses of intransigence to insist that France was heard, and by requiring all to see the splinters in their own eyes while overlooking the mote in France's eye, de Gaulle fashioned a policy style which could not fail to complicate and at times obfuscate France's relations with even her closest allies. The contention is further that, despite the passage of time which has seen both the fading of Gaullist rhetoric and the relative strengthening [at least in some areas] of French power, this style has persisted as an important part of the Gaullist legacy.

The path that carries Gaullist influence on defence policy to the late 1980s is long and complex and can be studied elsewhere[38]. The relevant waymarks in the present context are the lack of defence expertise of President Georges Pompidou and his choice of de Gaulle's close associates as his Prime Minister and Defence Minister; the role of the 1972 *Livre Blanc sur la Défense*, written largely under the direction of de Gaulle's Prime Minister Michel Debré, in conceptualising the defence debate almost wholly in Gaullist terms and in projecting into the future the Gaullist framework of thinking about defence, the tensions at the heart of Gaullist defence policy and the Gaullist responses to those tensions; and the pressure exerted on President Giscard d'Estaing by the French strategic community, the guardians and enforcers of Gaullist orthodoxy, to persuade him to backtrack on defence policy innovations.

When 23 years of right wing control of the *Elysée* came to an end in 1981 with the election of the socialist François Mitterrand the question was whether Gaullist influence on defence would continue. The defence policy of the *Parti Socialiste* had undergone profound changes between 1972, when the party manifesto *Changer la Vie* rejected the Gaullist model, and 1981 by which time most Gaullist ideas were accepted, but it remained to be seen if these changes would be observed in office[39]. In the event they were: Mitterrand did little to challenge the Gaullist framework nor to dismantle or decentralise presidential control of defence or foreign affairs declaring 'France's institutions were not made for me, but they suit me well enough'[40].

The elements of his defence policy were almost wholly consistent with Gaullist objectives. Overall defence spending rose from 104.4 billion Francs in 1981 to 158.0 billion in 1986 in line with previous trends[41]. Mitterrand directed the expansion and modernisation of French nuclear weapons in order to maintain independence and technical credibility, and confirmed his fidelity to Gaullist nuclear doctrine and strategy[42]. Franco-NATO relations remained distant, but the French left's antagonism to American hegemony and capitalism was kept in check by Mitterrand who understood the importance of the United States and NATO cohesion during the so-called 'Second Cold War' of the early 1980s and the continued need to anchor West Germany to NATO. In line with this Mitterrand took the opportunity of a speech to the Bundestag on 20 January 1983, the twentieth anniversary of the Franco-German *Elysée* treaty, to urge the German government to accept US Pershing II missiles[43]. In the process he forged an alliance with Helmut Kohl, the soon-to-be German Chancellor, which was to prove decisive in the construction of European defence. Outside Europe Mitterrand flirted briefly with *Tiers Mondialisme*, a more moral foreign policy, before reverting to Gaullist orthodoxy in the face of pressure from France's allies and from traditionalists within France on the basis that France could ill afford turbulence in French global policy at a time of potential crisis in Europe.

Mitterrand's role as a *status quo* player was confirmed for many on 10 July 1985 when French secret agents Dominique Prieur and Alain Mafart blew up the Greenpeace ship *Rainbow Warrior* in Auckland Harbour, New Zealand, killing the Portuguese photographer Fernando Pereira[44]. The conduct of Operation Satanic to sink the Greenpeace ship for the basest of national interest motives revealed something of the nature of the French state under Mitterrand and strengthened the perception that Mitterrand himself had become the 'most Gaullist of de Gaulle successors'[45].

Reflecting on 30 years of Gaullist dominance of defence and security issues between 1958 and 1988 it is evident that a 'consensus' of sorts emerged around the core ideas of de Gaulle which have been usefully summarised as:

> the absolute need for independence in decision-making, a refusal to accept subordination to the United States, the search for *grandeur* and *rang*, the primacy of the nation state and the importance of national defence[46].

To this definition needs to be added the Gaullist imperative of *Grand Ensembles*, the central importance to France of embedding French defence in a European context from which, as de Gaulle understood, French security could not be divorced, and through which French power and status could be enhanced.

Concealed within these ideas were the tensions which arose from the contradictions within Gaullism. How could national primacy be reconciled with European defence ambitions? Were France's European and global interests compatible? Did distance from NATO weaken French influence with the US and weaken the French role in Europe? Could French nuclear doctrine be reconciled with NATO and with the security of neighbouring states? The consensus thus centred not only on Gaullist ideas but also on Gaullist problematics.

This consensus cannot be understood as complete cross-party and body politic support within France for all elements of Gaullist defence policy. There has consistently been opposition to elements of Gaullist thinking on defence from across the political spectrum and even the most popular of Gaullist policies such as the possession of nuclear weapons have seldom attracted more than 70 per cent public support[47]. The consensus can better be approached as a broad agreement of political parties, the strategic community and the electorate around the core sub-set of Gaullist ideas described above. These ideas, however, turn out to share certain characteristics with a desert mirage: distinct and clearly defined at a distance, they prove to be elusive and ethereal on closer inspection.

Within the broad consensus around Gaullist ideas there has in fact been precious little agreement about the details of policy, the relationship of means and ends and even in certain cases about how to read Gaullist thinking. In part this was because de Gaulle did not resolve the tensions inherent in his ideas, in part it was because changing circumstances after 1958 put Gaullist ideas under strain, and in part it was because those like the Socialists and Communists who broadened participation in the consensus at a later date also broadened the perspectives and debates within the consensus.

With this in mind, arguably a useful way to understand the consensus is as the widespread agreement amongst political elites, the strategic community and the body politic that the defence debate should take place within the Gaullist framework of thinking about defence rather than across or outside it. An important and obvious consequence of this has been the stifling of debate and creativity in defence policy-making by limiting the discourse to issues within the Gaullist paradigm. In addition, as Anand Menon has helpfully observed, the consensus has also stifled debate within the Gaullist framework because maintaining fidelity to the consensus became something of an end in itself and conditioned French responses to events and technical innovations on the basis that France could not react to change in ways that risked breaking the consensus[48].

Reflecting the foregoing, an obvious and important question is how to account for the persistence of Gaullist thinking in defence and security matters. One reason might simply be that de Gaulle was brilliant and prescient in defining and implementing ideas from a range of alternatives which were of permanent, or at least persistent, relevance to French defence and security policy.

A second possibility is that, far from defining a new future for France based on far-reaching and correct insight, de Gaulle was in fact weaving together the warp of historical strands of French policy with the weft of prevailing geopolitical reality to create a policy determined rather more by lack of options than strategic foresight. It is certainly the case that many Gaullist ideas can be found in French history and that de Gaulle accepted many of the foreign and defence policy elements of the Fourth Republic. Moreover, it is hard to avoid the view that much of what de Gaulle did, though not all, was conditioned by the reality of the time and France's limited means to do anything else [France could not, for example, have long resisted decolonial-isation, nor could France have developed a nuclear arsenal to match that of the USA and USSR]. This explanation suggests that Gaullist ideas have persisted because there has been little else France could do given its limited means, and that de Gaulle's genius was to define the maximal room for manoeuvre within these limits.

A third explanation for the persistence of Gaullist ideas may be found in the 'vision' de Gaulle articulated for, and of, France. De Gaulle's ideas of *grandeur, rang* and independence – built on a highly idiosyncratic, romanti-cised, and selective reading of French history – offered the French a return to past glory and national importance which was immensely attractive in the post-Second World War era even if it bore little relation to France's actual situation. This view says that de Gaulle offered France the idea that it had only to believe once again in its greatness for that greatness to return, the continuing relevance of this being that deviation from Gaullist ideas means deviation from the belief in France's greatness and thus that that greatness would be lost. According to this view Gaullist ideas have persisted because the French have not been prepared to face their objective circum-stances and have preferred the projected image, some might say the delu-sion, offered by de Gaulle.

One can add the important observation of Philip Gordon that de Gaulle's ideas have persisted because of the pure historical force of the man and the observation that having twice been seen as the saviour of France [in 1944 and 1958] none of his successors have felt able to deviate from his policies because to do so would be tantamount to saying that they knew more about the long-term defence and security needs of France than the General[49].

Perhaps the most persuasive explanation is to be found in the policy-making process itself and the institutional framework established by de Gaulle in the Constitution of the Fifth Republic. In centralising policy in foreign and defence matters in Presidential hands, in curtailing the role of the govern-ment and parliament, and in situating the President at the centre of a polit-ical-military-technical network of advisers and councils, de Gaulle created an arrangement of simultaneously immense institutional inertia and momentum within which successive presidential incumbents have found relatively little room for manoeuvre but, using which, successive presidents have been able to exercise great power. Gaullism may have persisted above all because the institutions and process of defence policy-making have served to maintain fidelity to the Gaullist paradigm and to counter any attempt to move away from that paradigm. To understand this more fully it is important to consider the defence policy process itself.

The defence policy process

A good starting point for exploring the French defence policy process is the constitution of the Fifth Republic enacted on 4 October 1958. Although the idea has persisted that this constitution, created by de Gaulle, centralises power in the hands of the French president in the '*domaine réservé*' [reserved domain] of foreign and defence policy, this is not formally the position. The constitution provides for the distribution of authority in a series of articles which, when considered together, appear to create some ambiguities about who exactly determines and controls French defence policy. Article 15 states that 'the President of the Republic shall be the commander-in-chief of the armed forces. He shall preside over the higher national defence councils and committees'[50]. This authority is evidently tempered by Article 20 which states that '[T]he Government shall determine and conduct the policy of the nation. It shall have at its disposal the civil service and the armed forces'[51] and by Article 21 which notes that '[T]he Prime Minister shall direct the operation of Government. He shall be responsible for national defence'[52]. The constitution thus establishes the central dynamic of defence policy-making – relations between the President, Prime Minister and government – but fails to define the nature or the power of the respective functions. Understanding the latter consequently depends both on a fuller understanding of the formal relationship between the centres of power and more particularly on an understanding of the workings of this relationship in practice.

A closer reading of the text of the constitution reveals that the President has the dominant role, being simultaneously the head of the executive, the commander-in-chief of the armed forces, and the chairman of the National Defence Committee. Moreover, the authority of the President to appoint and dismiss the Prime Minister [Article 8], to appoint the civil and military posts of the state [Article 13], and to dissolve – subject to caveat – the *Assemblée Nationale* [Article 12] appears to put presidential primacy beyond doubt. Presidential dominance is strengthened further by decrees of 7 January 1959, 18 July 1962 and 14 January 1964 which widened his powers and, in certain respects, reduced the ambiguity in the constitution.

Presidential authority is also underpinned by a number of other factors flowing from the constitutional position. These include the status conferred on the president by direct suffrage and the power resulting from a long term of office [a presidential term was five years between 1945 and 1974 and has been seven years since 1974, and presidents may serve a maximum of two terms]. They include also what has been termed the 'personalisation'[53] of defence policy arising because the president has a uniquely powerful role in the control and use of nuclear weapons [though this is ostensibly qualified by the insistence of Article 35 of the constitution that it is parliament that would authorise a declaration of war] and because the president is the principal figure representing France in relations with other states.

Successive presidents have determined the main lines of defence policy as de Gaulle did in giving priority to nuclear weapons and withdrawing from NATO[54] and as Giscard d'Estaing did in taking a more Atlanticist stance in the mid-1970s[55]; have shaped or guided French doctrine and strategy both at the nuclear level [for example the shift towards a more flexible doctrine under Giscard d'Estaing[56], or Mitterrand's resistance to a revision of nuclear strategy in the post-Cold War era[57]] and at the conventional level [as when Pompidou resisted the restructuring of the armed forces urged by the military chiefs[58]]; and have exercised authority through *ad hoc* interventions to ensure compliance with broad policy objectives and to take French policy in new directions, as for example in Mitterrand's intervention in 1992 which placed a moratorium on nuclear weapons testing despite political and military opposition[59].

The key point to emerge from this evidence, elements of which will be discussed in more detail below, is that despite ambiguity and uncertainty in the formal position of the president with respect to defence policy, the practice of the past, at least since 1958, has established norms in which presidential authority is typically decisive and pervasive.

This at least is the position where there is political harmonisation between the president, the prime minister and government, because in such a situation the distribution of power established by the constitution is largely untested. To date the Fifth Republic has yielded at least two exceptions to this, the first, and less important, being when differences on defence issues open up between the president and government despite both being from the same political stable. Such a situation obtained during the presidency of Giscard d'Estaing who, having adopted a more Atlanticist approach in the first part of his term of office, came into increasing tension with Gaullists in government and Parliament on whose coalition support he was dependent. The president was persuaded into a series of compromises and eventually revised the major thrust of his policy back towards a more orthodox Gaullist position[60].

The second, and far more important point, arises in periods of 'cohabitation', that is in periods when the president and government are from opposite political parties. In such a situation the distribution of power specified by the constitution and the lack of clarity about function is tested in novel ways. Moreover, given that the last decade or so has yielded three such periods of cohabitation [1986–88, 1993–95 and 1997 to date] it is tempting to suggest that this pattern may be set to become the norm as the seven-year presidential term of office appears to be increasingly out of synch with a volatile electorate. It is thus worthwhile giving further consideration to the problem with the caveat that – in general – periods of political homogeneity between president and government can be expected to mitigate, though not necessarily eliminate, the issues of central tension in a period of cohabitation.

The evidence from the three periods of cohabitation since 1986 [the latter of which is unfolding at the time of writing] indicates that the exercise of power and authority in matters of defence depends crucially on the relative political power of the players which is itself shaped by a range of other contextual and institutional factors. The first period of political cohabitation between 1986 and 1988 was a novelty and presidential-government relations were marked by clashes shaped by respective uncertainties about the boundaries of responsibility for defence under the constitution. The situation was not helped by the personal animosity between the protagonists, President Mitterrand and his cohabitation Prime Minister Jacques Chirac, nor by the latter's 'bulldozer' approach to politics.

Chirac sought a commensurate role in defence and foreign affairs, not least to enhance his future presidential credentials, and established a presidential-style coterie of defence advisers, placing as many of his own people as possible in the foreign and defence ministries. Where the opportunities arose Chirac took operational decisions, occasionally recklessly so, amongst the most conspicuous of which were deals with terrorists to release French hostages, the release of the Rainbow Warrior bomber Dominique Prieur from custody in the South Pacific, and the ordering of military operations against separatists in *Nouvelle-Calédoine*[61].

Mitterrand for his part acted to retain as much presidential control as possible over defence and foreign affairs, for example vetoing Chirac's original nominees for foreign and defence minister, Jean Lecanuet and François Léotard respectively, both of whom were considerable political figures in their own right and could have strengthened the government challenge to the *domaine réservé*. Notwithstanding a few dramatic exceptions to the contrary, in general the popularity of Mitterrand [he was re-elected in 1988], the presidential ambitions of the Prime Minister Jacques Chirac, which meant that Chirac did not wish to see the authority of the office of President weakened, and the broad agreement of political left and right on core issues of defence meant that Mitterrand's dominance of defence matters was decisive though hardly unchallenged[62].

The period between 1993 and 1995 were constitutionally Mitterrand's last two years of office and a period of declining personal popularity and health. This relatively weak position allowed the government of Eduoard Balladur more room for manoeuvre and shifted the balance between President and government, prompting the French press to write of a *'domaine partagé'* [shared domain] in defence and foreign policy. Nevertheless government scope was still tempered, *inter alia*, by Mitterrand's political skill, Balladur's presidential ambitions, and by general agreement on defence issues evidenced by the publication of the 1994 *Livre Blanc sur la Défense*, which was written from within the government but which had presidential consent. When pressed by Balladur's government on issues such as the 1992 nuclear weapons testing moratorium or attempts to revise nuclear strategy Mitterrand reasserted his authority and his will prevailed[63]. Cohabitation between 1993 and 1995 was more skilfully managed by all parties than in 1986–88, evidence that a *modus vivendi* was emerging which, as far as possible, allowed difference but avoided open division.

The period since the election in 1997 of a Socialist government under Lionel Jospin has been more difficult for the Gaullist President Jacques Chirac. Because of Chirac's personal unpopularity[64], and his relative isolation within his own political coalition which effectively weakened his own power base, Chirac since 1997 has looked, at least to his critics, to be weak and drifting[65]. The government of Lionel Jospin has intervened in matters of defence policy to an arguably unprecedented extent on issues such as the management of defence, arms industries, and the maintenance of the nuclear weapons stockpile, and Chirac has appeared unwilling or unable to respond[66]. This does not, of course, preclude the possibility that changing political fortunes will eventually restore Chirac's eroded position.

The point of this brief discussion here of issues which receive greater attention later is to highlight the idea that the exercise of political authority over defence policy is shaped both by the formal role of the key players set out in the constitution [and in some cases by subsequent amendments and decrees[67]] and by the political fortunes of the players, with a key variable being the relative political strength of the President.

The interplay between the President and government, though a dominant element, is but part of the defence policy process as a whole. While it is not possible here to explore this fully it is useful to set out, as Figure 1.1 does, the main players in the process and to say something of their respective roles.

The President sits at the head of the apparatus of defence policy and takes the main defence decisions in three fora which he chairs: the *Conseil de Ministres*, the *Conseil de Défense* and the *Comité de Défense*. The *Conseil de Défense* is the highest authority in defence policy-making and not only directs all policy in matters of defence, but also serves as the War Cabinet. It comprises the President, the Prime Minister, the Defence Minister, the Foreign Minister, the Interior Minister, the Finance Minister, the *Secrétaire Général de la Défense Nationale*, the Chief of the armed forces and the three service heads of the armed forces, and may be supplemented by others, for example from the CEA or intelligence community, subject to the issues under discussion. In matters of defence the President has formal advisers in the *Chef d'Etat-Major Particulier du Président de la République* and defence *Conseillers* and

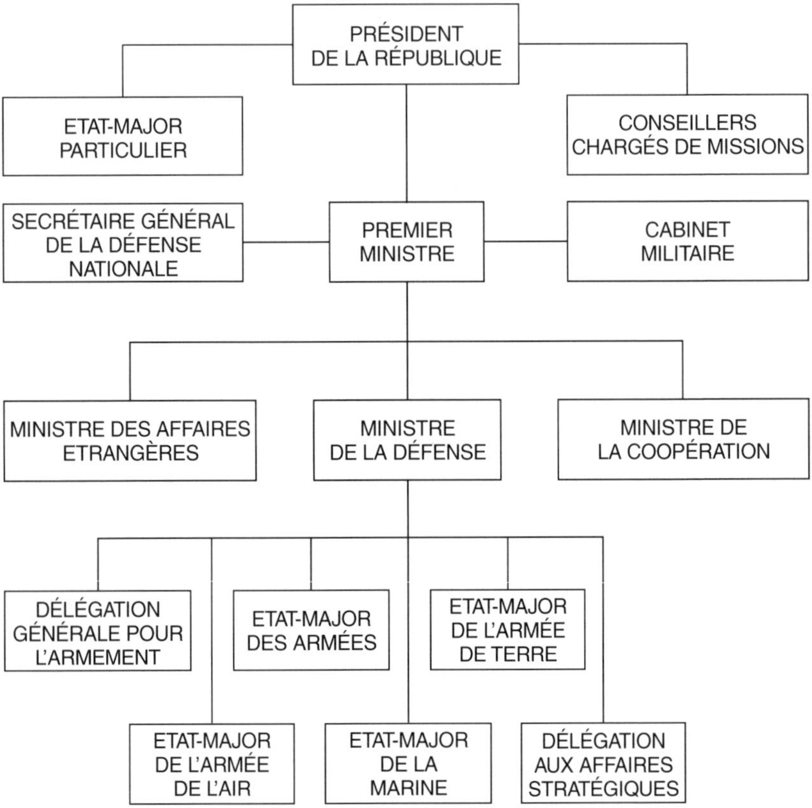

Figure 1.1   The Organisation of Defence.
Source   Défense: Annuaire des Relations Internationales de Défense (Association Française Frères d' Armes, 1994).

Chargés de Missions and may draw on personal advisers without a formal role.

The key role of the government is the implementation of the decisions decided upon by the *Conseils* and *Comités* chaired by the President. In relation to defence the Prime Minister, aided by his own defence advisers and *Cabinet Militaire*, implements decisions through the *Secrétariat Général de la Défense Nationale* [SGDN], though by the decree of 25 January 1978 the SGDN also liaises directly with the President. The primary role of the SGDN is interministerial co-ordination of the implementation of defence decisions, particularly between the Foreign Ministry and the Defence Ministry. Outside the Defence Ministry each Minister is responsible for the preparation and execution of defence measures to be carried out by their respective departments and, to facilitate this, each Minister is advised by a senior Defence Ministry civil servant.

In terms of input into policy-making at the highest level the Defence Minister is limited by the willingness of the President and Prime Minister to engage. Within the scope allowed by the President [and Prime Minister] the Defence Minister exercises wide-ranging authority over the implementation of policy in terms of fighting in government for a share of the public spending cake, threat assessment, strategy, decision-taking on research and development, procurement, defence industries, force levels and so forth.

Within the Defence Ministry the Defence Minister has his own *Cabinet Civil et Militaire* of advisers and directly controls DICOD. The armed forces are represented at the highest level by the *Chef d'Etat Major des Armées* [CEMA], France's most senior military officer, and by the respective heads of the armed forces, the *Chef d'Etat Major de l'Armée de Terre* [Army], the *Chef d'Etat Major de l'Armée de l'Air* [Airforce] and the *Chef d'Etat Major de la Marine* [Navy], the four officers who attend the *Conseil de Défense*. Under the constitution of the Fifth Republic, de Gaulle reasserted political control over the military in the wake of the Algerian crisis, and the armed forces, *La Grande Muette* [literally 'The Great Mute'], have exercised influence less publicly as a result.

The intellectual backbone of the Defence Ministry is provided by the *Délégation aux Affaires Stratégiques* [DAS], tasked in June 1992 with conducting research and analysis relevant to French defence in the long term. DAS is thus responsible for the intellectual coherence and cohesion of defence policy and plays a major part in the production of defence policy documentation.

Defence Ministry oversight of arms production and procurement is provided by the *Délégation Générale pour l'Armement* [DGA], tasked in a semi-autonomous manner with carrying out research and analysis on future military equipment needs, maintaining technical edge, contributing to the planning and programming of defence spending and co-ordinating state policy on arms production and procurement. The DGA also has a role in the production of nuclear weapon delivery systems, though the warheads and bombs themselves are the responsibility of the CEA and its nuclear weapons production element, the *Direction des Applications Militaires* [DAM].

Finally the Defence Ministry is served by elements of the intelligence community, the *Direction Générale de la Sécurité Extérieure* [DGSE] which in relation to defence policy has a key role in threat assessment, and the *Direction du Renseignement Militaire* [DRM] under the direct authority of the CEMA who co-ordinates the activities of French defence attachés around the world and thus also has a role in threat assessment[68].

During the periods of presidential ascendancy which have been the norm in the Cold War era, decision-making authority has consequently been centralised in a relatively small group subject to little outside influence or scrutiny. The extent of presidential dominance at these times can be judged from the oft repeated claim that the Foreign and Defence Ministers usually function as little more than senior civil servants to the President with little room for autonomy of action. This is not to suggest that neither minister has any authority, but rather to note that both are more dependent than other ministers on presidential indulgence and both are rather more likely to find their decisions over-ridden, their objectives thwarted, or their landscapes transformed by unilateral presidential initiatives.

In periods of cohabitation the position of the Defence Minister and Defence Ministry [and similarly the Foreign Minister and Ministry] is not necessarily any better. The President, as Mitterrand did in 1986, can veto government choices for office and thus influence the appointment of individual defence ministers. The incumbent can then become marginalised in the government precisely because he may be seen as leaning [however slightly] towards the President. The Defence Minister may also be caught in the tension between the President and Prime Minister, particularly where the latter tries to accrue status through the exercise of power in the *domaine réservé*. Sometimes, though, a defence minister is empowered by a close working relationship with a Prime Minister and may openly challenge the President as, for example, in the mid-1990s when François Léotard contradicted and rebuked Mitterrand in his final years of office over nuclear weapons policy decisions[69].

Defence policy itself is articulated and controlled through three key documents. These are the *Livre Blanc sur la Défense* [the Defence White Paper], the *Loi de Programmation Militaire* [literally the military planning law, known for convenience as the *Loi-Militaire*], and the defence budget element of the annual *Loi de Finance* [Financial law], the *Budget pour la Défense*.

The *Livre Blanc sur la Défense* [known as the *Livre Blanc*] is in many respects the most substantial and revealing of official defence documentation. It sets out in some detail an official assessment of the nature of contemporary international security and France's place and concerns within it, describes at length the major elements and objectives of French defence policy, its armed forces and principal defence industries, and examines related issues such as defence relations with other states and relations between the French armed forces and the French people.

The purpose of the *Livre Blanc* is largely one of explanation and prescription though it sets no budget or other precise constraints on defence policy. It identifies key issues which are of central and lasting concern to the defence and security of France and describes the French responses to those issues over typically the forthcoming 15 to 20 years[70], and thus to some extent it both describes the consensus within which the defence debate takes place and prescribes the ideas and trends shaping defence policy into the future. That there have been only two *Livres Blancs* since the beginning of the Fifth Republic – in 1972 and 1994 – is testimony to the stability of French defence policy and also to the broad scope and long-term relevance of the documents.

The Defence Ministry also prepares the periodic *Projet de Loi Relatif à la Programmation Militaire* [the formal title of the *Loi de Programmation Militaire*]. This is a document which typically covers a five-year period [the most recent being 1997–2002] and attempts to set a framework for defence spending, procurement and force levels, and includes detailed information about the disposition of the nuclear forces, the air force, army, navy and gendarmarie[71]. The *Loi de Programmation Militaire* is intended as the key prescriptive and explanatory document which shapes and steers defence policy over the mid-term. Its role, however, is undermined by two factors. Firstly, because of the political volatility of the French political system, particularly over the past decade or so, *Lois de Programmation Militaire* have often been superseded before the end of the framework term. Thus, for example, the 1995–2000 *Loi de Programmation Militaire* published in June 1994 following the *Livre Blanc* was superseded within two years by the 1997–2002 *Loi de Programmation Militaire* published in May 1996. Secondly, the spending plans, and thus the research and development procurement and force level projections set out in the *Loi de Programmation Militaire* are routinely altered by the annual *Budget pour la Défense*, although these seldom threaten the major programmes set out in *Loi-Militaire*.

The *Loi de Programmation Militaire* can thus be seen as the liaison between the longer-term objectives expressed in the *Livre Blanc* and the annual defence budgets, the better to ensure that the defence budgets maintain a general fidelity with longer-term objectives. The annual defence budgets themselves, presented under the annual *Loi de Finance*, are prepared by the Defence Ministry in consultation with government as part of overall government spending. These provide detailed costings and explanations of defence spending for the forthcoming year setting out the salary, equipment, pensions and related budgets[72].

In 1996 these three levels of defence documentation were amended by a novel document entitled *Une Défense Nouvelle 1997–2015*. This was produced through the Defence Ministry but was essentially a presidential document. It set out arguably the most substantial revision of French defence policy since the 1966 withdrawal from NATO's integrated military command and, while it is explored in some detail in Chapter 3, it is important to note here that it substantially revised the 1994 *Livre Blanc* and cut right across the 1995–2000 *Loi-Militaire* and the defence element of the 1997 *Loi de Finance*, eloquent testimony to the authority and scope of presidential intervention[73].

In any democratic society the issue which quickly surfaces [or ought to] when examining the policy-making of an executive is where exactly are the checks and balances on the exercise of executive power? In the case of France, and particularly in the areas of foreign and defence policy in which the President exercises such authority, the answer is that the checks and balances are few and those that exist are relatively weak.

Each French president draws democratic comfort from the fact of direct suffrage but the seven-year term of office and the range of issues on which the French electorate elect their ruler must qualify the extent to which executive policy-making on defence is checked or balanced by the public will. That said, the supposed consensus on defence policy and the strong and persistent support of French people for particular elements of policy such as the possession of nuclear weapons arguably functions as a buffer against extreme policy shifts from the *status quo*, even if such policy shifts were objectively desirable.

Perhaps the most important nominal check on the executive is that of parliament which is given a limited balancing role by the constitution. The two chambers of Parliament – the *Assemblée Nationale* and the *Sénat* – are in fact rather weak in large measure because of the constraints imposed by de Gaulle through the constitution of the Fifth Republic, precisely because he was concerned to temper parliament's ability to interfere in his role and that of government. Government thus largely dominates Parliament and does so in many ways. Obviously the government is elected on the basis of parliamentary majority and thus can expect to command a majority in Parliament's lower house the *Assemblée Nationale*[74]. The prevalence of coalition politics in France however not infrequently renders parliamentary support a fickle creature. To temper any unruliness, the government controls the timetable and agenda of the house, imposes limits on the extent to which financial bills may be held up, can operate a fast-track process for sensitive or difficult bills, and limits the extent of parliamentary scrutiny by limiting committee oversight[75].

This does not mean that Parliament is wholly toothless. It remains the essential legislative body and thus must be factored into all policy-making by government. Parliament consequently functions as another buffer against major changes in defence policy because movement away from the broad consensus on core matters of defence is likely to face sustained parliamentary opposition. Parliament also debates the annual defence budgets [producing an annual set of analytic reports on the defence budget] and, when relevant, the *Loi-Militaire* and *Livre Blanc*, and can raise questions to embarrass or wring compromises from government or suggest amendments to bills. Parliament thus exercises influence directly in the chambers and indirectly in forcing governments to make policy cognisant of likely weaknesses which will be highlighted and critiqued by parliament. Parliament can also use a variety of procedural and process measures [such as the Constitutional Council and Issues of Confidence] to delay and harass government proposals.

Some of the hardest questions and sharpest critiques of government policy are faced in Parliament's committee structures. In the more important *Assemblée Nationale* the permanent *Commission de la Défense et des Forces Armées* provides the key scrutiny of defence policy with a similar, though less influential, body in the *Sénat*. It is in these fora that some of the deepest oversights of government policy takes place. Yet this cannot be understood as necessarily checking or balancing executive power. The committees are usually dominated by government figures who, although not members of the executive, are nevertheless very closely allied to government. Moreover, some individuals in key committee positions themselves become important players in the executive defence policy process. Ultimately neither these committees nor parliament as a whole have any real powers to stop or substantively curb government action. Presidential decisions such as Mitterrand's 1992 decision to suspend nuclear weapons testing or Chirac's 1996 decision to revise defence policy in *Une Nouvelle Défense 1997–2015* have typically been subject to very little parliamentary influence. With respect to the former, parliament had no role in the decision and had no power to alter the decision. With respect to the latter, Chirac's announcement gave rise to debates in both chambers only after the decision had been announced, clearly indicating the absence of parliamentary influence on the decision. Even one of the most fundamental roles given to Parliament in the constitution, that of declaring war [Article 35], has been side-stepped for years by presidents and governments who simply conducted military operations without ever declaring war[76].

It is evident then that the president and the government exercise considerable power and latitude in defence policy and that this is only weakly tempered by the limited checks and balances in the system. One final caveat needs to be added. While in theory each of de Gaulle's successors has matched his power in office, in practice in defence this has not been so because de Gaulle's successors have had to work within the constitution and conceptual framework determined by de Gaulle, rather than being in a position to set out their own. As such they have been constrained not only by the formal elements of the constitution but also by the Gaullist *esprit de défense* which imbues the strategic community. In sum the defence debate after de Gaulle has operated within a Gaullist paradigm, albeit a paradigm under strain because of its inherent tensions.

If one accepts this view, the door may open to some Kuhnian observations about the Gaullist paradigm. Firstly one would expect the strategic community to remain faithful to the paradigm in the face of growing contradictions and diminishing relevance because of the absence of an alternative paradigm and because the strategic community, 'socialised' in the norms of the paradigm, have a vested interest in its retention. Secondly, the contradictions and declining utility of the paradigm would eventually be expected to lead to its collapse and rejection, freeing the strategic community from its constraints. Thirdly, a post-Gaullist paradigm with rules and prescriptions better suited to the context and French means would be expected in due course to emerge. The progression between these stages, however, would not be expected to be smooth.

By the late 1980s, on the verge of great change, the Gaullist dominance of defence was already eroding but the geopolitical transformation of the end of the Cold War and subsequent events completely recast French defence policy and as such posed deep questions about the residual value of Gaullist defence ideas.

2
Adjusting to Change: French Defence Policy, 1989–94

Introduction

In 1988 the first cohabitation experiment of the Fifth Republic came to an end when in the elections of 24 April and 8 May François Mitterrand defeated his presidential rival Jacques Chirac by a comfortable 54 per cent of the vote to 46 per cent. The renewal of Mitterrand's mandate provides an appropriate point to take stock of French defence policy before charting the passage of France through the geopolitical turbulence of the following years.

As was argued in the previous chapter French adherence to Gaullism, for all its qualities, served as a constraint on defence and security policy-making. By accepting the rules and tools of Gaullism the French sacrificed creativity and innovative scope and were firmly prisoners of their own construct. Many who have commented on the period have argued that by 1988 the tensions, contradictions and strains in French defence policy had become intolerable and that difficult choices had become inevitable[1]. To understand this, it is useful to examine three mutually informing elements of policy: the Gaullist defence agenda, the budgetary problems, and the unresolved tensions and ambiguities within the defence consensus.

Gaullist thinking situated France at the hub of three conceptual concentric circles [France, Europe and the rest of the world] and Gaullism conditioned French policy in each of these and led to the development of specific roles – and hence specific military needs – for the French armed forces. France required forces for national defence, a role in the defence of Western Europe, and forces for global reach. Rank and independence required independent nuclear forces, conventional forces for national, European and global roles, and a defence industrial base to enable France to be virtually self-sufficient in nuclear and conventional arms[2].

By the late 1980s France was consequently undertaking nuclear weapons modernisation and expansion to keep pace with the other acknowledged nuclear powers and to respond to potentially constraining technologies [ASAT, anti-submarine warfare and BMD], maintaining and developing conventional forces for battlefield use in Europe, deploying and developing forces [particularly naval and army] to garrison and project power into the developing world, and seeking to maintain national military industrial autonomy. The demands made on France by these diverse military commitments became an acute problem by 1988 because France did not have the means to meet them as established levels and patterns of defence spending had become unsustainable.

Part of the explanation for this lay in the past. Like other Western states France had been subject to spiralling defence inflation and to the economic turbulence which followed the 1973 and 1979 oil crises. French problems though were exacerbated by the rigidity of Gaullism which set a demanding agenda but excluded realism [that is recognition of France's limited means as a medium power with great power ambitions] and creativity [in rejecting that which did not conform]. They were exacerbated also by the spending dislocation caused by the French nuclear deterrent, the sturdy child of Gaullism, which was responsible between 1960 and 1975 for the relative neglect of conventional forces and consequently for the conventional 'catch-up' problems which emerged after 1975[3].

Part of the explanation was contemporary. By 1988 the French economy was in some difficulty and economic policy was characterised by *rigeur*, that is by discipline and constraint to manage, *inter alia*, the budget deficit, unemployment running at around 10 per cent, generous social and welfare provision, damaging public sector strikes, the need for economic stability to meet European convergence criteria, and the consequences of Prime Minister Jacques Chirac's 'cohabitation' experiment between 1986 and 1988 with 'Thatcherite' popular capitalism[4]. Within this context the defence budget was, by 1988, subject to tremendous strain.

The third element which in turn greatly complicated the need to rethink French defence policy and to revise French defence budgets was a lack of political and intellectual agreement within the fast eroding defence consensus. Simplistic left-right divisions in France obscured substantive differences *within* each side of the political debate [for example right-wing Gaullists were rather more sympathetic to Mitterrand's Gaullist emphasis on 'pure' deterrence than to the Gaullist Jacques Chirac's ideas about nuclear 'flexibility']. A further tightening of the Gordian knot seems to have arisen as a consequence of Gorbachov's reforms in the Soviet Union and emerging East-West détente which overlaid the fault-lines within the French defence discourse with further splits which emerged on the question of whether to respond cautiously or optimistically to Soviet change. Within the ruling French socialist party this split was evident in tensions between those like the Defence Minister Jean-Pierre Chevènement who favoured caution and those like the Prime Minister Michel Rocard who argued that emerging détente was more substantive. Not surprisingly an important strand of this debate was played out in terms of the size and trajectory of the defence budget.

Faced with profound choices but faced also with divisions within his own party and within the wider body politic, and with uncertainty about the future shape of European security, Mitterrand decided to buy time by postponing major decisions in favour of relatively minor adjustments and not a little window dressing. The 1990–3 *Loi de Programmation Militaire*, announced on 18 May 1989, proposed that FF45 billion would be cut over the next four years from the budget anticipated by the previous 1987–91 *Loi-Militaire*, but it announced too that all the major projects – the nuclear modernisation, the Charles de Gaulle aircraft carrier, Leclerc tank, and Rafale fighter amongst them – were to be maintained. The circle was to be squared in the short term by project stretching, by purchase order delays, and by purchase quantity reductions, though multinational projects (such as the Tigre helicopter and the Helios satellite) were to remain unscathed [5]. A more substantive, if nevertheless limited, response was the restructuring and streamlining of the armed forces, also announced in 1989, under the *Armées 2000* initiative[6] aimed at 'improving the efficiency and operational readiness of [available] forces . . . and . . . their suitability for combined arms and inter-services operations'[7].

Under *Armées 2000* [and the related ORION infrastructure rationalisation programme] the awkward arrangements by which France was organised in six army, four airforce, three navy and six gendarmerie regions overseeing in turn 22 territorial areas were simplified into an arrangement, overseeing just ten defence areas, which divided France into just three integrated regions [the North East region headquartered at Metz, the Atlantic region headquartered at Bordeaux, and the Mediterranean region headquartered at Lyon]. Of these the Atlantic and Mediterranean also served as integrated regions with the navy, though the navy itself was headquartered at Brest and Toulon.

On the operational side, restructuring under *Armées 2000* reflected the deepening French commitment to West Germany and thereby to NATO forward defence, and in some respects was a working through of some of the defence clauses in the Elysée Treaty and joint Franco-German studies on the uses of France's 47 000 strong projectable *Force d'Action Rapide* [FAR]. Under *Armées 2000* the three Army Corps within the French First Army were simplified to a new *Corp de Manoeuvre Aero-Terrestre* [Airland battle manoeuvre corp] comprising two Army Corps headquartered respectively at Baden in West Germany and at Lille, together with the FAR headquartered in Paris under the direct control of the General Staff. Each of the two new Corps was strengthened to comprise five [rather than the previous three to four] divisions of which three within each Corps [rather than the previous two] were armoured. To facilitate closer air-land battle co-ordination in the central region, *Armées 2000* moved the French First Army headquarters from Strasbourg to Metz because the latter was also the home of the *Force Aérienne Tactique* [FATAC] responsible, *inter alia*, for tactical air support [and incidently also for airforce 'prestrategic' nuclear weapons][8].

With hindsight *Armées 2000* can be seen as an attempt to adapt French forces to US-style air-land battle concepts, to respond to budgetary pressures and to address French-European security tensions by rationalising and restructuring forces to better serve Franco-German defence co-operation. It is also evident now that *Armées 2000* was conceived of wholly in terms of the Cold War division of Europe and the need for forward defence on the inter-German border. Working within the existing defence budget, attracting little in the way of additional resources, and resulting in 8000 service and defence-related job redundancies[9], *Armées 2000* did little to address the core problems at the heart of defence policy. Moreover in being conceptually grounded in the Cold War it did not begin to prepare France for the upheavals which subsequently unfolded. On both counts *Armées 2000* hardly justified Chevènement's claim that it was bringing France 'step by step to the dawn of the next century and [giving] France a renewed Army'[10].

Years of upheaval

It is not necessary to rehearse in any detail the transformation of Europe which took place between 1989 and 1991. The principal events – the fall of the Berlin wall on 9 November 1989, the replacement of communist regimes across Eastern Europe, the reunification of Germany on 3 October 1990, and the collapse of the Soviet Union in December 1991 after the abortive August coup – are well known and have been subject to extensive analysis elsewhere. What matters here is how these events were seen in France and how they impacted on French defence policy. The key to answering both these questions can be found in reminding ourselves that France, despite de Gaulle's critique of the Yalta settlement, was through most of the Cold War a *status quo* power. French security objectives were well served by the NATO –Warsaw Pact balance both because the broad security structure in Europe anchored the United States to European continental defence to balance Soviet and to a lesser extent German power and because it weakened Germany by division and constrained the more powerful West Germany and orientated it to the West by integration into Western security arrangements. The old adage that NATO kept the US in, the Soviets out and the Germans down also functioned as a virtual résumé of French security concerns.

Furthermore the stability and seeming perpetuity of this arrangement had since de Gaulle, afforded France the possibility of semi-detachment and diplomatic autonomy which would arguably not have been possible had the situation been less stable or the US and West German commitment to the *status quo* less assured. In this respect it is instructive, as for example in the Euromissile crisis in the early 1980s, that in times of heightened tension or crisis within the alliance France throughout the Cold War consistently moved quickly to shore up NATO and the commitment of the US and Germany to it.

The events which unfolded from the summer of 1989 shook the very foundations of French security and defence policy because they placed in doubt each of the elements upon which policy had rested. Moreover although Europe was the epicentre of change it is important to note that the shock-waves which went out across the world impacted in turn on the French presence and the French role elsewhere in the world.

Before discussing the effect of these changes on French policy in more detail two observations are useful. The first is that as a *status quo* power committed to the glacis which had either solved or suspended historic French security concerns in Europe, France was reluctant to accept, let alone condone or support, change which in any way eroded its position. For years after 1989 France consequently functioned as a reactionary player opposing significant change and seeking to slow down, mitigate or qualify the consequences of change. The fact that in the process France was sidelined by the main thrust of events gives us some insight into France's role in European security at the time and also into the futility of paddling against a prevailing current.

The second observation is to suggest that part of the French resistance to change was rooted in the paradox that, despite de Gaulle's critique of Yalta, and indeed his seemingly prescient ideas about the nature of relations between peoples which allowed him to see ideological differences as temporary, these ideas found no expression in contemporary French debates and thus there was little intellectual preparation in France for a post-Yalta Europe.

As Mitterrand's second term unfolded in 1988, little hint of the profound changes of the next three years could be discerned. Developing his theme of warming to the Soviet Union and Eastern bloc in line with West German preferences Mitterrand undertook a series of visits to Eastern bloc countries. These began in the Soviet Union in November 1988, and took in Czechoslovakia in December 1988, Bulgaria in January 1989, and Poland in June 1989. They continued with visits to East Germany in December 1989, Hungary in January 1990, a newly democratic Czechoslovakia in September 1990, and a post-Ceaucescu Romania in April 1991[11]. This tour of course was interrupted by the events of October and November 1989 and a process, which began as an attempt to adjust to the implications of Gorbachov's new ideas, to position France beside West Germany and to build common European links to enhance the French role, became after November 1989 an exercise in damage limitation.

As the feet of those moving across Eastern Europe rubbed away the old divisions, Mitterrand chose the Franco-German summit of 3 November 1989, six days before the Berlin wall fell, to declare 'I am not afraid of [German] reunification'[12]. To many, Mitterrand's words served only to confirm the French preoccupation with arguably their least welcome outcome of the political turbulence sweeping Eastern Europe. Mitterrand for his part is understood to have believed that Gorbachov would not, and indeed could

not, allow German reunification, claiming that he had been told by Gorbachov 'the day German reunification is announced, a two-line communique
will announce that a Marshal is sitting in my chair'[13]. Mitterrand's mistake,
if such it was, appears to have been in believing that Gorbachov was in control of events. Gorbachov's mistake was simply in getting wrong the timing
of his prediction.

As the fact of swift reunification became evident, first through the elections in East Germany on 18 March 1990 which gave Helmut Kohl's CDU
cross-border German political support for reunification, and then through
the '2 + 4' process which brought the two Germanies into a forum with their
erstwhile conquerors to address the complicated questions surrounding
reunification [including borders, German membership of NATO, German
adherence to arms control agreements and armament constraints], German
single-mindedness and French marginalisation became increasingly
evident[14].

In the year following the fall of the Berlin wall, French historic anxiety
about German reunification and subsequent neutralism or, even worse, eastward drift seemed increasingly justified. Mitterrand responded by moving
swiftly to deepen political, economic and security ties with the new Germany. Until the autumn of 1990, however, much of this effort was stymied
as German attention not unnaturally focused on reunification. The pivotal
issue for France was assuring Germany's commitment to the schedule for
economic and political union defined by the Delors Committee Report of
April 1989. After spats earlier in the year, France and Germany reached a
compromise at the European Council meeting in October 1990, just a few
weeks after reunification which accepted but slightly delayed the European
Monetary Union [EMU] and European Political Union [EPU] processes. In
France this was sold as a compromise – German reunification for European
unification[15] – but this overstates French and indeed wider European influence. German reunification was dictated by the Germans, handled largely
bilaterally with the Soviet Union, and French influence was minimal. Similarly Kohl's commitment to the European project – if not to the more Fed-
eralist ambitions of Delors – was assured. The linkage between the two was
not so much a deal as the public expression of the acceptance of new norms.
It was important, though hardly conditional, that Germany had European
support for reunification; it was important, and certainly conditional, that
the EC had German support for the phasing of EMU and EPU. The deal, if
there was one, was within the EC rather than between the EC and Germany
in that Germany sought closer political union in exchange for the surrender
of the Deutschmark in the process of EMU. In all these cases Germany not
France was in the driving seat.

In terms of security, Franco-German initiatives in 1990 were embedded
within the ongoing discourse about political unity, a discourse which itself
qualifies any assessment about the nature of tensions between France and

Germany after 1989[16]. Beginning with the 19 April 1990 statement by Mitterrand and Kohl on EC Union[17], continuing at the 26 April Franco-German meeting prior to the 28 April EC summit in Dublin, and culminating in the 6 December 1990 letter which Mitterrand and Kohl jointly addressed to the Rome IGC[18], France and Germany put together a series of proposals to lay the foundations for building a European security and defence identity [ESDI]. In the French conception this new entity would be intergovernmental in nature, would be capable of co-operating closely across the range of military activities [though nuclear weapons were a special case], and would work closely with the United States though without subordination to it. The 6 December letter thus states with respect to an emerging European political entity: 'foreign policy will extend to all areas. It should include a real common security policy which will eventually lead to a common defence'. In addition the letter noted that the European Union would eventually have a 'clear organic relationship with the WEU' which in due course would 'fuse' with the European Union and determine its common security and, eventually, defence policy.

The letter however also contained the comment that the 'Atlantic alliance will be reinforced by the increase in the role and responsibilities of the Europeans'. This phrase is revealing because, by suggesting that ESDI was also about strengthening the European contribution to NATO, it identified the central if subtle difference of emphasis between France and Germany and also rather less subtle differences between France and the most Atlanticist states in the European Community. These differences turned on the precise relationship between NATO and a potential ESDI and centred on the role of the Western European Union in relation to NATO and to the European Community. The wider argument was essentially between those like the US, UK and Netherlands who favoured a more Atlanticist Western Europe in which the WEU and ESDI functioned as a means to strengthen the European pillar in NATO and in which the development of ESDI took place within NATO. In this formulation the WEU was not to be 'fused' to the EC but rather was to serve as a bridge between the EC and NATO. This concept of the 'bridge' was rather imprecise but was taken by some to mean that the links between the EC and WEU would give the EC a relationship with the WEU inside NATO serving as NATO's European pillar and thus eventually give the EC a role in policy-making albeit subject to US decision.

In contrast the French and Germans, the driving force of the European Union, together with some smaller states like Belgium and Luxembourg, favoured a more European solution in which the WEU and ESDI developed not within but outside or at least across NATO in a much closer 'organic' relationship to the EC, culminating in the eventual merger of the EC/EU and WEU. In this formulation the emergent WEU/ESDI would work closely with NATO but would not be subordinate within NATO to US hegemony. This did not mean the imminent disengagement of European states from

NATO but rather that European states would seek to strengthen the European pillar within NATO while building a new security entity outside it. While the debate remained conceptual the practical issues raised by these rather crude formulations, and the multitude of positions between them, could be ignored and tensions and contradictions smoothed over. Indeed it is arguable that the extremes of the debate were only tenable as long as the hard issues of decision and implementation were over the horizon.

The debate between Atlanticists and Europeanists can thus be seen as a debate about the future evolution of security in Europe. Those on the Atlanticist wing were arguing that NATO would continue to be the dominant security organisation in Europe and that a US continental presence would remain essential. Those on the Europeanist wing took a different view and argued that eventually the US could disengage and thus Western Europe [or even a wider Europe] would have to stand on its own feet. Given the long lead times for preparing for such a situation [in terms of equipment, organisation, strategy, and so forth] the building of an ESDI had to begin promptly and had to take place outside or at least in an autonomous relationship to NATO.

What complicates this picture still further and indeed what takes us to the heart of differences between France and Germany is the incompatibility of the two processes. Historically many Europeans, West Germany amongst them, have feared that strengthening European defence too much or challenging US hegemony within NATO risked provoking US disengagement. This concern explains, for example, European coolness towards French-led European security initiatives during the Cold War. In 1990 the Atlanticist/Europeanist dichotomy was informed by the same fear. Atlanticists worried that moving to create an ESDI independent of NATO would only precipitate US withdrawal. Europeanists worried that subordinating ESDI to NATO would both perpetuate US/NATO dominance and stymie the creation of a genuine ESDI against the day when the US did disengage and NATO folded. After 1989, although [West] Germany placed itself publicly in the Europeanist camp in line with the ongoing process of European political and economic union, it was evidently more concerned about the potential for US disengagement than France and subtle tensions between the two crystallised around this anxiety.

In order to understand this in more detail it is necessary to examine French thinking about NATO in the year or so after the fall of the Berlin wall. Contemporary sources are replete with French government affirmations of the importance and centrality of NATO[19] but none are more insightful than Mitterrand's speech at the *Forum de L'Ecole Superiere de Guerre* on 11 April 1991. In the speech Mitterrand noted:

> The defence of Western Europe, for the present and for many years ahead, can only be conceived of in a context of respect for the Atlantic

Alliance. It is not a matter of creating a defence organisation that will substitute for NATO. It is a matter of understanding the limits of the Atlantic Alliance and its military organisation... [which will] continue to play its major role in the maintenance of peace[20].

The precision in this language is fascinating for while Mitterrand affirms the importance of NATO he leaves a great deal of room for manoeuvre around NATO. A 'context of respect for the Atlantic Alliance' is not an acceptance of the dominance of NATO or of the singularity of NATO as the instrument of Western European defence. Arguing that it is 'not a matter of creating a defence organisation which will substitute for NATO', Mitterrand is allowing for an organisation which may arise alongside and may in certain respects compete with NATO. 'Understanding the limits of the Atlantic alliance and its military organisation...' is very much a static view which seems to preclude the possibility of the development of NATO, while the suggestion that NATO will 'continue to play a major role in the maintenance of peace' does not exclude other organisations with equally important roles.

Arguably the most insightful of these comments is that which sees NATO in static terms. Allied to the earlier observation that Europeanists saw the US as eventually disengaging and NATO folding, this comment suggests that Mitterrand was content to have NATO remain as a powerful if functionally limited collective defence organisation [US in, Soviets out, Germans down], but deeply reluctant to see NATO adapt and redefine itself in new ways and to new roles which would make it as dominant in the new security context as it had been in the old. Resisting the evolution of NATO in unwelcome ways [that is ways which enhanced NATO's relevance to the new context] both created space for French-led security initiatives and held out the prospect – barring Soviet revanchism – that NATO would become an anachronism and would eventually wither and die. Philip Gordon has insightfully observed that the French debate in the early 1990s was not a 'debate about the existence of the alliance [but] a debate about NATO's organisation, content and scope'[21]. This should not however be read as French willingness to participate in debates about the evolution of NATO but rather as a French attempt to enhance the European role within NATO while pinning the Alliance to its limited collective defence role.

The events of 1989 and 1990 did not therefore prompt a rapprochement between France and NATO but rather cast France as the reactionary nay-sayer to NATO's efforts to reinvent itself. The skill of French diplomacy was to wrap this opposition in the cloak of prudence and intellectual openness which becomes evident when one considers the French attitude to NATO's early attempts to extend its areas of competence and its geographic scope.

With respect to NATO's areas of competence France consistently argued that NATO should confine itself to military rather than political or other issues and that even in the military sphere NATO should remain an exclusive collective defence organisation[22]. In line with this, France consistently promoted other organisations as the preferred instruments of pan-European security progress. Thus the French pushed the CSCE as the forum for collective security and arms control and the WEU for emergent European collective defence. In the political realm Mitterrand even went so far as to propose on 13 December 1989 a European Confederation – an inclusive pan-European forum which specifically excluded the North Americans – to take forward discourse about what Gorbachov had termed the 'common European home'. The European Confederation idea was intended to bring the Eastern bloc states into a dialogue with the European Community, enhancing the former's prospects of eventual EC/EU membership, although its subtext was about embedding [West] Germany in additional European structures and providing a context of constraint for West German engagement with the Eastern bloc[23]. In the event the Confederation idea quickly began to sink, weighed down by US anger at exclusion, West German anxiety about its challenge to NATO and the CSCE, and Eastern European concern about the potential erosion of their emergent relationship with the United States. It was eventually scuttled by Mitterrand's own hand when he declared in June 1991 that Eastern bloc countries were probably decades away from EC membership, thereby removing what little attraction the European Confederation idea had[24].

With respect to NATO's geographic scope, French opposition centred on two issues: firstly the evolution of NATO's area of military competence to extend links, security guarantees, or even eventual membership to former Warsaw Pact states; and secondly the extension of NATO's geographic theatre to include roles and missions outside the area described by Article 6 of the North Atlantic Treaty[25]. In relation to the former, France opposed the establishment of links between NATO and the new East European democracies and was consequently highly critical of the North Atlantic Co-operation Council [NACC] announced in December 1991 on the grounds that the NACC conflicted with the CSCE, raised Eastern European expectations about a deepening security relationship with NATO, and was inconsistent in excluding 'neutral' European states such as Austria and Sweden.

In relation to the latter, the possibility that NATO forces might project their power outside the NATO theatre, and the French critiques of such an option, were both profoundly shaped by the Persian Gulf War of 1990/1. At the same time the Gulf War exposed differences between the US and Europeans, between Europeans themselves and between France and Germany, which in due course were to complicate still further the Atlanticist/Europeanist debates about the future of European security. As a consequence it is essential that we consider the conflict and its impact on French defence and security policy.

France and the 1990/1 Gulf War

Were it the case that the Gods played dice with human fate, it is doubtful they could have conjured a more revealing test of emergent security relations in Europe, nor indeed of French defence policy, than the 1990/1 Gulf War. Iraq's invasion and occupation of Kuwait on 2 August 1990 sparked the first post-Cold War international crisis thereby becoming a defining and seminal event. By the time the Gulf War ended formally on 28 February 1991, a great deal more was known about the nature of the post-Cold War international security system, not least about the place within it of the United States, NATO, Europeans and the United Nations. Even more importantly for our purposes, the Gulf War unsparingly exposed the limitations and anachronism of the French armed forces and of French defence policy more broadly, provoking in turn a rapid process of reflection and change in France.

From the outset Iraq's invasion of Kuwait placed France in a political and diplomatic dilemma and exposed divisions within the ruling socialist government and in the wider French body politic. As a permanent member of the UN Security Council, with a long established pro-Arab strand to foreign policy, and with important strategic interests in the region – not least the purchase of oil and the sale of arms – France could not stand aside from the conflict. At the same time the Gaullist requirement for independence and the maintenance of a special relationship with Arab countries demanded a policy which in scope and emphasis was distinct from that of the United States. Thus in the weeks after invasion, as the US-led military build-up unfolded, Mitterrand walked an uneasy tightrope between support for the UN demands for Iraqi withdrawal and distance from Washington. At the same time the US military build-up provoked domestic opposition within France to the US action and the possibility of French participation in it. This criticism was evident from within the *Parti Socialiste* motivated by anti-Americanism, pro-Arabism or neutrality, most notably from the defence minister Jean-Pierre Chevènement himself, from the *Parti Communist*, and from others – including Gaullists – opposed to foreign adventurism[26].

Mitterrand's decision to shift decisively behind the United States, while nevertheless continuing until the 11th hour to seek a peaceful settlement, seems to have been determined by a number of factors including the desire to avoid marginalisation and to have a role in a potential post-conflict settlement. Hubert Védrine, Mitterrand's close foreign policy adviser, records in his memoirs French policy on this issue being driven by the relentless US build-up which convinced Mitterrand that the United States would inevitably use force, by the decision of the British on 14 September to send an armoured brigade, and '*surtout*' by the violation, also on 14 September, by Iraqi troops of the diplomatic immunity of the French Amassador's residence in Baghdad. On hearing of the latter, Védrine quotes Mitterrand as angrily declaiming: *'C'est inacceptable! Ça, c'est la guerre! Ils nous cherchent? Ils vont me trouver!'*[27]. The next day France began the deployment of ground troops into the Gulf War theatre.

This deployment, however, continued to be conditioned by domestic political issues. Mitterrand's desire to hold his own party together, particularly those around Chevènement, tempered French activities. From the outset Mitterrand took the decision, reflecting public opinion and political constraints, not to send conscripts into the combat theatre[28]. This was a fateful decision given that conscripts were deployed across the armed forces rather than concentrated in particular units or roles. As one analyst noted: 'not only was the overall percentage of conscripts in the French armed forces high [55 per cent for the army, 26 per cent for the navy, and 37 per cent for the airforce] but several key units including the FAR . . . were made up largely of non-professional soldiers who could not be sent abroad [to fight]'[29].

As a consequence of the domestic political context and the conscript decision, the initial French deployments were limited and tightly circumscribed. Despite total armed forces of more than 450 000 personnel, an army of over 280 000 and the existence of the FAR which was supposed to be specifically earmarked for 'out-of-area' operations, France at first was able only to deploy around 5000 troops to the region, of whom half were Foreign Legionnaires. Moreover French efforts – led by Chevènement – to maintain distance from the United States meant the French troops were deployed some 150 km away from the Iraqi/Saudi and Kuwait/Saudi front lines and away from US and British deployments to distance France from war preparations. Similarly French troops initially dealt only with Saudi commanders rather than the US commanders directing the build-up and developing strategy.

Mitterrand's position finally hardened after the passing on 29 November 1990 of UN Resolution 678 which set the deadline for unconditional Iraqi withdrawal by 15 January 1991 and authorised the use of 'necessary means', taken by everyone to mean force, in the event of non-compliance. It was a countdown to war. The issue for France thus became the nature of French participation in the conflict: whether to be left scowling behind the lines or whether to take a more active role consistent with national perceptions of *rang*. Mitterrand chose the latter and in so doing set in motion several processes which were logically inevitable. The first was an increase in the scale of the French deployments, the second was closer co-operation with the United States, the third was the heightening of domestic political tensions.

After considering options Mitterrand elected on 8 December to boost French force levels in the region to 19 000 personnel[30]. This figure comprised the 6th Light Armoured *'Daguet'*[31] division of 12 000 troops [of whom around 9500 were combat troops and the rest logistic and support], 2400 naval personnel, 1160 airforce staff and a reserve force of 3400 based in Djibouti. In terms of equipment the initial deployment were supplemented until France, by 11 January 1991, had deployed 40 AMX-30 tanks, 650 armoured vehicles [of which 100 were AMX-10RC armoured cars], 18 155 mm artillery pieces, 120 helicopters [70 attack] and Crotale, Mistral and Stinger ground-to-air missile systems. The French airforce eventually deployed more than 60 aircraft, comprising 14 Mirage 2000C, 8 Mirage F-1C and 28 Jaguar A combat aircraft, 5 C-135FR refuellers, 6 Mirage F-1CR reconnaissance and *Sarigue* electronic surveillance aircraft. The French naval contribution eventually comprised the destroyers *Du Chayla*, *Jean de Vienne* and *La Motte-Picquet* and the frigate/escort vessels *Commandant Bory*, *Doudart de Lagree*, *Dupleix*, *Premier-Maitre-L'Her*, and *Protêt*, together with a number of support vessels. Of these only the destroyer *Jean de Vienne* took part directly in offensive operations in support of Desert Storm and Desert Sabre.

Although in many respects a considerable force, the French deployments to the Gulf were nevertheless dwarfed by those of the United States [in excess of 515 000 troops, 2000 tanks, 1300 combat aircraft and six aircraft carrier fleets], considerably weaker than those of the United Kingdom [36 000 personnel of whom 29 000 were ground troops, 160 tanks, 300 armoured vehicles, 76 155 mm artillery pieces, 80 combat aircraft and 23 naval vessels] and numerically inferior to those of Saudi Arabia [67 000 personnel, 550 tanks, 1840 armoured vehicles, 500 artillery pieces and 140 combat aircraft], Egypt [35 600 personnel, 450 tanks and 11 artillery pieces] and Syria [20 800 personnel, 300 tanks and 54 artillery pieces][32].

In relation to the 'Desert Storm' air war, France contributed less than 4 per cent of the total number of combat aircraft, and even more tellingly was able to conduct only around 1 per cent of the allied combat sorties[33]. The reason for the latter was that French Mirage aircraft were virtually indistinguishable from Iraqi Mirage aircraft and thus could not be used, the remaining Jaguar A aircraft had poor avionics and thus could not operate effectively in the dark or in poor weather, and finally the aircraft lacked compatible IFF ['Identification Friend or Foe', an aircraft recognition system] and thus could not work closely with other allied planes.

In relation to the 'Desert Sabre' land war, the role of the *Daguet* division was in the end a minor triumph of necessity and circumstance. For political reasons Mitterrand wanted a clearly defined and distinct role for the French forces, one which would have a specific objective to reduce the risks of being sucked into a messy open-ended commitment, and one in which the French role was clearly visible. For military reasons – the small size of the force, its lack of interoperability with coalition forces and the lightness of its armament – the allied coalition also wanted the French division to have a commensurate and essentially self-contained mission. Reflecting both, the *Daguet* division was given the task, strengthened by a brigade of the US 82nd Airborne division, of executing a far left flank manoeuvre, largely across open desert, to first attack and destroy Iraqi forces at As-Salman and then to seize control of its airport and transport intersection ['*noeud routier*'], the purpose of this manoeuvre being to facilitate the eastward swing of the

US XVIIIth Corps, to guard the allied westernmost flank, and to ensure the logistic lines of support[34].

By all accounts the Daguet mission was a great success. French and 82nd Airborne forces rapidly overwhelmed the Iraqis and within 36 hours of the mission beginning at 03.30 on 24 January the objectives were achieved and part of the French force, the 6th *Régiment Etranger de Génie* together with the 101st US Assault force, had pressed on northward to As-Samawah on the Euphrates to bisect one of the main roads linking Baghdad to Basra and thus to Kuwait[35]. In the event French and American forces managed to interoperate successfully[36], though according to one analyst this owed more to the 'professional qualities of [French forces] and their gift of improvisation than to staff work in France itself'[37].

Although the French made much of the fact that their forces had come closer than any of the allies to Baghdad and although Mitterrand argued in his first speech after the liberation of Kuwait that France had 'upheld its role and its rank' in the Gulf War[38], it is hard to objectively avoid the conclusion that the French role was marginal. The small size and lightness of the armour of *Daguet* meant it could not engage the bulk of Iraqi heavy armour as the US and British had. *Daguet* had a support role which could have been fulfilled by a small proportion [numerically around 2 per cent] of the US forces deployed had the French not been available. Moreover French combat effectively came to an end on 26 January while the main allied forces fought on until 28 January and French forces at no time came nearer than 200 km to the main combat theatre around Basra and Kuwait. This is not to question the quality or courage of French troops [two of whom were killed and 35 injured], but it is to ask that the *Daguet* mission be placed in its proper perspective[39]. In all but an important political sense the French contribution was both geographically and metaphorically peripheral to the outcome of the Gulf War and, if France did indeed uphold its role and rank, both were patently diminished.

A second outcome of Mitterrand's decision to take a larger role in the war was the need to work much more closely with the United States during the war. In order to do so Mitterrand was compelled to break Gaullist taboos in for example allowing US B-52 aircraft to overfly French airspace en route to the Persian Gulf and in allowing the US to base refuelling aircraft on French soil at Istres, the first basing of US aircraft on French soil since the withdrawal from NATO in 1966[40]. Most controversially Mitterrand allowed French forces to be placed under coalition command and control, thereby appearing to erode the Gaullist *verities* of national independence and autonomy, provoking further domestic debate[41].

A closer look at this issue reveals that the notion of French operational subordination is simplistic. Hubert Védrine offered a useful insight in writing: 'the debate [about French subordination] was ... obscured ... by a confusion between autonomy of decision and autonomy in the execution. The first is essential, the second in the context of a coalition is absurd'[42]. In his memoirs of the Gulf War the French CEMA at the time, General Maurice Schmitt, argued with respect to the same point that it was necessary to distinguish between three levels of command and control: strategy, operational control and tactics[43]. Applying these ideas to the French role in the Gulf War clears up the ambiguity. With respect to the French decision to participate in the war, Mitterrand retained full autonomy of decision over French participation in the US-devised strategy. In other words Mitterrand could have simply refused French participation in the war, or on seeing the US strategy and the French role within it, he could again have said no to French participation. Having agreed to participate and to participate specifically in the US-led strategy, Mitterand necessarily ceded operational control to the coalition commanders under General Norman Schwartzkopf. To attempt to micromanage French activities within an overall multinational operational plan could have risked the whole plan and would, as Védrine noted, have been absurd. The operational plan to evict Iraqi forces from Kuwait in turn gave the *Daguet* division a reasonably autonomous role and thus tactical control [that is exactly what the division was to do to fulfil its mission] was ceded in turn to the *Daguet* commander General Michel Roquejeoffre.

It was consequently at the operational control level that issues arose about French subordination but these need to be seen within the wider context of French strategic decision and French tactical autonomy. Even at the operational control level French participation was 'governed' by the Ailleret-Lemnitzer accords. These accords provided guidelines about French participation in multinational operations with NATO member states – including with respect to elements like missions and duration – and specified that while French forces remained under French command they could nevertheless accept and execute allied orders. It is thus clear that even having ceded operational control to the coalition Mitterrand could have pulled the French out at any point or redirected them if necessary, albeit at the risk of the wider operation. French 'subordination' was consequently rather less to do with command and control issues than it was to do with the relative impotence of the French military contribution.

It is worth making two other points. The first is the notion, explicit in the Ailleret-Lemntizer accords, that French subordination to coalition authority during the Gulf War was an unusual event of limited duration which neither established a new norm, nor indeed necessarily had any formal implications for the future of French participation in multinational operations. The second is that, while the bulk of French forces came under coalition control as part of the coalition's XVIII Corps, the brigade of the US 82nd Airborne was under French command, providing at least a sop to some of the critics of Mitterrand's decision.

The third consequence of Mitterrand's decision-making after November 1990 was a heightening of political tension within France as the extent and nature of the French role became evident. With hindsight the period between November and March 1991 offers rich insight into French defence decision-making and in particular the presidential role.

The decision to move closer to the United States, accept coalition command, and prepare for war deepened the strains between Mitterrand and Chevènement and in turn between their respective camps within the *Parti Socialiste*. Unable to accept the general thrust of French policy Chevènement offered his resignation shortly after the 8 December decision to boost the French contribution and again on 7 January as the war became imminent. On both occasions Mitterrand refused to accept it preferring to keep Chevènement within his executive to placate the anti-war faction within the *Parti Socialiste* and to maintain a dialogue with Iraq[44] which might yet have realised a peaceful settlement. At the same time Chevènement's open hostility to the war and French participation in combat became increasingly uncomfortable and Mitterrand moved to marginalise him by using his Elysée aides, notably Admiral Jacques Lanxade, as a direct link to the CEMA Maurice Schmitt to circumvent Chevènement and his Defence Ministry staff and by making several public statements to clarify the government's position and override the pronouncements of Chevènement.[45] After enduring what can only have been a personally difficult period[46] Chevénement's resignation was finally accepted on 29 January.

The French role in the Gulf War also provoked opposition in the wider body politic, particularly from the political parties of the extreme left and right and from public opinion[47]. Anxious to avoid tripping up in the *Assemblée Nationale*, Mitterrand's government scheduled the main debate about the Gulf War for 16 January, the day after the UN Resolution 678 deadline expired and the day before the air strikes began, and put to the *Assemblée* not a decision about a declaration of war [which under Article 35 of the constitution was a parliamentary prerogative] but a motion of support on the basis that French action was part of UN resolution enforcement and was thus a police action and not a war. In the event Mitterrand gathered the public support of the former President Giscard d'Estang and the Gaullist champion Jacques Chirac to forge a solid political consensus, public opinion swung to a two-thirds majority in support of the war, and the vote in the *Assemblée Nationale* was carried by a huge majority of 523 in favour to 43 against [with 2 abstentions][48].

All this said, the Gulf War was still in many respects a watershed for French defence policy in terms of both military and wider political issues. The speed and manner of the allied coalition victory were in themselves to pose entirely new questions for France, not least about how to respond to a newly potent and clearly hegemonic United States. The rapid, decisive and, from an allied point of view at least, largely casualty-free outcome to a conflict which involved in excess of one million combatants appeared to erase the last traces of 'Vietnam Syndrome' from the American military psyche and seemed to validate everything from US power projection capability and air-land battle concepts, through emerging military technologies, to the very utility of military power itself in the post-Cold War context. The lessons for France from all of this were deep and complex.

The core questions asked of France were firstly why had France not been able to project stronger forces, commensurate with its self-image, into the region. This question had added frisson given the size of the French defence budget and armed forces, the existence of the FAR, French global interests, and France's historic propensity to project force overseas in support of pro-French regimes in *Francophonie*. A second question was why, once deployed, were French forces so dependent on the United States across virtually every mission area. A third and consequent question was what should France do about these problems.

The answer to the first question was multi-faceted but each facet was clearly linked to the rigidity of Gaullism. France had been limited in its capacity to deploy forces into the Gulf theatre for three reasons: the lack of suitable conventional forces, the lack of power projection capability, and the lack of professional soldiers.

The lack of suitable conventional forces was itself explained by a number of factors. One was the nuclear emphasis in French defence policy which at least until the mid-1970s left French conventional forces neglected and, even after Giscard's boost to conventional defence spending, subject to a further lack of priority under Mitterrand as a consequence of his nuclear modernisation. As a result of this neglect, much French equipment was out of date [for example the aircraft carrier *Clemenceau* carried Crusader aircraft, designed in the 1950s, which were so outmoded they could not be deployed in the Gulf] or poorly equipped [for example the Jaguar avionics discussed above].

Another factor was the composition of conventional forces. The emphasis on the defence of *La France Métropolitaine*, reticence about a formal commitment to NATO or a forward defence of Western Europe, and the limited scale of out-of-area operations had each informed the blend of conventional forces rendering them largely structured for static national defence with only limited forces suitable for projection, despite the FAR and despite the restructuring under *Armées 2000*. An additional point was the emphasis on independence in the defence industrial sector which acted against the French purchase of more advanced, specialised or interoperable military hardware from elsewhere, thereby confining France to the limits of her own technologies and industrial base[49].

The second problem for France was the lack of power projection capability. French forces were configured primarily to project relatively limited and lightly armed forces into the *Francophonie* region. None of the interventions in Africa for example between 1962 and 1995 involved more than 3500 troops and often many of these forces were already garrisoned in close proximity[50].

France lacked the capability to project numerically large forces or heavy equipment [such as tanks] into distant theatres. The deployments France eventually undertook in the Gulf took months to conduct and were dependent *inter alia* on the requisitioning of civilian aircraft and ships and the leasing of transport aircraft from the United States. In the words of one commentator, these difficulties 'underscored the gap between France's global rhetoric and the country's lift, logistical and force projection capabilities'[51].

It was the lack of professional soldiers however which was finally decisive in limiting the French contribution to the Gulf War. Had Mitterrand been able to send conscripts the foregoing problems would not have been as important. The British after all had similar problems with ageing equipment and lack of power projection capability but with a professional army their contribution had been far more substantial[52].

The second military issue raised by the French Gulf War deployment was the extent to which France had been dependent on the United States within the theatre. Undoubtedly the greatest dependence was that on US satellite and other intelligence-gathering systems. Despite French dominance of European space programmes and France's national Arianne launch system, France [and indeed the other Western European states] was severely deficient in intelligence-gathering capability, and acutely so in the Gulf region. With the single exception of the civilian SPOT satellite [*Satellite Pour l'Observation de la Terre*] which had some limited utility in the Gulf War[53], France was almost wholly dependent on the United States for theatre intelligence[54]. The French defence minister Pierre Joxe summed up the situation in a speech before the *Assemblée Nationale* on 6 June 1991:

> the weakness of our [intelligence-gathering] means prevented us from having access to information in an autonomous and comprehensive fashion. Without intelligence from the allies, that is the United States, *we were almost blind* [emphasis added]. To leave our systems in their present state of insufficiency and dependence would be to weaken considerably our present and future defence effort[55].

France had little choice but to rely on US intelligence and thus to be guided by the collection, analysis and presentation of data the US wished France [and the other coalition allies] to see. This gave the United States in another sense control of almost everything that happened on the allied side in the war.

French dependence was not however limited to intelligence. The 'semi-detachment' of France from NATO meant that French forces were not able to interoperate or integrate smoothly with the United States in the way in which, for example, the British were. To overcome the worst problems, the French were given some compatible IFF systems to assist the recognition of coalition aircraft, communication systems to facilitate multinational communications and NAVSTAR navigation equipment to facilitate manoeuvre[56]. In addition the autonomous role given to the *Daguet* division was itself designed to reduce the French requirement for interoperability.

The answer to the question of what France should and could do about the military problems exposed by the Gulf War was the subject of intense debate inside France, a debate necessarily situated within the wider debate about post-Cold War defence policy. This debate seems to have been premised on the assumption that the Gulf War was more than an aberration and that it was likely to be but the first of many interventions by Western states as the axis of international security swung away from East-West issues[57]. As such the central lessons of the Gulf War were important in the general evolution of post-Cold War defence policy. Arguably the Gulf War contributed three main lessons to the wider discourse, reflecting the three main problems in the French deployment. Conventional force weakness required a rethink of the conventional-nuclear relationship, the modernisation and blend of conventional forces and of power-projection capabilities[58]. Lack of professional soldiers required a rethink as a minimum about the mix of professionals and conscripts and, in the minds of many, a deeper analysis of the utility of conscription itself[59]. And the intelligence weakness required France to focus attention on both national and European intelligence needs[60].

The second set of lessons to be drawn from the Gulf War were political. To begin with, the Franco-German relationship at the heart of European security was subject to novel pressures. In contrast to France, the newly unified Germany played only a very minor military role in the Gulf War, deploying an Alphajet squadron of 18 aircraft and 11 Hawk and Roland batteries to Turkey [and naval minesweepers to the Persian Gulf] within NATO's ACEMF principally in case of an Iraqi threat to the major NATO base at Incirlik. The main German contribution was through financial contributions to the war which totalled more than $9 billion, together with some logistical support[61]. The reasons why the German military role was minimal – including preoccupation with the unification process taking place as the Gulf crisis unfolded, and political resistance to an interpretation of the constitution [Basic Law] which could have allowed Germany to contribute forces directly[62] – are here less important than the implications of the differing responses of France and Germany for the Franco-German security relationship. One issue was whether the impotence of Germany (by choice or circumstance) cast doubt on the value of the Franco-German security relationship and by extension on the emerging ESDI. Atlanticists within the Alliance were certainly quick to draw the inference that German inaction revealed the weakness of the core relationship at the heart of the ESDI concept[63]. This general critique was given added weight by the fact that, aside from the UK and France, no other Western European state contributed ground forces to the war and, although Belgium, Denmark, Germany, Greece, Italy, the Netherlands,

Norway and Spain provided naval vessels in the theatre, none took part in offensive operations during the war itself. Western Europe's single combat contribution to the war, aside from that of the British and French, was the involvement of eight Italian aircraft in the air war[64]. It was consequently hard to argue with Jacques Delors' own assessment that with respect to the war 'Europe was rather ineffectual'[65].

A war which exposed the weakness of Western European collective defence did not however signal the immediate demise of ESDI, rather it strengthened the reasons and resolve of those promoting it. For Jacques Delors, speaking in March 1991, 'the limitations of the European Community in the Gulf... [were]... another argument for moving towards a form of political union embracing a common foreign and security policy'[66] and in certain respects the war provided a blueprint for the kind of military capability Europe needed to become a credible military player.

A second question asked of the Franco-German security relationship by the Gulf War was whether the war exposed fundamental differences between France and Germany which, despite the intentions of politicians, might in the end drive the security relationship apart. On the one hand France was a UN Security Council member with global interests and a foreign policy orientation which included the developing world and the Mediterranean region in particular. By contrast Germany had few international interests or obligations, a foreign policy centred on West and Central Europe and historical, legal and political obstacles to greater international engagement. Despite Franco-German policy overlap, the 'pulls' of these interests were arguably in different directions. While the Cold War dominated the security agenda of the two states, these underlying differences were less important but when the Cold War ended they became increasingly important and in the Gulf War they became highly visible[67].

One question after the war thus became the extent to which the two states could continue to harmonise security interests. After 1989, as we have seen, France fell increasingly in step with German policy towards the former Warsaw Pact states but, as the Gulf War made clear, it was rather less certain that Germany could or would move towards French conceptions of extra-European security interests. The suggestion that Germany exercised rather more 'pull' on France than the reverse raises the subsidiary question of whether French interests will eventually be subsumed to those of Germany or, perhaps more realistically, whether French interests will become increasingly centred around those areas on which there can be Franco-German convergence and agreement, resulting perhaps in the erosion of France's wider interests.

The point is complicated by a second lesson from the Gulf War, namely the importance of multilateralism. The United States could in principle have responded unilaterally to the invasion of Kuwait but instead it built a multinational coalition within the purview of the United Nations and UN Security Council and situated the coalition response within a framework of UN resolutions and international law. In so doing, the US enhanced the role and status of the UN and of international law and enhanced the importance of international co-operation in the regulation of international conflict. In seeking military, political and financial support for the war [with respect to the latter the US received more than $50 billion in contributions primarily from Kuwait, Saudi Arabia, Germany and Japan[68]] the US signalled the importance of multilateralism, leaving smaller states like France to conclude that international status was going to flow increasingly from the capacity to contribute to multilateral initiatives on behalf of the wider, if imprecise, international community.

After the Gulf War it became evident that to uphold its *rang* in a context of disparate and geographically varied threats to international or regional stability, France could find itself involved in conflicts in many different parts of the world and in different *ad hoc* coalitions[69]. This appears to place France in a particularly tricky position. Within Europe its interests risk being curbed by the need to build European coalitions, particularly with Germany, in the pursuit of ESDI while outside Europe the demands of rank require participation in multilateral world order roles.

The Gulf War had important implications also for France's relations with NATO and the United States. In underlining US hegemony and enhancing the role of NATO – not least because the alternative organisations were either impotent or irrelevant to the conflict – the Gulf War and NATO's prompt adaptations afterwards forced a French rethink of its initial responses to NATO after 1989. The emergence of a new 'out-of-area' role for NATO, which became formal on 28/9 May 1991 when the Defence Planning Committee [DPC] announced the new force structure and the creation of the NATO Rapid Reaction Corps [RRC] under permanent British command, effectively dashed any French hopes that NATO could be confined to its Cold War functions[70]. Moreover the RRC with Belgian, British, Dutch and German troops was interpreted in France as a direct threat to French ambitions for the creation of a European Army, while NATO's 'out-of-area' role seemed likely to complicate Franco-NATO relations still further by linking relations between the two with France's interests in the 'South'[71], a complication which did not exist as long as NATO confined itself to the Article 6 theatre. It was not irrelevant to what followed either that the tensions in French intra- and extra-European security were reflected in NATO's adaptation to the end of the Cold War and its need to shape up to non-Article 5 threats while retaining a core collective defence role.

The implications were not however wholly negative for France. French co-operation with the US during the Gulf War had to a point reassured Western European states that European defence co-operation need not necessarily be anti-American. Similarly converging views about the nature of post-Cold War security in France and the United States, together with a proven French willingness and capability to offer military support to the US in the right circumstances, opened the door to closer relations between the two and, by extension, between France and NATO provided the right balance of interests could be struck, particularly in the European theatre. Furthermore, 'out of area' roles for NATO held out the prospect that military integration, the *bête noir* of France's relations with NATO, would be abandoned because, as Frédéric Bozo observed, in the new context what mattered was 'not the integration of forces but [the flexibility] for each country...to act together – or not – when the time comes'[72]. If this assessment was correct, NATO's adaptation could open the door to much closer Franco-NATO relations.

The period between the end of the Gulf War and the agreement of the Maastricht Treaty in December 1991 was in many respects seminal both in terms of the construction of post-Cold War security architecture in Europe and in terms of France's place and role within it. At the North Atlantic Council [NAC] meeting in Copenhagen on 6/7 June 1991, these elements became more perceptible. The Copenhagen Declaration which followed the meeting reaffirmed NATO's role as consistent with the Treaty of Washington and the 1990 London Declaration but in part IV noted: 'The affirmation of a European security and defence identity will show that Europeans are ready to assume a greater responsibility for their security [which] will help to reinforce transatlantic solidarity'[73], a statement which for the first time openly endorsed NATO acceptance of increased European influence in security matters. The statement did nothing, however, to resolve the question at the heart of European security, namely what was to be the nature of relations between NATO, the WEU and the EC/EU.

In response to British and Italian initiatives over the summer to try to locate the WEU within the Alliance and to delay progress towards a common defence for the EC/EU, the French and Germans responded on 14 October 1991 by restating the thrust of their December 1990 letter to the Rome IGC that European political union would include a common defence and that the 'WEU [would be] an integral part of the process of European union'[74] and by announcing the creation of a European Corps, the Eurocorps, and inviting other WEU partners to offer troops to it.

Although presented by Germany as a logical outworking of the process of deepening bilateral military ties between France and Germany and – because of Germany's place in NATO and the RRC – as bringing France more closely into a relationship with the Alliance, in France the Eurocorps was seen as the progenitor of a European army and as 'a symbol...of common defence in the making, of European military integration *outside* [emphasis added] the structure of NATO'[75]. The Eurocorps initiative was important to France in other respects as well: it provided the pretext, announced on 31 October 1991, to reverse the June 1990 decision to withdraw all French forces from German soil in the wake of the '2 + 4' agreements; and by the retention of one of three armoured divisions in Germany [under the *Armées 2000* restructuring] gave substantive form to France's commitment to Germany and gave France a symbolically important European role for its forces. In a reciprocal move France invited German officers to be indefinitely based at the Eurocorps headquarters in Strasbourg, a hugely resonant step[76]. The Eurocorps was also the first time since 1958 that France had accepted the integration of any of its military forces with those of another state, a further example of the French recognition that it had to cede a degree of national autonomy for the prize of German engagement and constraint.

In the aftermath of the Eurocorps announcement the 7–8 November NATO meeting in Rome was suffused with difficulties. US officials wanted to know if the Eurocorps meant Europe was intent on creating competitive structures to NATO, but appear to have been reassured [not least by Germany] that it was not. NATO moved ahead with the North Atlantic Co-operation Council [NACC] which although critiqued by France was welcomed by the Germans as complementary to the engagement of Eastern Europe by the EC as part of the emerging CFSP. For all their rancour it was not lost on the French that the NACC was one more framework within which German engagement with Eastern Europe would be embedded and thus the seeds of French acceptance of the NACC were present from the outset. NATO also announced its New Strategic Concept which, while confirming French concerns about NATO's continued adaptation to the post-Cold War context, at least had the virtue of accepting a role for European forces albeit without explaining what that role or its relationship with NATO should be.

The Maastricht meetings of 9–10 December 1991 to amend the Single European Act and agree a plan for European political and economic union thus arrived with the core issues about the EC/EU, WEU and NATO unresolved. The wording of the treaty with respect to defence and security matters consequently used ambiguous and inclusive language to smooth over the difficulties and present a united position to the outside world. Thus in Article B, Title 1 the Treaty declares one of its objectives to be 'the implementation of a common foreign and security policy including the eventual framing of a common defence policy, which *might* [emphasis added] in time lead to a common defence'. With respect to the matter of relations between the EU and the WEU, the treaty in Article J 4.2 notes that the EU can 'request the WEU which is an *integral part of the development of the Union* [emphasis added] . . . to elaborate and implement decisions on actions of the Union which have defence implications'. In a separate declaration attached to the Treaty on 10 December 1991 by the WEU [made necessary by the differing memberships of EU and WEU], the WEU stated its intention to 'develop WEU as the defence component of the EU and as a means to strengthen the European pillar of the Alliance'[77]. It sought also to facilitate the evolution of relations between the EU, WEU and NATO by offering WEU membership to non-WEU EU members [Denmark, Greece and Iceland] and by offering associate membership of WEU, allowing 'full participation', to European non-EU members of NATO [Iceland, Norway, Turkey].

The implications of the Maastricht Treaty and appended declarations for French defence and security policy were mixed. On the one hand Maastricht agreed a logic of deepening European defence co-operation which 'might' in time lead to a common defence, and it established that the WEU was an integral part of the European Union and that it would develop as the defence component of the EU. But on the other, it explicitly stated that the WEU was to be developed as a means to strengthen the European pillar of the Atlantic Alliance, a formula of words which situated the WEU much closer to the Atlanticists' conception of a 'bridge' between NATO and the EU than the French concept of 'fusing' with the EU outside the NATO purview. The latter point was underlined by two other details. Firstly Maastricht had agreed that the evolution of CFSP would take place within a distinct and separate intergovernmental framework rather than being fully integrated within the structures and processes of the EU itself. Secondly in describing the relationship between the EU and WEU, Maastricht had explicitly noted that the EU was to '*request* [emphasis added] the WEU ... to elaborate decisions on actions of the Union which have defence implications', rather than the French preferred phrase that the EU would '*instruct* the WEU ... etc'. This was much more than semantics because the latter implied a subordinate role for the WEU as the policy instrument of the EU and was thus much closer to the 'fusing' of the two. The use of the word to 'request' quite clearly excluded that interpretation.

In the period between the agreement of the Maastricht Treaty, its appended declarations on 9–10 December 1991 and its formal signing on 7 February 1992, the Soviet Union collapsed in late December following an abortive coup against Gorbachov in August diminishing still further the potential threat from the East. A year or so earlier this event might have deprived NATO of a vital prop of support [not least in the US Congress] but by the end of 1991 the agenda had moved on, NATO had a new strategic concept, a new force structure and new roles in force projection and security stabilisation within the NACC framework. The collapse of the Soviet Union was consequently little more than a blip in terms of the discussions about European security architecture even if it was to further complicate relations between the West and the former Warsaw Pact states.

By early 1992 the conceptual and semantic debates about Western European security were increasingly overshadowed by the worsening situation in Yugoslavia which in June 1991 had crossed from crisis into war. In much the same way as the Gulf War had subjected the internally focused and often petty Western debates about security to a blast of heat from the real world, so the bloody break-up of Yugoslavia was to impact on these debates from the middle of 1991. As with the Gulf War, the wars in Yugoslavia both exposed and influenced relations between France and the US, NATO and the other Western European states and impacted in turn on defence and security debates in France.

France and the wars in Yugoslavia

Yugoslavia was in certain respects an artificial state-nation created from the outside in the aftermath of the First World War around the entity of Serbia. In the formation of Yugoslavia five disparate kingdoms were brought together – Bosnia, Croatia [which included much of Dalmatia], Montenegro, Serbia and Slovenia [the area described by modern Macedonia was assimilated by Serbia in 1913] notwithstanding ethnic and cultural distinction, four main languages [Serbo-Croat, Albanian, Macedonian and Slovenian] and three main religious groupings [Greek Orthodox, Catholic and Muslim]. During the Second World War Yugoslavia was made a Nazi state dominated by Croatians who in turn fought a brutal civil war with the communist resistance comprising mainly Serbs. For much of the Cold War Yugoslavia was held together by the communist domination of political power, until 1980 under Tito, and by external pressure from both the West and from the Soviet bloc from whom Tito split. Following Tito's death in 1980 the cohesion of the Federal Republic of Yugoslavia began to weaken. After 1989 the demise of communism and the spread of new political freedoms in Eastern Europe precipitated the break-up of Yugoslavia, catalysed in the eyes of most by the nationalist policies of the Serbian president Slobodan Milosevic. With most of the republics in Yugoslavia agreed on separate destinies, the principal focus of conflict was not secession itself but rather the modalities of secession, and in particular the question of whether the right of self-determination was to be exercised by the respective republics or by the peoples populating ethnic territories which cut across existing borders[78].

Fuelled by a history of past atrocities and injustices and shaped by ethnic, cultural, religious and language differences, the wars in Yugoslavia escalated in ferocity. The first war of Slovenian secession in the summer of 1991 was short and relatively constrained. The second war of Croatian secession from the summer to the autumn of 1991 was also short but much more brutal given the presence of a large Serbian population in Croatia. The third war for Bosnian secession from 1992 through to 1995 was for the most part a confused and savage conflict between Croats, Serbs and Muslims. As the EU, NATO and WEU spoke of a new era of peace and stability in Europe, Yugoslavia descended into a nightmare of murderous 'ethnic cleansing' unparalleled in Europe since the Nazi holocaust[79].

The shooting in Yugoslavia began little more than three months after the guns fell silent in the Gulf War. Few of the lessons of the Gulf War had by then been assimilated and fewer still of the organisational and procedural changes in France, NATO and elsewhere had been implemented. National rather than collective policy-making continued to dominate the

Western responses and once again France and Germany quickly found themselves in distinctly different positions. After maintaining a reasonably coherent position within the European Community around the idea of a 'single and democratic' Yugoslavia resolving its problems by peaceful and democratic means, the European consensus quickly fell apart following the outbreak of fighting. From September 1991 Germany began to argue publicly for the prompt recognition of the breakaway republics of Slovenia and Croatia. In early December, frustrated by the lack of EC progress on the issue and by Serbian and JNA action which included the destruction of Vukovar and the shelling of Dubrovnik, Germany announced that it would recognise both breakaway republics before the end of the year. For several weeks the EC sought to reach a compromise, eventually proposing to offer recognition from 15 January 1992 to any republics meeting specific criteria. This idea was promptly rejected by Germany which went ahead with recognition on 23 December 1991, leaving France and the rest of the EC with little alternative but to follow suit in line with their own schedule.

The German decision was widely seen as a watershed. It raised the prospect of a newly assertive Germany flexing its foreign policy muscles within Europe, it suggested that national rather than collective instincts were reasserting themselves as a driving factor in German foreign policy [particularly after Kohl's single-mindedness with respect to German reunification], and it was widely attacked as precipitating the subsequent descent of the rest of Yugoslavia into chaos as both Bosnia-Hercegovina and Macedonia in turn sought recognition fearing a Serbian dominated future in a residual federation[80]. The reasons why Germany pushed so strongly for recognition have been widely analysed and a near-consensus view appears to be that a number of predisposing structural issues combined with a particularly strong set of domestic pressures forced Kohl's hand. Amongst the former was the idea that a German people only just reunited after 50 years as distinct states could hardly argue against the rights of self-determination for nations or in favour of the inviolability of existing borders. Similarly, historical and religious links to Croatia and Slovenia predisposed Germany towards a solution in line with their wishes, a view reinforced by the EC's general antipathy towards the Serbs. Likewise a Germany already absorbing large numbers of refugees from the former Warsaw Pact states feared that more would flow from sustained conflict in Yugoslavia and evidently hoped that recognition would lead to stabilisation[81].

German domestic political pressures seem however to have been decisive in pushing Kohl to prompt recognition. Hubert Védrine records a particularly insightful exchange between Mitterrand and Kohl at the Franco-German summit in Bonn on 15 November 1991:

Kohl: I am going . . . to be obliged to recognise Croatia [and Slovenia]

Mitterrand: Don't do it, it will be a mistake
Kohl: *without doubt* [emphasis added], but the pressure at home is very strong...my party, my political allies, the Church, the press, not counting 500 000 Croats who live in Germany, everyone [is] push[ing][82].

France approached Yugoslavia, and hence the question of recognition, from a very different perspective and French leaders were subject to very different pressures. To begin with, France remained a highly state-centric society which largely accepted the Gaullist idea of the state as the ultimate focus of human loyalty and the most appropriate locus of power for good governance. It consequently preferred solutions centred on Belgrade rather than in the respective republics of Yugoslavia which seemed potentially subject to cascading divisions. This preference was reinforced by French antipathy towards the self-determination of peoples both because France had problems of its own with nationalist groups within France [not least the FLNC in Corsica] and because the integrity of *Francophonie* depended on the peoples of the DOM-TOMs remaining loyal to the French state.

A second issue which logically flowed from the first was French anxiety that the assertion of the principle of self-determination in Europe through the break-up of Yugoslavia could precipitate or accelerate similar processes elsewhere in Europe, particularly but not exclusively within the former Warsaw Pact, threatening European stability and, until December 1991, the Gorbachov reforms. Maintaining the cohesion of Yugoslavia was consequently seen as placing a brake on similar processes elsewhere.

In France historical ties to Yugoslavia played differently to those of Germany. France had historically good relations with Serbia going back to the Napoleonic era and continued through two world wars, a point stressed publicly by Mitterrand when the Yugoslavian Prime Minister Ante Markovic visited Paris in May 1991[83]. Finally French thinking was also influenced by a concern that the recognition of Croatia and Slovenia closely linked to Germany would strengthen German influence in *Mitteleuropa* and pull Germany even more strongly towards a Eastward orientation.

In terms of domestic politics in France there was little pressure on the government in any particular direction other than support for resolving the conflict through peaceful and initially collective means. Muted French Catholicism, the absence of substantial immigrant populations from any of the protagonists, insulation from East European refugee flows, and the low-key political debate about the crisis within France meant that the French government experienced little of the pressure or urgency impacting on German policy-makers in the second half of 1991.

From such different perspectives Franco-German policy towards Yugoslavia was unavoidably subject to tension[84]. Arguably what smarted most with the Germans was the willingness of France, its supposedly closest ally, to publicly scapegoat it for consequent events in Yugoslavia while France itself eschewed any responsibility for Western policy failures. Given that there were at least two important respects in which French policy played a role in the decisive events in the break-up of Yugoslavia, France's attacks on German policy seemed once again to be motivated by latent anti-German sentiment. Firstly by insisting initially that Yugoslavia stay together at virtually any cost, France [and even more so the United Kingdom] effectively closed the door on more creative processes which might have managed the transition from the unitary state more peacefully. Secondly by trading on Franco-Serbian friendship to seek to position itself as an intermediary between Serbia and the wider international community in the hope of exerting influence in the region and perhaps even brokering a peace [very much paralleling French manoeuvres in the run-up to the Gulf War when France sought to trade on its special relationship with Iraq], France strengthened the hand of Serbian nationalists and undoubtedly weakened the cohesion of collective European, transatlantic, and United Nations responses which could have constrained Serbian aggression.

In any event the issue of external responsibility for events in Yugoslavia moved quickly to the periphery as both France and Germany faced up to the new reality on the ground, centred on the vicious war in Bosnia which followed Bosnian independence in March 1992. For France several new elements were now impacting on policy. Firstly the *de facto* break-up of Yugoslavia required a policy adjustment to the new situation. Secondly the agreement at Maastricht and its French-led elements on CFSP and European defence required a more cohesive Franco-German response if ESDI was not to be stillborn. Thirdly the pro-Serbian position of France began to erode in the face of growing evidence about Serbian aggression and growing public and political disquiet in France about Mitterrand's line. Finally France's initial approach of seeking a solution through European fora had clearly failed[85]. The European Community, despite Jacques Poos' announcement of the 'hour of Europe' and his confident assertion that 'if there is any problem which Europe can solve it is Yugoslavia'[86], had proven impotent in the face of continued fighting. The CSCE, the pan-European common security fora, proved to be similarly impotent, paralysed by the divergent interests of members and the lack of suitable instruments to respond to a series of hot wars. NATO, which France sought to exclude anyway, was initially content to maintain a low profile in part constrained by its treaty's tightly defined theatre of operations and *casus foederis* and by an American reluctance to get drawn into an open-ended civil war [the ghosts of Vietnam had perhaps not after all been fully exorcised], and in part by a gleefully accepted American opportunity to sit back and watch the architects of 'European defence' founder in the face of their first real collective test. Finally France's efforts to promote the WEU, plausibly because it did not face the same constraints as NATO, were hampered by the WEU's limitations as a functional military force [it had for example no command structure and no dedicated assets], by wrangling about its role, and by unresolved questions about who exactly could or would provide troops. The whole idea of a WEU 'interposition' force was finally torpedoed on 19 September 1991 by the British, whose participation in any force would have been essential[87].

In contrast to the foregoing, France found the United Nations, in which it was coincidentally president of the Security Council in the latter half of 1991, to be willing to play a role in Yugoslavia and French attention thus shifted from a European to a UN-centred approach. Within the UN France had a role which was more than nominally the equal of the great powers, German influence was largely excluded, and within the UN Security Council France could continue the old game of manoeuvre between the influences of the United States and the Soviet Union/Russia. After the collapse of the Soviet Union the Soviet veto on UN involvement in Yugoslavia ended and UN intervention on the ground became practical because of the Croatian cease-fire brokered by the US Special envoy Cyrus Vance in late December and early January. France took a lead role in agreeing UN resolution 743 on 21 February 1992 which agreed to the creation and deployment to Croatia of up to 13 000 United Nation Protection Forces [UNPROFOR, known in France as FORPRONU, *La Force de Protection des Nations Unies*] and France provided the largest single contribution of forces to UNPROFOR I deploying 2752 personnel out of an initial UN total of just over 10 000[88].

The outbreak of fighting in Bosnia in April 1992 and in particular the bombardment of Sarajevo by Bosnian Serb forces seems to have had a profound impact on French society and Mitterrand's position of 'balance' towards the Serbs to preserve a 'special relationship' came under growing domestic pressure. On 8 June and 26 June the French played a decisive role in the UN passing respectively resolutions 758 and 761 which agreed an additional UN force of up to 1100 UN peacekeeping troops [*casques bleues*] to open and protect Sarajevo airport to allow the flow of humanitarian aid to those caught up in the war. After relieving an initial Canadian deployment, France made much of the fact that it again was contributing the largest proportion of armed forces to the tri-national deployment [French, Egyptian, Ukranian] while the British, Germans and Americans continued to refuse to participate[89].

The UN however was unwilling to inject the force into Sarajevo without the agreement of the warring parties and in frustration Mitterrand made a surprise personal visit to Sarajevo on 28 June 1992. This visit was dismissed from some political quarters in France as a stunt ['*très showbiz*'[90]] designed to boost Mitterrand's flagging personal rating and that of his party, and critiqued by some European allies as both an empty '*beau geste*' motivated by delusions of *grandeur*[91] and for undermining precisely the European solidarity Mitterrand himself had called for at the Lisbon conference a few days earlier, but it was both a more complex and more important '*geste*' than these critiques allowed.

As Serbia's only friend in Western Europe, Mitterrand's visit to Sarajevo was as much an ultimatum as a humanitarian appeal, threatening by his presence the erosion if not the loss of French support if the siege continued. It is a measure of the importance of this link to the Serbs that the siege was promptly lifted, allowing both the dramatic arrival of French aid transport aircraft to crown the moment and the subsequent arrival of UN forces. Behind Mitterrand's action lay a complex set of factors. In choosing the 28 June for the visit, the anniversary date of the assassination also in Sarajevo of Archduke Franz Ferdinand which sparked the First World War, Mitterrand was reminding everyone of the risks if the Balkan conflicts went out of control. At the same time Mitterrand was seeking in a single action to respond to a number of competing pressures. Firstly a breakthrough would demonstrate French leadership and placate those at home critical of Mitterrand's line. Secondly, a French-led success would be evidence that Europe could play a positive role in the resolution of conflict in its own backyard, a point Mitterrand hoped would strengthen French support for the Maastricht treaties in the crucial run-up to the French referendum on the issue on 20 September. Thirdly a French-led 'European' and UN success could also condition US involvement in the conflict and stay the use of force against Serbia being increasingly advocated by the United States itself coming under domestic pressure to act in the wake of the siege of Sarajevo and mounting evidence of atrocities in Bosnia mainly, but not exclusively, perpetrated by Serbs[92].

The lifting of the siege of Sarajevo at the end of June proved not to be the turning point Mitterrand had hoped for and the limits of French influence with Serbia, if not yet of French tolerance of Serbia, were quickly exposed. Throughout the second half of 1992 UN passivity and in particular the logjam between American and Franco-British diplomacy – with the former favouring the 'lift and strike' policy of lifting the arms embargo against the Muslims in Bosnia and using airpower against Serbs to correct the balance of forces, and the latter opposing this strongly on the basis that lifting the embargo would fuel the war and air strikes would provoke Serbian retaliation against UN forces on the ground – failed to provide any constraint on Serbian aggression and human rights violations. The French responded to critics by rebuking those like the United States and Germany who called for the use of force yet were unwilling to put troops on the ground. The French were nevertheless complicit in failing to find a working solution and in consistently blocking or watering down EC and UN measures against the Serbs which allowed the latter to pursue their ambitions virtually unimpeded[93]. The price for UN and great power vacillation, in which France played no small part, was paid at the 'concentration and extermination' camps at Omarska, Prijedor, and elsewhere[94].

From July, principally because of domestic and international reaction to Serb atrocities, French policy began to shift towards the acceptance of some action against Serbians, albeit limited and conditional. Following the London peace conference on 26–8 August 1992, France supported UN resolution 776 adopted on 14 September, which accepted NATO's offer of forces to defend aid convoys. Because it marked the formal entry of NATO into the conflict on the ground, French acceptance of 776 in effect signalled the demise of French hopes for a lead role for the WEU and the continued marginalisation of NATO. In response to UN resolution 776, France sent 1350 troops as part of the NATO forces in UNPROFOR II deployments which began in October 1992[95], a deployment which saw British combat troops arrive for the first time [the British contributed only a medical unit to UNPROFOR I] and the contribution of a US field hospital unit[96].

France then supported UN resolution 781 to monitor flights over Bosnia but resisted the US push for the imposition of a no-fly zone to curb the use of Serbian air power. By the end of 1992 the French position had hardened against Serbia and seemed to be converging with that of the United States. Opinion polls showed that in late 1992 more than two-thirds of French people were in favour of military intervention[97], a point of some significance just a few months before major legislative elections. By late 1992 the French political class and intellectuals were also virtually united in advocating a stronger French line[98]. In addition Mitterrand faced two novelties. Serbian successes in Bosnia supported by the Serbianised Yugoslav National Army [JNA] raised the prospect of further aggression in Kosovo [where there had been some sporadic violence in 1992] and Macedonia. Unless checked, Serbian aggression risked expanding the war and potentially embroiling Albania and Bulgaria and perhaps even Greece and Turkey. Moreover Serbian successes in Bosnia had been achieved principally at the expense of the Muslims who had suffered the brunt of atrocities and who by the spring of 1993 were pinned back to enclaves within Bosnia surrounded by Serbs. The French 'special relationship' with Serbia thus came into conflict with the French 'special relationship' with the Islamic world, forcing Mitterrand into a new calculation.

French reluctance to enforce a no-fly zone came to an end when Bosnian Serbs rejected the Vance–Owen plan in March 1993 [and by referendum in May] and in response to continued Serbian operations in eastern Bosnia and around Sarajevo. At the same time the incoming administration of Bill Clinton raised French expectations that the US would lead new initiatives to resolve or at least stabilise the conflict. On 31 March France voted to accept UN resolution 816 agreeing the enforcement of the no-fly zone, a decision which brought NATO into the front-line as the only organisation capable of enforcing the no-fly zone. France offered aircraft to the NATO's Operation Deny Flight to avoid marginalisation and succeeded in retaining UN as opposed to NATO political control of operations.

The no-fly zone did little to hamper Serbian operations in Bosnia and by May the position of Muslim enclaves was critical. France was one of the key states in the UN which pushed for the designation of Muslim enclaves as 'safe areas' and this was accepted in UN resolution 824 of 6 May and resolution 836 of 4 June which created six 'security zones' in Bosnia. The French promptly sent additional troops to protect the 'safe areas' and by late June had more than 5300 personnel on the ground in former Yugoslavia [2239 in UNPROFOR I and 3096 in UNPROFOR II], more than one quarter of total UN forces committed, a figure which was to rise still further to more than 6700 little more than a year later[99].

The Clinton administration proved no more willing that the Bush administration to put US troops on the ground and over the next 12 months wrangling between the US, UK, France and Germany, and the continued passivity of the United Nations, brought the end of conflict no nearer[100]. Not even international revulsion at the deaths of 70 people in a mortar attack on a Sarajevo market on 5 February 1994[101] was enough to prompt the West to decisive action. Subsequent ultimata to the Serbs around Sarajevo and the enclaves in Bosnia proved largely ineffective. In this context the presence of large numbers of French troops on the ground generated both powerful domestic criticisms of French policy in former Yugoslavia and powerful pressure for the withdrawal of French troops. By February 1994 France had spent a year stuck between *Scylla* and *Charybdis*, unwilling to pull troops out and leave the Muslims and people of Sarajevo to their fate, but increasingly also unable to justify the cost in French lives [18 killed and 260 injured[102]] of remaining in a static and destructive context[103]. Ironically France, which for so long had resisted the use of force and a NATO presence and lead in Yugoslavia, was by the spring of 1994 arguing repeatedly for the US and NATO to take a lead and to act decisively against the Serbs.

At this point it is necessary to leave the discussion of the French role in the former Yugoslavia to reflect on the wider defence debate between 1991 and 1994. Before doing so it is, however, useful to explore the 'lessons' of the experience of Yugoslavia for French defence and security policy. Firstly, in low-level humanitarian intervention of the type conducted by UN forces in UNPROFOR I and II, France had found a role for which it was well suited and one which allowed France to 'punch above its weight' by being willing and able to act where others were not. The wars in Yugoslavia also appeared to confirm the continued primacy of national interest in foreign and defence policy-making, a point made clear by the failure of the EC/EU to reach a cohesive position on anything substantive and by the failure of the Franco-German partnership to function as the core of an European security response. Finally, European impotence in the face of the worsening conflict had lead to the deepening involvement of the US and NATO as the only organisation with sufficient military clout to make a difference in the theatre. From the spring of 1993 France was forced to accept the primacy of NATO within Bosnia and to shelve more optimistic expectations about the WEU and CFSP. Yugoslavia, like the Gulf War, seemed a signpost of how far Europe had yet to travel to create an ESDI rather than a context within which such an identity could be forged.

The end of the Cold War, the Gulf War and the ongoing wars in Yugoslavia thus changed the geopolitical landscape, multilateralised the French defence and foreign policy framework, recast Franco-German, Franco-NATO and European security relations, and exposed the limitations of French military power. All these issues were to impact on the defence policy debates which unfolded from the early 1990s to adapt French defence to the new era.

3
The Reform of Defence

Introduction

The publication of a defence white paper in March 1994 was evidence that the post-Cold War adaptation of French defence policy had reached a point of intellectual maturity. By 1994 there was considerable agreement about the nature of the post-Cold War security context, the lessons of the Gulf War, the defence and security implications of the Maastricht treaties, the lessons of the conflicts in the former Yugoslavia, the nature of a new French 'international mission', the impact of technical-military developments, and the broad aims and objectives of French defence and security policy. This did not mean that a new consensus had appeared – there remained sharp differences in particular about Franco-NATO relations, nuclear weapons and the defence budget – but it did mean that the post-Cold War *'champ de bataille'* of the defence debate was becoming clear and that debates inside France were increasingly centring on the detail rather than the broad lines of policy, on means rather than ends, and on the pace rather than the trajectory of change.

The purpose of this chapter is firstly to examine, against the background of changes outlined in the previous chapter, how defence policy was reformed between 1991 and 1994 in the process which culminated in the publication of the 1994 *Livre Blanc*. The second half of the chapter then discusses the transition from the Mitterrand to the Chirac presidency and brings the reform of defence up to date by examining Chirac's impact on defence policy, his attempt to resolve the main tensions which remained at the heart of the white paper and the consequences for defence of the election of Lionel Jospin's Socialist government in the spring of 1997 which opened France's third period of political cohabitation in little more than a decade.

The defence debate in France 1991–4

To understand the reform of defence it is firstly necessary to say something about the French analysis of the new security context. Much of this begins with the view that since the end of the Cold War France itself, for arguably the first time in its history, faces no direct threats to its frontiers. In the new era France is surrounded by formally allied states [with the exception of the Swiss] each of whom in turn face no direct threats to their respective frontiers [Germany is for example closer to North Africa than to Russia][1]. Because of this, indirect, latent or future threats have replaced present and direct threats as key determinants of policy with a resultant loss of certainty and predictability and crucially a shift in focus from the present to the future.

France shares much of the Western analysis that Europe in the post-Cold War era is bounded by an eastern and southern flank of instability. To the 'East' the dominant issues are the strategic vacuum left by the collapse of the Soviet Union and the potential for Russian revanchism, political and economic instability, nationalism and the potential for civil war for which the break-up of Yugoslavia provides the precedent. To the 'South', primarily shorthand for the Mahgreb and Middle East, the issues are centred around population growth and the political, economic and social weakness of the southern states to deal with its consequences; access to strategic resources [particularly oil]; religious [Islamic] fundamentalism; the proliferation of weapons of mass destruction; and transnational threats such as terrorism and drugs[2]. The absence of direct threats to the French state and the shared perceptions of peripheral threats, combined with the integrational implications of the European project, have eroded the French conceptual distinction between national and European interests, thereby rendering irrelevant many of the old debates about national sanctuarisation[3].

French global interests outside Europe, defined by Jacques Lanxade as: 'the exercise of sovereignty over the DOM-TOMs; . . . obligations to countries with whom [France] is linked by co-operation and defence accords, notably in Africa; the security of [French] residents abroad (more than 1.5 million); and . . . the responsibilities . . . of [being] a permanent member of the United Nations Security Council'[4], are determined to be subject to essentially the same range of threats as those on Europe's eastern and southern flanks and in the new context there is consequently greater overlap than ever between the two European and global elements of French foreign policy.

As a result the old 'three circles' construct is being eroded in a context in which the distinction between France and Europe is diminishing and threats to Europe and those in the wider international system are converging. A novel and officially endorsed formula now proposes that the international security *désordre* may be resolving to one dominated not by east–west nor north–south issues but rather by issues arising from a global stratification into *zones riches* [rich zones], *zones tampons* [buffer zones] and *zones misérables* [poor or wretched zones][5]. This in turn has paved the way for a conceptual realignment of French military security interests based not on geography but on the thematic nature of threats to French interests.

In facing up to these threats two features of the new context are important in understanding French defence. One is the emergence of what may be termed a 'multilateral imperative', the need to work with others in pursuit of security objectives rather than seeking uniquely national responses. This has arisen because of limited French means to act unilaterally [evident in the Gulf War], because of the transnational nature of newly acknowledged threats [such as proliferation, terrorism, drugs] which demand a multinational response, and because of the legitimisation in France of multinational fora in the post-Cold War era.

A second feature is the emergence of a novel dichotomy in French policy. France has increasingly taken a foreign policy line centred around international law and multinational fora for global problem-solving. This is not to deny that French national interest no longer shapes policy, but it is to argue that since the end of the Cold War France has given preference where practical to a legalistic, multilateral and diplomatic approach to security. At the same time there is also a recognition in France that these approaches alone are not sufficient to guarantee security [a principle also illustrated by the Gulf War] and that consequently military force remains an important policy tool in the new context[6].

The dichotomy arises because collective and legalistic approaches to security and the exercise of military power may to a certain extent be contradictory. The promotion of the observance of international law, the acceptance of shared norms, and the utility of international fora risk being undermined by a failure or unwillingness to observe these norms, and by the pursuit of national rather than common or collective interests [both points with considerable relevance to French adaptation to multilateralism]. Moreover, the pursuit of security by non-military means may be undermined by military planning and operations which strike at the very elements of trust, tolerance, mutuality, and even fairness, necessary to facilitate diplomatic, economic and political progress.

With these issues in mind it is useful to take the discussion of French defence adaptation forward by first looking at France's relations with NATO. For a number of reasons on both sides Franco-NATO relations improved markedly between 1990 and 1994. Firstly when the Cold War ended Western Europe, along with the rest of the world, was the subject of a strategic reappraisal by the United States. Western Europe remained an important security concern for the US but, in the absence of direct threats, it no longer justified the attention or resources it had received from the United States during the Cold War. Indeed as US strategic interests shifted towards the stabilisation of the former Soviet Union and, in the wake of the Gulf War, to the Middle East, Western Europe increasingly became defined in terms of its relationship to those issues[7]. As such it became a conduit through which the United States could pursue its wider strategic interests rather than an object of security in its own right, and increasingly the main utility of NATO for the US has become the supply of potential partners for US-led world order roles[8]. A certain Franco-American rapprochement became inevitable once France showed itself, as it did in the Gulf War and in Bosnia, to be a willing and capable partner in some of those ventures.

Secondly, as a result of arms control deals such as the CFE and unilateral measures such as the elimination of US short and medium-range land-based missiles from the European theatre, the US military presence in Europe was by 1994 substantially reduced [US troop levels in Europe fell, for example, from 349 000 in 1989 to 160 000 in 1994, while US nuclear weapons based in Europe fell from around 7000 in 1989 to perhaps as few as 500 in 1994[9]]. Even allowing for the downsizing of European militaries, a *de facto* adjustment in the relative military clout within Europe had taken place, with symbolic if not immediately practical ramifications.

Finally the budget-sensitive Clinton administration which took office in January 1993 was openly committed to the idea of greater European burden-sharing and thus supportive of European efforts to take a greater share of responsibility for their own security with the important caveats that this did not undermine either the centrality of NATO or the importance of the US within it.

France for its part was also subject to a set of changes and pressures between 1990 and 1994. As we have seen, France quickly had to come to terms with its failure to confine NATO to its traditional functions and thus was forced to adapt to NATO's changing role and its continued dominance of the European security agenda.

Secondly, France had to come to terms with the stalling of its 'European' ambitions for Western European security arising from the lukewarm reception for the Maastricht treaties in France and the impotence of the EU and WEU [and of the CSCE] in responding to the Gulf and Yugoslavian Wars. Few European states were prepared to pay more than lip-service to the idea of ESDI [some like the UK initially opposed it, others like Germany supported it but could not accept a challenge to or the displacement of NATO] and rapidly falling defence budgets in Western European states made progress towards a functionally autonomous ESDI virtually impossible in the short to medium term.

Finally, France had renewed its interest in retaining a US continental presence not only for historic reasons *vis-à-vis* Russia and Germany but also because US reluctance to engage in Bosnia initially raised the spectre of US disengagement [which has always sent France rushing into American arms], while the subsequent unfolding of the war convinced the French that the US and NATO alone had the muscle to resolve it.

Mitterrand had consequently agreed in December 1992, largely at Chancellor Kohl's request, that the Eurocorps could be placed under NATO's operational command [subject to certain caveats[10]] in line with French ambitions for strengthening the European pillar of NATO. He had also agreed to facilitate French participation in NATO military staff preparations for UN implementation of the Bosnian settlement because it was essential for French participation in the mission[11]. Mitterrand, however, remained circumspect about NATO, adhering to Gaullist norms in the face of pressures for engagement which were questioning the very validity of those norms.

The pace of French engagement with NATO increased following the election in April and May 1993 of Eduoard Balladur's centre-right government, an election which also condemned Mitterrand to a second period of cohabitation for the final two years of his term of office. The position of the centre-right towards NATO had been clear for some time. In widely quoted remarks the RPR leader Jacques Chirac picked up several of the pressures bearing on France with respect to NATO and drew from them a clear conclusion about France's European security ambitions:

> With respect to Europe, we are forced to note that the substantial reduction in the American military presence has not stimulated any decisive European process, far from it. Several of our partners have even begun considerably to reduce their armed forces and are placing themselves even more than ever under American protection ... through NATO.
>
> I conclude that if France wants to play a determining role in the creation of a European defence entity, it must take into account this state of mind of its partners and reconsider ... the form of its relations with NATO. *It is clear ... that the necessary rebalancing of relations within the Atlantic Alliance, relying on existing European organisations such as the WEU, can only take place from the inside [and] not against the United States, but in agreement with it* [emphasis added][12].

With a centre-right government in office, these views began to impact on policy and to contribute to an observable shift in the centre of gravity of French thinking. As one analyst noted: 'while during the initial post-Cold War period France saw NATO as at best irrelevant to the development of ESDI and, at worst, as a major obstacle to it, the French view shifted towards recognising NATO as a necessary instrument in the building of ESDI'[13].

In June 1993 France agreed that the separate NATO and WEU naval task forces enforcing the embargo against Serbia could be combined under the operational control of NATO's integrated military command, subject in turn to the political control of the North Atlantic and WEU councils. In so doing, France appeared to have adjusted to the primacy of NATO in accepting the need to consult with the United States prior to European crisis initiatives and in accepting the principle that uniquely European operations would only take place if the United States did not wish to participate. This was important in two respects: firstly it established a priority – NATO first, WEU second – in responding to crises and, secondly, it amounted to an implicit acceptance of a US veto on European initiatives. The formulation gave Europe theoretical autonomy if the US did not wish to act but, since the US and NATO alone had the hardware and muscle to act decisively, a no from the US was effectively a no to all but the most low-level operations[14].

That French policy was able to shift so far while Mitterrand remained President requires some explanation. It is important to reflect that in his second period of cohabitation from 1993–95 Mitterrand was not as strong as in his first. Despite the primacy of presidential office and his long years of experience, Mitterrand's position was eroded by his political unpopularity [and that of his party], by speculation about his health, war record and private life, and by the fact that he was in the twilight phase of his political career. The *domaine réservé* gave way from the spring of 1993 to a *domaine partagé* [shared domain][15], identifying an adjustment if not quite an equalisation of power between the President and government.

This 'adjustment' had several important consequences. One was the enhancement of the government's autonomy and thus the assertion of a more Atlanticist agenda, albeit one not infrequently circumscribed by Mitterrand. A second was the 'politicisation' of foreign and defence policy. In periods of political harmonisation between the presidency and government, the consistency and stability of policy allows the foreign and defence ministries to have a stronger role in policy formulation and implementation. At times of cohabitation – particularly those more balanced between president and government – foreign and defence policy becomes an axis of tension between the president and government. As a consequence the debates and details of policy are 'politicised' and the roles of the respective ministries are diminished[16]. From 1993–5 this was important because the reactionary and Gaullist Foreign Ministry in particular was less able to assert an orthodox brake on the trend in Franco-NATO relations.

The importance of even this limited rapprochement is hard to underestimate. French co-operation with the US and NATO acted like a lubricant on European defence co-operation. Closer French-US relations reduced the tensions in German policy, which historically oscillated between Paris and Washington, encouraging Germany to step up the tempo of European defence construction. French engagement with NATO also reduced the anxiety that French security ambitions were a threat to NATO, paving the way for enhanced European co-operation within NATO. Finally, US assent to European defence co-operation persuaded even the reticent British to participate for fear of being marginalised. The lesson for France was profound: while building ESDI outside NATO was a non-starter, within NATO and with US consent it was at least feasible.

There were also lessons arising directly from French participation in NATO operations in Bosnia. In order to co-operate with NATO in peacekeeping, France had to exercise a presence in the relevant NATO meetings in which the issues were discussed. Because it was at first prohibited from participation in formal committees, France participated instead in *ad hoc* committees set up to avoid the problems raised by French non-integration in NATO. Pushed by the requirements of engagement France soon established new military links with the three NATO MSCs reporting to SACEUR, and improved liaison between the French military and SHAPE while pressure grew for high-level French participation in NATO's Military Committee and the Defence Planning Committee[17].

The need to work with NATO in the former Yugoslavia was thus driving the extent of French rapprochement with NATO with two important consequences. The first was that France found itself playing an enhanced role in the evolution of NATO because NATO was itself being shaped by events in Yugoslavia. France was first drawn deeper into NATO deliberations by the Yugoslavian wars and once there, perhaps unexpectedly, found itself influential. The second consequence was the further convergence of French and US policy to the point where Franco-American co-operation [and disagreement] was effectively driving NATO policy in Bosnia. This was important because it demonstrated to the US that France was a mover and shaker amongst Western European states, a player willing and able to act and crucially perhaps the only state capable of mobilising other EU states around collective action. In seeking allies to police the post-Cold War order, it was not lost on the United States that 'both the United States and France define security policy as a fundamental interest, that both conceive of national defence in global terms, and [that both] maintain militaries capable of intervention all over the globe ([albeit] on a different scale)'[18]. French action in Yugoslavia and elsewhere in turn undoubtedly also influenced US thinking about the potential for ESDI as a partner in regional and perhaps global security.

These strands came together in the seminal January 1994 NATO summit in Brussels at which NATO announced support for ESDI and the creation of the Combined Joint Task Force [CJTF] concept to in due course enhance the operational reality of ESDI by making NATO assets available, under certain circumstances, to the WEU for military operations[19]. For France it was 'the best NATO summit ever… [giving] wholehearted backing to two major French policy goals: the development of ESDI through the EU, and strengthening the European pillar of the Atlantic Alliance through the WEU'[20].

NATO's second major initiative announced at the summit, the Partnership for Peace [PfP] process for dialogue and engagement with the former Warsaw Pact states, was less welcome in Paris and underlined another issue driving Franco-NATO rapprochement: non-participation in wider NATO deliberations did not spare France the consequences of decisions taken by NATO over which France exercised no influence[21].

France did however play a leading role in the third major initiative of the summit: the establishment of a NATO policy framework to respond to biological, chemical and nuclear weapons proliferation, issues which overlapped neatly with French preoccupations. The French approach to proliferation in the new context was consistent with broader lines of policy: an emphasis on legalistic and multinational problem-solving coupled with the pursuit of measures to provide national and collective military options in the event other approaches failed.

From the early 1990s France dropped years of opposition to international arms control processes and came into a number of fora for the control of weapons of mass destruction. Thus France signed the Nuclear Non-Proliferation Treaty [NPT] in 1992 in time to be influential in its renegotiation in 1995, and came in to the Comprehensive Test Ban Treaty [CTBT] framework in order to participate in the negotiations which began in January 1994 to end the physical testing of nuclear weapons[22]. France renewed its enthusiasm for the biological and chemical weapons conventions as 'genuinely inclusive and meaningful processes' and for the Missile Technology Control Regime [MT-CR][23]. Mitterrand also took a lead in creating a context for nuclear arms control success through the French nuclear weapons test moratorium announced on 8 April 1992 by Mitterrand's Prime Minister Pierre Bérégovoy[24].

Under Mitterrand France seems to have become persuaded of the importance and utility of multinational fora in the new context, in part because processes were no longer dominated by the US and Soviet Union, in part because post-Cold War openness and trust between former rivals opened the door to improved verification of agreements, and in part because the new fora were both more inclusive and less unequal than in the past. In the new context the Moch-de Gaulle principles which had circumscribed French participation in arms control for decades seemed to be largely met, although processes which still abrogated these historic French concerns – such as the US-Soviet/Russian START treaties – continued to be eschewed by France[25].

At the same time France saw a renewed opportunity to exert international influence through international arms control not least because of France's leading role as a nuclear weapons state, a principal supplier of nuclear technology on the international market [with unfortunate past consequences for the very proliferation issues which France was now seeking to address], and as a state with 'special interests' in some of the key regions in which weapons of mass destruction were proliferating, notably the Maghreb and Middle East. France consequently pushed NATO towards a collective response to these issues, again converging with US global interests[26].

Nonetheless, uncertainties around diplomatic and political approaches to containing proliferation provoked both nuclear and conventional responses in France to the threat. Understanding these responses is best done in the context of wider analysis of respectively nuclear and conventional weapons policy in France between 1990 and 1994.

There was no serious debate in France following the end of the Cold War about whether to keep or relinquish nuclear weapons. The need to retain a nuclear arsenal was seen as self-evident for a variety of reasons including the latent threat of Russian revanchism, the proliferation threat, the political utility of nuclear weapons, the centrality of an independent nuclear deterrent as a cornerstone of French defence policy, the huge disparity between the size of the French arsenal and that of the Americans and Russians, and the broad political and public support for retention. Political leaders and defence intellectuals of left and right were consequently quick to reaffirm the basic tenets of nuclear policy – *dissuasion, independence, suffisance* – which had guided French thinking since de Gaulle[27].

There was nevertheless an intense and far-reaching debate about the adaptation of nuclear weapons in the new context and in this respect few elements of policy were left unexamined. As we have seen, the Gulf War of 1990/1 provoked a rethink of the nuclear-conventional mix of French armed forces but as early as 1988, the start of Mitterrand's second term, the nuclear priority had come under scrutiny in the response to Gorbachov's reforms and as the cost of nuclear modernisation butted against budgetary constraints. The statistics reveal that nuclear spending began a downturn from 1988 falling the following year from 17.51 to 17.29 per cent of the defence budget, reversing a pattern of year-on-year increase evident since 1975[28]. The 1990–93 *Loi de Programmation Militaire* announced in June 1989 set the framework for nuclear spending at around 16 per cent of the defence budget [in fact by 1992 it had fallen to 15.37 per cent] but kept broadly on track the ongoing modernisation plans[29]. These comprised the phased replacement of the *Redoutable*-class and *Inflexible*-class ballistic missile submarines with the *Triomphant*-class vessels, the upgrading of the submarine ballistic missiles themselves with the M-45 and by the year 2000 the M-5, the implementation of the ASMP and *Hadès* programmes [with three rather than five Mirage 2000N squadrons and fewer *Hadès* missiles than originally planned], and the modernisation of France's nuclear command and control system [the transmission systems ASTARTE and Ramses and the related systems Jupiter, SYDEREC, Syracuse II, Telemac, and Transfost][30].

Downward adjustments in the nuclear budget and projected nuclear force expansion were accompanied by two important debates about nuclear policy. The first centred on the adaptation of nuclear strategy to the changed context, the second on how to exploit the possession of nuclear weapons by France in the construction of ESDI.

The adaptation of strategy was driven in large measure by the events in the 1990/1 Gulf War and the discovery afterwards by the UN weapons inspectors of advanced Iraqi biological and chemical weapons programmes[31]. The implication of wider proliferation threats to France and French interests around the world forced a new calculation in Paris based on the need to deter [or rather to *dissuade*] political elites in developing nations who may appear subjectively 'irrational' or who may not make the same cost-benefit analyses the West had assumed of Cold War adversaries. The brutal dictatorship of Saddam Hussein served as the seminal [though by no means only] example of this problem. Hussein had butchered and gassed many of his own people, sent tens of thousands to their deaths in two costly wars within a decade, and had fired ballistic missiles against a nuclear-armed neighbour. It was by no means self-evident that such an individual [or elite group] would be influenced by a general deterrent threat or that counter-city or counter-value targeting alone would exercise *dissuasion*.

In the aftermath of the Gulf War, Western analysis asked what kind of threats might deter Hussein or someone like him and provided the answer that the ability to destroy his weapons of mass destruction, his command and control infrastructure, the political apparatus which enabled him to remain in power, and [implicitly] the ability to terminate his life, would constitute a powerful deterrent threat. What flowed from this was the idea that nuclear weapons should be flexible and capable of being used with control and discrimination against very precise targets. By these means the powerful states might deter 'crazy' or 'irrational' leaderships, an idea summed up in France as *'dissuasion du fort au fou'* [the dissuasion by the strong of the mad][32].

Given that *dissuasion* rests on an assumption of adversarial rationality, predictability and control, the debate about how to respond to the emergent threat of proliferation was fundamental. Between 1991 and 1994 a divide opened up in Paris between what David Yost termed, by his own admission somewhat crudely, the supporters of a 'more operational' and 'less operational' response to this core question[33]. The former favoured the development of precision nuclear weapons and strategies to allow nuclear use and flexible targeting, the latter favoured traditional nuclear hardware and concepts[34].

In the new context both the political left and right were willing to look again at the logic of the arguments. By 1992 even the Socialist Defence Minister Pierre Joxe [arguably influenced by the military-technical community] was suggesting that alongside its traditional nuclear role France 'should perhaps develop more flexible...arms, creating *dissuasion* by the precision of the strike more than by the threat of a general nuclear exchange'[35]. In his insightful study *Vive la Bombe* Pascal Boniface argued that this line was evident also in the annex of the 1992–94 *Loi de Programmation Militaire* and that for a while the political left looked as though it might be about to shift policy towards a counter-force position[36]. In his follow-up study *Contre le Révisionisme Nucléaire* Boniface both charted and undoubtedly influenced the movement of the left away from this line. In the study Boniface critiqued the assumption that leaderships in the developing world are necessarily irrational, attacked the premises of *'dissuasion du fort au fou'*, and argued that the policy of *'dissuasion du faible au fort'* by which France had expected to deter the much stronger Soviet Union would serve well equally in reverse – *'dissuasion du fort au faible'* – in allowing France to deter weaker states[37]. Underpinning these views was the belief that existing concepts in nuclear policy and existing and scheduled weapons in the nuclear arsenal would continue to meet French needs in the new context.

Many amongst the centre-right however did not share Boniface's analysis and, following the April-May 1993 Balladur victory, the 'more operational'/'less operational' debate moved to the centre of the cohabitation discourse. On the 'more operational' side, arguments emerged for the development of low-yield, 'special effects' [i.e., neutron] and precision nuclear weapons, while those on the 'less operational' side opposed these kinds of developments.

Rather tellingly such weapons did not subsequently materialise. In his final year of office Mitterrand made a decisive and hugely important statement on the nuclear forces on 5 May 1994, an intervention termed by one commentator 'a Presidential *Livre Blanc*'[38], which confirmed the fidelity of France to traditional *dissuasion* concepts and which slapped down notions of 'flexibility' and 'operational utility' for French nuclear weapons[39]. By the time Chirac came into office a year later, when the 'more operational' line might have been expected to reassert itself, the political and strategic debate had moved on.

There is an important corollary to this analysis however which becomes evident when attention is switched from the political to the military level and to the evolution of French nuclear forces, particularly after the spring of 1996. This corollary suggests that the idea of nuclear flexibility was not killed off by Mitterrand but in fact has emerged in a revised form as an important element in French nuclear policy. This idea and the issues related to it are explored in Chapter 5.

France also took significant steps forward during this period in promoting the idea and functional reality of a European nuclear deterrent to underpin the construction of ESDI. The idea was not exactly new: despite French critiques of the validity of the US nuclear guarantee to Western Europe, French defence intellectuals, if not always French policy-makers, had a long history of advancing the idea that the French deterrent had a utility outside France's national interests alone. The debates in the 1970s and 1980s about '*sanctuarization élargie*' were arguably the clearest expression of this, and Mitterrand's offer in 1986 of consultation with Germany, subject to caveats, on nuclear use clearly demonstrated the political value of the idea, if not at the time its practicality.

Arguably the event which shifted the debate closer to the policy mainstream was the signing of the INF treaty in December 1987. The removal of US short and intermediate-range land-based missiles from Europe had two immediate effects. The first was to raise again questions in Western Europe about the US nuclear guarantee and the strategy of flexible response. The second – at least in the minds of the French – was to 'upgrade the French nuclear deterrent, since the US withdrawal [left] France as the only European power with a comprehensive "prestrategic" and strategic nuclear armament'[40]. Both elements played to a certain extent in a degree of Franco-British rapprochement on nuclear matters[41]. As early as March 1987 the French had put out feelers to the UK about enhancing bilateral nuclear co-operation and both countries had agreed to work together on 'the evaluation of enemy defences that our nuclear forces have to cross'[42]. In December 1987 the French Defence Minister André Giraud proposed that France and Britain might take the process forward by consulting on nuclear targeting but the idea was rebuffed by the British Prime Minister Margaret Thatcher, although an agreement promptly emerged in January 1988 to enhance bilateral military co-operation generally and to allow French SSBNs to visit British ports[43].

December 1987 also saw an agreement to study the joint development of a tactical nuclear air to surface missile [TASM] to replace the UK's ageing WE177 and France's ASMP, both of which were potentially vulnerable to Soviet air-defence[44]. These discussions were complicated by the fact that the UK was engaged in parallel discussions with the US about an American missile, and by the profoundly important consequences which would have flowed for the UK and Europe had the UK switched its nuclear partner for the system.

The notion that a fledgling co-ordinated European deterrent was emerging through Franco-British co-operation was patently premature. The British were deeply unwilling to weaken the transatlantic relationship or NATO and thus discussions with France were low-key and slow-paced. The idea that a European deterrent could emerge even if it was mutually desirable was itself critiqued in the late 1980s on the grounds that French and British force levels were too low to match the Soviet Union alone, that there existed no means to manage and control the co-ordination of uniquely European nuclear forces, that the creation of such means posed seemingly insurmountable political and strategic problems, and that there was in any case 'no vacancy' for the job of nuclear guarantor in Western Europe while the US and NATO remained in place[45].

The events of 1989–91 set these issues in a completely different context and created for many analysts a clear watershed in French thinking by the end of 1991 following the signing of the Maastricht treaty. According to one analyst, French policy on a European nuclear deterrent can be characterised as belonging to two 'radically different periods': that prior to 1992 'when those responsible for French policy categorically refused to provide French nuclear arms for European defence' [notwithstanding French overtures to West Germany] and the period from January 1992 when Mitterrand stated that the question of a European nuclear deterrent would become one of the major issues in the development of a joint European defence, the latter amounting to a 'major strategic revolution' in French thinking[46].

This 'revolution' was brought about by a range of factors. US denuclearisation and the large-scale withdrawal of US forces following the end of the Cold War raised doubts about the long-term reliability of the US nuclear guarantee and thus faced Western Europe with the question of how to meet its deterrent needs in the medium to long term. Uncertainty about the US in this respect at least created a less hostile climate for consideration of European nuclear ideas, albeit tempered by the traditional refrain that Europe should not act in ways which might precipitate that which it feared. A related risk was that German unification and the erosion or withdrawal of the US nuclear guarantee might tempt the Germans down the nuclear path[47], particularly in a context in which Russia continued to deploy large numbers of nuclear weapons. France was acutely anxious to forestall this possibility by ensuring that Germany felt neither 'vulnerable to Russian nuclear coercion or preoccupied with an apparent lack of nuclear protection'[48].

France was also persuaded by the idea that in the end a European political entity, implicit for many in the Maastricht treaty, will have to have an autonomous nuclear deterrent if it is to be credible in a context in which other global players – the US, Russia, and China – possess such a capability[49]. With an autonomous nuclear capability and autonomous ballistic missile capability [unlike the British], the argument went, France alone was placed to provide the necessary nuclear weapons for a uniquely European arsenal.

The final point appears to be one of legitimacy, that is a French concern that French nuclear weapons could emerge as an obstacle to closer European defence and security harmonisation, particularly if gaps between the nuclear haves and have-nots persisted or widened around the need for, and role of, European nuclear weapons. France consequently had an interest in finding ways to promote consensus and cohesion around a shared perception of nuclear forces in Europe.

In the immediate aftermath of the signing of the Maastricht Treaty, at arguably the maximal moment in terms of overt French optimism about ESDI, senior French political leaders began to articulate their expectations for the Europeanisation of the French nuclear deterrent. Mitterrand's January 1992 statement on the issue was followed in quick succession by Defence Minister Joxe's assertion that French and British nuclear weapons could be combined and that European co-operation in the nuclear domain was achievable, and by a Defence Ministry statement that the coexistence of a purely national deterrent and close European political union were logically questionable[50].

In addition to the rhetorical promotion of Europeanising nuclear weapons, France took a more pragmatic line in seeking to begin building functional nuclear relationships with individual European allies. The most obvious candidate for this approach was the British, Europe's only other nuclear power, with whom France already had an ongoing process of nuclear dialogue.

The renewed Franco-British nuclear dialogue begun in 1987 continued throughout the late 1980s and in 1990 meetings between the respective national premiers and defence ministers[51] addressed the issues of bilateral nuclear dialogue and co-operation against the background of wider military co-operation. The process culminated in October 1992 with first an explicit call from the French Prime Minister Pierre Bérégovoy to co-ordinate nuclear policy as a step towards the creation of a European deterrent[52] and then the disclosure at the NATO Gleneagles summit that closer Franco-British nuclear co-operation had been given US approval[53]. In the wake of Gleneagles, France and Britain established a Joint Commission on Nuclear Policy and Doctrine [JCNPD] and at their bilateral summit the following year on 26 July 1993 the two countries agreed to make the Commission a permanent body[54].

The establishment of the JCNPD needs to be seen as being conditioned by two external factors: one the unification of Germany which enhanced the mutual value of France and Britain to each other as a counter-balance to the perceived power of the emergent German state, the other enhanced French-US nuclear dialogue which formed the second side of the French-British-American nuclear triangle and brought NATO's three nuclear states into an increasingly important new relationship.

On the conventional side, the fruits of the wide-ranging defence debates of the early 1990s also became evident through the 1992–94 *Loi de Programmation Militaire*. The *Loi-Militaire* was remarkable in several respects. It was the first made available in modular form, the details appearing in six module papers dealing respectively with: the nuclear deterrent; space, communications and intelligence; air-naval operations; air-land operations; security and support; and, preparation for the future. The *Loi* also appeared by an unusual political route. Anticipated in 1991, the *Loi* failed to materialise and instead important defence decisions through 1991 were taken by the government under presidential authority without reference to parliament[55]. As a consequence of the *Loi*'s non-appearance, the annual defence budget debate in November 1991 [for the 1992 budget] was conducted in something of a vacuum and the budget itself was pushed through using a *'vote réservé'* formula which in effect passed the budget without parliamentary vote[56]. When the *Loi* finally appeared in July 1992, its budgetary prescriptive framework thus covered only two years, far less than the usual five or six years of a *Loi-Militaire*, though it sought to project trends through to 1997[57].

The most important aspect of the 1992–4 *Loi-Militaire*, however, is that the policy-makers' view of the strategic context, the threats to France and French interests, and the place and role of French armed forces, had begun to shift to a post-Cold War agenda. This was evident in the slowing down of the nuclear modernisation, discussed above, which was projected to see nuclear spending fall overall by 6.6 per cent by 1997, and by the emphasis on space, intelligence and communications [to rise by 3 per cent], and multi-service operations [air-naval operations equipment was to rise by 5.5 per cent and air-land by 1.9 per cent][58].

To understand the extent of this shift it is worthwhile explaining briefly the main projects under each of the modular umbrellas subject to budgetary growth. With respect to 'space, communications and intelligence' [*l'espace, la communication, le renseignement*] the *Loi* envisaged a major programme to expand and enhance French capabilities. The spending through to 1997 was intended to support initial programme development which would then continue through to 2010 and beyond. These programmes fell into three main strands. The first, telecommunications, was concerned with the development of military satellite communications and the exploitation of civilian satellites to extend force connectivity. The centrepiece of this proposal was the Syracuse II-Telecom 2 and, from 2000, the Syracuse III-Telecom 3 satellites, eventually anticipating a multinational European Military Satellite Communication system [EUMILSATCOM] beyond 2010. The second, and arguably most important, strand was the development of dedicated space-based intelligence assets beginning with the Franco-Italian-Spanish Helios I optical imaging satellite due for launch in 1994, and followed from 2000 by the Helios II optical and infra-red system, the Franco-German radar observation satellite Osiris [later Horus], and the Cerise and Zenon signals intelligence systems[59]. Under the intelligence rubric the *Loi* also included provision for a new generation Sarigue intelligence gathering aircraft fleet[60]. The third strand concerned itself with 'enhancing operational coherence' by exploiting civilian satellites such as the SPOT observation system, Météosat and DMSP meteorological systems, ERS and Topex-Poseidon oceanographic systems, and accessing allied state technologies, in particular the US Navstar/GPS network[61].

With respect to air–naval operations, the *Loi* maintained or boosted spending on the *Charles de Gaulle* aircraft carrier, projected at the time into service in 1998, and anticipated the construction of a second carrier from 1997, both intended to deploy the nuclear-capable Rafale aircraft. It included also spending on the multinational NH-90 helicopter [with Germany, Holland and Italy], primarily for anti-submarine and tactical transport roles, and the development of other elements of the surface naval fleet.

On the air-land side the *Loi* cut back 'Cold War' programmes including the AMX30B2 heavy tank which was terminated and the Leclerc heavy tank which had its initial purchase order of around 1400 cut by almost half to around 800. By contrast, projects for multi-service force projection were upheld or increased including AMX10RC light tanks and VBL armoured cars, the former capable of being carried on a Boeing 737, as well as combat helicopters, anti-tank and air-defence missiles[62].

In the important area of long-range transport aircraft the *Loi* was, however, circumspect, providing only for a commitment to examine a replacement for the ageing DC-8 and Transall aircraft after 1994 if European initiatives for a Future Long-range Aircraft – FLA] were not advanced by that date.

All this said, there was still considerable Cold War programme inertia in the system and, despite the shift towards 'out of area' force projection and intelligence, the changes overall were still set in a framework of organisational adjustment consistent with the *Armées 2000/ORION* programmes. The

most important of these changes, in part flowing from budgetary constraint, was the projected reduction in the size of the army from 280 000
to 225 000 by 1997, a 20 per cent cut, the deepest single reduction in force
levels since the end of the Algerian war.

A leading defence commentator noted however that the 1992–94 *Loi-Militaire*
left three central contradictions of French defence policy unresolved: how
to reconcile an independent defence policy with a foreign policy increasingly
co-ordinated with Europe; how to find, within a mixed conscript-professional
army still orientated towards national territorial defence, the means to rapidly
deploy substantial numbers of professional soldiers beyond French frontiers;
and how to modernise French forces as a whole against a background of
other allied states reducing their defence spending and raising the public
expectation for a 'peace dividend' in France?[63].

The 1994 *Livre Blanc sur la Défense*

In May 1993, within a matter of weeks of election, the new centre-right government of Eduoard Balladur set up a commission, under the chairmanship
of Marceau Long, vice-president of the *Conseil d'Etat*, of 20 members drawn
from the Prime Minister's office, the Ministries of Defence, Foreign Affairs,
Budget, Interior, Research, Industry, Co-operation, the Atomic Energy Commission [CEA], permanent secretariat and six invited 'personalities'[64], to examine defence policy and prepare a defence white paper for publication in
1994[65]. Both the timing and evident urgency of these developments were
curious given that a two-year period of cohabitation had just begun, presidential elections loomed in 1995 and no obvious event or political, economic,
or strategic pressure triggered the process.

To the question of why a defence white paper was published in 1994, the
Livre Blanc itself offered three answers: 'to better understand our times and
the role of defence in our country. To place the politics of defence in a long-
term perspective . . . [and to] . . . explain defence to the French people and
seek their support for it'[66]. Of these the first and third argued that the intention was to explain to the French public the changes since 1989 in defence
policy which had not been clearly articulated but which had been implicit
in every development after *Armées 2000/ORION*. The second pointed to a
prescriptive or at least framework-setting role in 'placing defence in a long-
term' perspective, a perspective which was intended to cover the period
through to 2010[67]. Neither answered the specific question of why it was
published in 1994 rather than waiting, for example, a year until a new president could approach the issues with an uncontested political mandate.

One explanation for this, widely credited in Paris, suggests that the *Livre
Blanc* appeared when it did because the new Prime Minister Eduoard Balladur
had presidential ambitions himself and wished to give his name unambiguously to a major national document, so clearly within the presidential purview,

in order to enhance his own credentials[68]. This point is useful but must be qualified by the fact that the *Livre Blanc* was begun in May 1993, well before Balladur's presidential ambitions were evident. Another part of the answer perhaps is that the *Livre Blanc* appeared in 1994 because by then the framework of post-Cold War French defence policy was clear, the main lines of policy evident, and the context of cohabitation was conducive to the production of a consensual document which would need cross-party support to see daylight and which thus would not be expected to be overturned by whichever party took power in 1995. Arguably the degree of fidelity to the *Livre Blanc* of both the RPR government of 1995–97 and the Socialist government since 1997 bears this out.

The *Livre Blanc* has four parts dealing respectively with the new strategic context and France's military security objectives within it, strategy and force predisposition and requirements, defence resources, and the relationship between defence and society. The first part describes in some detail a strategic context in which there are no direct threats to French frontiers, a new relationship is evident amongst the world's major powers, and in which the threats to France and French interests arise primarily from proliferation, regional crises and instability, and the 'new vulnerabilities' of drug-trafficking, religious extremism and terrorism. Within this context French defence policy objectives are set out both in terms of direct national interest and in terms of playing a role in multinational fora – in Europe, NATO, the UN and through bilateral defence accords [mainly in Africa] and multilateral arms control – in pursuit of national interest. Arguably the key formalisations presented in this section are the wish to enhance the role of the UN in international security, and the decision, reflecting Franco-NATO rapprochement, to have the Defence Minister attend NATO's NAC [in its DPC incarnation] and the CEMA NATO's Military Committee on a case-by-case basis and subject to Presidential and Prime Ministerial assent.

In the second part, French strategy is set out affirming traditional concepts of *dissuasion* while articulating a new strategy for the use of conventional forces centred around three new roles: *prévention*, that is anticipating and defusing potentially dangerous situations, pre-empting the use of force, and seeking to confine crises and conflicts to their lowest level; *action*, enhancing inter-service and multilateral operational capability, maintaining conventional technological advantage, and developing the forces necessary to intervene in a range of operations from Gulf-War type situations, through peacemaking and peacekeeping to limited military interventions for other purposes; and, *protection*, the protection of France from emergent threats – primarily proliferation – through expanded conventional roles centred primarily around air defence and anti-missile defence[69].

To demonstrate these traditional and new roles, six scenarios are presented and for each the function of nuclear and conventional forces is illustrated, reflecting the strategic positions set out above and the French force predisposition or requirements in terms of intelligence-gathering assets, nuclear and conventional forces, and overall organisation. The scenarios are:

(1)   regional conflicts which do not threaten French interests;
(2)   regional conflicts which could threaten French interests in Europe, the Maghreb or the Middle East;
(3)   threats to the DOM-TOMs;
(4)   the implementation of bilateral defence treaties or accords [mainly in Africa];
(5)   operations in support of peace and international law; and
(6)   the re-emergence of a major threat to Western Europe [primarily Russian revanchism], considered 'an improbable scenario [over] the next twenty years, but one that cannot be ignored because it constitutes a "lethal risk"'[70].

The key development here is the assertion of a role for French nuclear weapons in the second, third and last of these scenarios. While the third and last fall within the traditional purview of *dissuasion*, the nature of the second scenario and the statement that 'a deterrent manoeuvre, adapted to [each] context, may accompany our decision to intervene' clearly implied a novel projected nuclear role.

A second main emphasis is on the need to boost conventional forces, picking up the strands evident in the 1992–94 *Loi-Militaire*, stressing the requirements for intelligence assets for European autonomy and command and control upgrades to facilitate inter-service and multinational operations.

The third part focuses on personnel within the overall issue of resources, confirming the choice of *une armée mixte* [i.e., professional and conscript soldiers] and discussing the issues around conscription while nevertheless arguing that conscription 'must be retained'[71]. This section also contains important ideas about the evolution of French defence industries arguing for increased European co-operation, the revision of state-defence industry relations, the enhanced exploitation of dual civil-military technologies, and the continued importance of defence exports, arguably the most important assertion being that France would have to give up its self-sufficiency in arms developments [by 1994 a fiction anyway], a point reinforced in June 1994 by Henri Conze, the head of the DGA, France's defence procurement agency[72].

The fourth and final part of the *Livre Blanc* considers defence and society, examining relations between nation and army, civil defence, the defence economy, and defence and public opinion. The latter, subject to the usual caveats about opinion surveys, offered some interesting evidence indicating that the end of the Cold War had barely impacted on French support for the armed forces or for conscription [both running at around 70 per cent][73]. It was evident from all this that the white paper was a consensus document which set out areas and issues of broad conceptual and policy agreement but which also glossed over or omitted more contentious issues, such as nuclear strategy, nuclear weapons testing, conscription, and Franco-NATO relations.

In April 1994 the government published the *Loi de Programmation Militaire* 1995–2000 to implement the broad policy lines set out in the *Livre Blanc* and to support the principal procurement programmes identified. Like the *Livre Blanc* this *Loi-Militaire* was also contextualised by the imminence of the presidential elections. It provided for a year on year increase in the defence procurement budget of 0.5 per cent annually. While growth of this scale was well below predicted inflation [2.2 to 2.6 per cent] it nevertheless continued to buck the Western trend of deep defence procurement cuts since 1989. This level of growth was, even so, projected to be insufficient to meet the range of defence needs particularly with spiralling equipment costs [Henri Conze in a contemporaneous interview stated that while the procurement budget over the next 15 years would remain roughly constant, in real terms equipment costs would double over the same term[74]].

It is clear with hindsight, and was probably clear at the time, that the *Livre Blanc* and 1995–2000 *Loi-Militaire* were articulating a 'wish-list' in insisting on nuclear modernisation, the retention of conscription, and a virtually across-the-board conventional modernisation within the projected budget. Spending ambitions were clearly at odds with financial realities but in 1994 neither of the main political parties were willing to admit this, let alone cut or abandon programmes in line with these realities. Neither side wanted to be seen to be responsible for large-scale job cuts in the run-up to a presidential election and thus defence industry restructuring, ongoing project cutbacks and terminations, and new project production decisions – including the FLA, a second aircraft carrier, NH-90 helicopter, and the *Arme de Précision a Très Grande Portée* [APTGP] missile – were postponed until after the elections and in some cases until 1997.

During the election campaign there was, as a result, a 'conspiracy of silence' on defence matters[75]. Given that all roads pointed to inevitable force reductions, programme cuts, and industrial downsizing in the defence area – an assessment made clear in François Heisbourg's *Les Volontaires de l'An 2000*[76] – there was little incentive to bring these implicit job cuts into a campaign centred on solving the problems of unemployment.

From Mitterrand to Chirac

While steering clear of the more problematic defence issues in the run-up to the election, presidential candidate Jacques Chirac was openly critical of France's position in Bosnia, arguing that the paralysis caused by an unwillingness either to act or withdraw was exacting an unacceptable toll in French lives[77] [by July 1995 43 French UN troops had been killed in Bosnia[78]]. He also openly favoured a resumption of nuclear weapons testing, ending Mitterrand's three year moratorium, on the grounds that the moratorium was undermining the credibility of the French nuclear arsenal[79].

In the first round of elections on 23 April 1995 the right-wing vote split allowing a Jospin victory of 23.3 per cent to 20.8 per cent for Chirac and 18.5 per cent for Balladur, eliminating the latter. The right came together in the second round on 7 May to deliver victory to Chirac by 52 per cent to Jospin's 48 per cent. On assuming office Chirac made a number of moves which signalled his intent. He pointedly visited the grave of General de Gaulle at Colomby-les-Deux-Eglises as part of his inaugural ceremony and had de Gaulle's old desk taken out of storage and reinstalled in his Elysée palace offices[80]. Chirac had made no secret of his admiration for de Gaulle's strength of leadership, his emphasis on the primacy of French interests, and his willingness to take internationally unpopular decisions often in the face [from the French perspective] of hypocritical Anglo-Saxon criticism, and through these gestures he appeared to many to be trying to assume de Gaulle's mantle. The day after taking office Chirac made a visit to Germany to reassure Chancellor Kohl of continuity on the Franco-German relationship[81].

With respect to Bosnia, Chirac chose the offensive, promptly denouncing the Bosnian Serbs as the principal aggressors in phraseology which went even further than that of the exasperated Mitterrand in revising traditional French links with Serbia[82]. Chirac was instrumental in changing the mandate of the UN Protection Force [UNPROFOR] in Bosnia, beginning with the NATO defence minister's agreement in early June 1995 to set up a Rapid Reaction Force – comprising French, British and Dutch forces – under UN command. Armed with helicopters, anti-tank weapons and artillery, this force 'beefed up' the UN deployments in Bosnia and widened UN military options. While the introduction of this force was too late to prevent the Serbs overrunning the 'safe areas' of Srebrenica on 11 July and Zepa on 25 July, its presence did begin to turn the tide of events. After Srebrenica and Zepa, peacekeeping forces were redeployed away from Serbian harm, the threat of NATO airstrikes was extended to the remaining 'safe areas' of Bihac, Sarajevo and Tuzla, and in a key move British and French artillery took up positions on Mount Igman overlooking Sarajevo on 23 and 24 July[83]. With the vulnerability of UN peacekeepers diminished, the way was clear for the UN and NATO to step up military action.

Through the summer of 1995 the Serbian position on the ground was weakened following the Croatian victory in Krajina and Muslim/Croat federation territorial gains in Bosnia. On 28 August 1995 a second market place slaughter in Sarajevo, which killed 37 and was widely blamed on the Serbs, provoked NATO to respond, pushed by the French, with Operation Deliberate Force. NATO airstikes began on 30 August and, after a short pause between 1 and 5 September, continued through to 14 September. In all, 800 missions were flown by NATO aircraft [including French Mirage 2000Ds flying from

Cervia in Italy[84]] against Serbian artillery, communications, missile, radar and tank assets in Serbian-controlled Bosnia. Arguably, the airstrikes were instrumental in pushing the Serbs to the negotiating table. Militarily weakened by the airstrikes, Serbian forces withdrew from around Sarajevo and, more importantly, were subject to opportunist assaults from Bosnian government and Croat forces from 11 September which forced Bosnian Serbs out of 20 per cent of the Bosnian territory they held. The balance of forces on the ground had shifted decisively and the Serbs sought a quick accommodation to preserve what they had. A two-month cease-fire which began on 12 October provided the breathing-space for proximity talks in Dayton, Ohio, and paved the way for the subsequent agreement initialled on 21 November and signed formally in Paris at the Elysée Palace in mid-December[85].

French delight at the Dayton outcome was quickly soured by stories which emerged from late December concerning Serbian ill-treatment of French armed forces personnel and Franco-Serbian deals which allegedly traded the safety of UN peacekeepers for 'safe areas'. The former related to 11 French soldiers taken hostage at Vrbanja who were subject to brutal treatment by the Serbs[86] and to two pilots, José Souvignet and Frédéric Chiffot, shot down during Operation Deliberate Force, also allegedly tortured by Bosnian Serbs[87]. Both cases were accompanied by allegations of deal-making between France and Serbia involving the freeing of the French personnel and the suppression of the extent of Serbian mistreatment. More seriously, allegations surfaced in early 1996 that France had traded the safety of UN hostages in return for turning a blind eye to Serbian actions in 'safe areas' and staying NATO's hand. The specific allegations centred on claims that the French UN Commander Bernard Janvier and 'French officers under orders from President Chirac' were instrumental in agreeing with Bosnian Serb commander Ratko Mladic that NATO airstrikes would not be conducted against Serbian forces assaulting Srebrenica if UN hostages – mainly Dutch and French – were released. These allegations were strongly denied in Paris on the grounds, *inter alia*, that UN and NATO decisions were collective rather than French and thus France was not in a position to deal-make bilaterally, and on the basis that France had throughout the summer been the most vocal in calling for NATO military action against Serbia. Whatever the truth of it, the Serbs did overrun Srebrenica, NATO at the time conducted only two belated airstrikes against Serbian forces, and all UN hostages were released[88].

The Dayton Peace Accord accommodated competing interests by maintaining the legal status of Bosnia as a unitary state but in practice partitioning it into the Bosnian-Croat Federation and the Bosnian Serb Republika Sprska. To implement the accord, UNPROFOR was superseded by the NATO Implementation Force [IFOR, or *la Force d'Interposition de l'OTAN*] mandated for a year to monitor and enforce compliance with the military aspects of the Dayton Agreement. France contributed 7500 personnel to the IFOR deployment set at 60 000 troops [50 000 NATO and 10 000 non-NATO], including for the first time 20 000 US combat troops[89].

In a remarkable eulogy in the right-wing French newspaper *Figaro* in late December 1995 the RPR secretary-general Jean-François Mancel heaped praise on Chirac for his 'decisive role' in ending the Bosnian conflict. After lauding Alain Juppé's part between 1993 and 1995 in establishing Operation Deny Flight, creating the Bosnian 'safe havens' and setting up the contact group of nations, Mancel moved on to trumpet Chirac's achievements. Chirac, he argued, had 'known how to impose the creation of the [French-British-Dutch] RRF to put an end to the impotence of UNPROFOR' and had known too 'how to convince the Americans to get more deeply involved . . . leading in turn to Operation Deliberate Force and [subsequently] to the peace negotiations'[90]. Embedded within this rhetoric, perhaps unwittingly, were some deeper realities: the acceptance of the impotence of Europe without the United States and the admission that Chirac's triumph, if such it was, lay less in what France itself had been able to do, than in what France had been able to persuade the United States to do[91].

If Bosnia brought Chirac plaudits, his second policy initiative – the resumption of nuclear weapons testing – marked for many the beginning of his political decline and by December 1995 the storm around the testing decision was largely drowning out the French role in the Dayton Peace Accords.

Anyone reading the runes in 1994 and the spring of 1995 could have foreseen Chirac's decision. He was himself publicly supportive of a testing resumption and close defence advisors like Jacques Baumel and Pierre Lellouche had both critiqued Mitterrand's moratorium decision and called repeatedly for a resumption[92]. In an effort to defuse tension on the issue between Mitterrand and the Balladur government during the cohabitation period, two 'independent' committees had been established to examine the resumption question. The first of these, the Lanxade group under the chairmanship of Admiral Jacques Lanxade, was assembled in July 1993. It reported privately to Mitterrand and Balladur on 4 October 1993 that the moratorium could continue but that a number of tests would be imperative after the 1995 presidential election[93]. The second committee was established on 28 October by the *Assemblée Nationale* and comprised five *députés* [including Baumel and Lellouche] under the chairmanship of Réné Galy-Dejean. The Galy-Dejean group produced a 60-page report in six weeks and published on 15 December the recommendation that a test resumption was essential, that between 10 and 20 tests would be needed to guarantee the credibility of the French nuclear deterrent, and that data should be gathered for the PALEN programme [*Préparation à la Limitation des Essais Nucléaires*] computer simulation project to reduce reliance on physical testing[94].

Two other factors also probably played in Chirac's decision. Chirac's political style and his admiration for de Gaulle attracted him to the grand gesture, particularly one so centred on French national interests. Secondly mutual animosity between Chirac and Mitterrand also fed into the decision. Chirac had twice been defeated by Mitterrand in the presidential elections of 1981 and 1988 and had spent two difficult years from 1986 to 1988 in political cohabitation with Mitterrand, during which Mitterrand had acted to try to neutralise Chirac as a presidential rival. There was consequently little love lost between the two. In May 1994 Mitterrand made a long statement about the future of the French nuclear deterrent in which he tried to hamstring his successor by arguing that the next French president would not be able to resume testing because of the CTBT and NPT processes and because of international pressure, particularly from regional states[95]. Subsequent statements from the political opposition rejected the suggestion that an incumbent president could tie the hands of his successor and many senior figures expressed open hostility to, as they saw it, Mitterrand's blatant attempt to engineer future policy[96]. It cannot have been irrelevant to Chirac that his grand Gaullist gesture to resume testing would at the same time be one in the eye for Mitterrand.

With all these pressures bearing on him, Chirac announced on 13 June 1995 that France would conduct up to eight nuclear weapons tests in the South Pacific between September 1995 and May 1996[97]. If the announcement of eight tests [rather than the 10 to 20 recommended by the Galy-Dejean group] was intended to signal French restraint, it failed to convince the international community. Squeezed by the NPT negotiations taking place through the summer of 1995 and the CTBT process due for completion in 1996, Chirac had a relatively small window of opportunity for the tests. The year 1995 was nevertheless a particularly bad year for the French to abrogate the international moratorium because it was both the 50th anniversary of the nuclear bombing of Hiroshima and Nagasaki and the tenth anniversary of the sinking of the Greenpeace ship *Rainbow Warrior*. This context was bound to amplify the international reaction to the French decision.

Following the announcement, Chirac and France reaped a whirlwind of international protest not only from expected regional states such as Australia and New Zealand, but also from Japan, from across Europe [including a protest inside the European Parliament as Chirac rose to address it], and even privately from the Germans piqued at not having been consulted on the decision[98]. Perhaps more seriously for Chirac, French public opposition to the tests rose to around 60 per cent[99] and his domestic popularity began to tumble. The United Kingdom alone appeared to stand by France, supportive both because its own nuclear position closely resembled that of France and because of the deepening nuclear ties between the two through the JCNPD[100].

Many on the right in France rallied to Chirac's defence claiming, amongst other things, that the international outcry was orchestrated by the perfidious Anglo-Saxons to prise France out of the South Pacific and, more credibly, that the furore was a storm that would blow over as states sought to rebuild relations with France once the tests had ended[101]. On one level the latter point proved valid and three years after the tests ended it is difficult to see the lasting impact of the French tests on Franco-German, Franco-Japanese or even Franco-Australian relations. But neither Chirac nor France sailed through the storm unscathed.

To begin with, Chirac paid a heavy domestic price for his decision and while it would be wrong to suggest that the tests were the only factor in his political decline the test furore nevertheless marked a turning point in his fortunes and the end of his Presidential honeymoon. Secondly the protests impacted on both French nuclear policy and French arms control diplomacy. In the face of the political storm Chirac cut the number of tests from eight to six [these took place on 5 September, 1 October, 27 October, 21 November, 27 December, 1975 and 27 January 1996] in order to curtail French difficulties[102]. This figure was well below the number of tests wanted by the military-technical community and, while France confidently asserted that the tests conducted were all successful, it can be doubted that just six tests, to confirm the credibility of the arsenal, collect data for PALEN and facilitate the production of a 'robust' follow-on to the TN-75 [the so-called TNN], fully resolved France's technical difficulties[103].

Even more importantly, in response to the furore Chirac was either bounced or persuaded into a series of maximal decisions with respect to French testing and arms control. He announced in August 1995 that France was committed to a 'zero-option' CTBT agreement, which prohibited low-level sub-kiloton fissile testing, in order to demonstrate a French lead in the process[104]. As Pascal Boniface has pointed out, this decision was critical because it closed the door on further developments of the kind of specialised nuclear weapons required for certain concepts of the 'more operational' nuclear doctrine favoured by many on the political right. As such 'the proponents of a resumption of nuclear testing on 13 June won a pyrrhic victory because the price they paid was a definite end to all [physical] testing and [thus] the freezing of the nuclear doctrine'[105].

A third impact of the tests was on French-European nuclear relations. The resumption decision, taken without consultation, was widely seen in Europe as French unilateralism, putting national before collective interests. The clumsy French offer on 31 August 1995 of an extended French nuclear guarantee for Europe and the claim that the testing decision was made in the interests of wider European security were widely derided by France's allies and set back the careful and patient advancement of ideas about the Europeanisation of the French nuclear deterrent.

Finally the testing resumption stirred up anti-French sentiments in the South Pacific DOM-TOMs and may yet be seen to have played a part in undermining French authority not only in the region but in the DOM-TOMs and *Francophonie* more widely[106].

President Chirac's third defence-related initiative was to set up on 11 July 1995 a Strategic Committee [*Comité Stratégique*] chaired by Defence Minister Charles Millon to conduct a strategic review of defence policy[107]. In addition to Millon the Strategic Committee comprised 12 individuals: the *Chef d'Etat-Major des Armées* Jean-Philippe Douin, the *Délégué-Général pour l'Armement* Henri Conze, the *Secrétaire-Général pour l'Adminstration* François Roussely, the *Chef d'Etat-Major de l'Armée de Terre* [Army] Amedée Monchal, the *Chef d'Etat-Major de la Marine* [Navy] Jean-Charles Lefebvre, the *Chef d'Etat-Major de l'Armée de l'Air* [Air Force] Jean Rannou, the *Directeur Général de la Gendarmerie Nationale* Bernard Prevost, the *Chef du Contrôle Général des Armées* Jean-Claude Roqueplo, the *Chef d'Etat-Major Particulier du Président de la République* Jean-Luc Delaunay, the *Directeur Chargé des Affaires Stratégiques* Jean-Claude Mallet, a *Chargé de Mission auprès du Premiere Ministre* Bruno Racine, and the *Secrétaire Général de la Défense Nationale* Jean Picq[108].

This committee is strikingly different in composition from the one which drew up the 1994 *Livre Blanc*. Heavy with senior military personnel it was clearly intended to help smooth over military opposition to change by co-opting the military brass into the decision-making process. At the same time the absence of Foreign and Financial Ministry representation indicates the limits of the committee's remit. The Strategic Committee was asked to reorganise the armed forces and defence industry cost effectively and to do so as far as possible within the wider framework established by the white paper. It had no formal decision-making power but was intended to feed into the ongoing process of debate within the *Conseil de Défense* which met intensively on the issue between September 1995 and February 1996. To this end the Strategic Committee's work was shared amongst five working groups dealing respectively with the future of the nuclear deterrent, the adaptation of conventional forces, the format of the armed forces and the role of professionalisation, the industrial politics of defence, and the modernisation of the management of defence. Overshadowing all was the view that the 1995–2000 *Loi de Programmation Militaire* was way too expensive and that the committee should examine ways of meeting defence needs while cutting defence spending.

As the committee deliberated the government cut the projected defence budget set by the 1995–2000 *Loi*. In July FF8.5 billion was cut from projected expenditure and a further FF3.5 billion was cut in December[109]. In all the procurement budget fell from a projected FF102.9 billion to FF91 billion, a fall of 11 per cent[110].

The deliberations of the *Conseil de Défense* and the *Comité Stratégique* became public knowledge on 22 February 1996 when President Chirac announced first in a TV interview carried jointly by France's premier channels TF1 and France2, and the day after in an address to 500 armed forces personnel at the *Ecole Militaire*, sweeping defence reforms under a Presidential initiative entitled *Une Défense Nouvelle 1997–2015*. Heralded by *Le Monde* as '*une véritable revolution*', the announcement prefigured what for many was the most far-reaching shake-up of the French armed forces since the end of the Algerian War[111].

Une Défense Nouvelle

The clearest public statement of the substance of the defence reforms announced by Chirac is to be found in a *dossier d'information* published by the Ministry of Defence through the armed forces public relations service SIRPA [*Service d'Information et de Relations Publiques des Armées*, now DICOD]. In all, six major changes were announced: the professionalisation of the armed forces; a reduction in the overall size of the armed forces of around 30 per cent; a reduction in the defence budget to a stable level well below that of the 1995–2000 *Loi de Programmation Militaire*; a simplification of the nuclear arsenal to a nuclear dyad based on the SNLE ballistic missile submarines and the ASMP stand-off missile [and follow-on ASMP-*amélioré*]; a redefinition of the functions of the armed forces; and the restructuring of the French defence industries. Each of these elements requires elaboration.

The professionalisation and downsizing of the armed forces are best understood together. The end of conscription which had existed in a similar form since 1905 ended almost a century of policy continuity but more importantly it severed a relationship between citizen and army with historic resonances back to the French revolution and the '*levée en masse*' at the Battle of Valmy. Those critical of professionalisation both before and after 1996 reached into this history for much of their argument and bewailed the isolation of the armed forces from the body politic and the loss of conscription as a social leveller and as the forge of a shared national identity and sense of individual citizenship[112]. Behind these anxieties lay the more practical problems of the cost of professionalisation [put by some critics as high as FF25 billion[113]], the impact on unemployment, the problems of future recruitment, and the concern that professionalisation would speed the subordination of French forces to NATO.

These anxieties are qualified somewhat by looking simultaneously at the downsizing planned for the armed forces. Table 3.1 shows the force level reductions determined by *Une Défense Nouvelle*. Overall the size of the armed forces was planned to fall from 577 360 to 434 000 but taking civilians from these numbers reveals a cut from 502 460 to 352 700 military personnel. Removing the Gendarmerie and those engaged in '*services communs*'[114] from the latter figures leaves just 244 500 operational combat forces [including an army of just 136 000], a reduction of 37 per cent from the comparable 1995 figure of 392 100 operational combat forces [including an army of 239 100].

Prior to the decision to professionalise the armed forces, conscripts did not of course comprise all or even the majority of armed force personnel. Figures for 1996, the year of change, show that the armed forces comprised a

Table 3.1:   The New Model Army in 2015

	Situation 1995	Horizon 2015
Army	239 000 military 32 000 civilian 271 000 total	136 000 military 34 000 civilian 170 000 total
	9 Divisions, 129 Regiments 927 heavy tanks 350 light tanks 340 helicopters	Around 85 Regiments organised in 4 main forces 420 heavy tanks 350 light tanks Around 180 helicopters
Navy	63 800 military 6 600 civilian 70 400 total	45 500 military 11 000 civilian 56 500 total
	101 ships [excl. SSBN] Total Tonnage: 314 000 – 2 Aircraft Carrier groups – 15 main frigates – 6 nuclear attack submarines – 7 diesel attack submarines – 33 maritime patrol aircraft	81 ships [excl. SSBN] Total Tonnage: 234 000 – 1 or 2 Aircraft Carrier Groups – 12 main frigates – 6 nuclear attack submarines – 0 diesel attack submarines – 22 maritime patrol aircraft
Air Force	89 200 military 4 900 civilian 94 100 total	63 000 military 7 000 civilian 70 000 total
	405 combat aircraft 86 transport aircraft 11 refuellers 101 helicopters	300 combat aircraft 52 transport aircraft 16 refuellers 84 helicopters
Gendarmerie	92 230 military 1 220 civilian 93 450 total	95 600 military 2 300 civilian 97 900 total
Services communs	18 130 military 29 780 civilian 47 910 total	12 600 military 27 000 civilian 39 600 total
Totals	502 460 military 74 900 civilian 577 360 total	352 700 military 81 300 civilian 434 000 total

Source   Loi De Programmation Militaire 1997–2002, Ministére de la Défense, May 1996.

total of 573 081 personnel of whom 297 836 were professional [52 per cent], 201 498 were conscripts or volunteers[115] [35 per cent] with a further 73 747 civilians [13 per cent][116]. Set against the planned force levels for 2002 these figures make interesting reading. *Une Défense Nouvelle* anticipated 434 000 personnel overall comprising approximately 326 000 professional [75 per cent], 27 000 volunteers [6 per cent] and 81 000 civilians [19 per cent][117]. In absolute terms professionalisation was thus going to require a rise in professional personnel from 297 836 to 326 000, an increase of about 28 000 or 9 per cent. Alarmist predictions about recruitment problems were consequently rejected. Similarly, exaggerated figures for the costs of professionalisation were also rejected on the basis that the direct cost of a downsized professional army would be about the same as that of a larger mixed force and that savings would in fact flow from the need to supply and equip a considerably smaller force[118].

Without itself specifying figures *Une Défense Nouvelle* set out cuts in overall defence spending which were subsequently detailed in the new 1997–2002 *Loi de Programmation Militaire* which was published in May 1996. The 1997–2002 *Loi* was the first of a three-stage *'plannification'* intended to complete the defence reforms by 2015, with a second phase from 2003–8 and a third phase from 2009–14. Following the first phase downsizing and professionalisation of the armed forces and the restructuring of the defence industries, the second phase would oversee an enhancement of the power and effectiveness of the French armed forces as new military hardware was delivered and became operational. The third phase would thereafter see France achieve a 'new equilibrium' of military power in a more 'Euroatlantic' defence context[119].

For phase one the new *Loi-Militaire* earmarked FF516 billion for defence equipment over the six-year [inclusive] period, around FF86 billion per annum. The annual figure for 1997 thus represented a cut of FF1.2 billion from the budget delivered in 1996 of FF87.2 billion and was set at fully FF18 billion below the FF104 billion planned for 1997 by the superseded 1995–2000 *Loi-Militaire*. Overall the 1997 defence budget comprised just 2.9 per cent of French GDP, the lowest figure since the Second World War[120]. Of the FF516 billion set for 1997–2002, FF389.5 billion [75 per cent] was planned for conventional forces, FF105.8 billion for nuclear weapons [21 per cent] and FF20.7 billion [4 per cent] for space-related military programmes.

On the conventional side the 136 000 combat personnel of the army of 2015 [in place from 2002] will be organised around four new highly mobile formations of 15 000 troops each, comprising a heavy armoured group, a mechanical group, a rapid armoured intervention group and an infantry assault group. In principle this ought to allow France to rapidly project up to 60 000 troops abroad, but in his speech to the *Ecole Militaire* Chirac suggested that the intention was to be able to project up to 30 000 troops into one distant combat theatre while simultaneously projecting a brigade [3–5000 troops] into one or two other theatres[121]. Perhaps against the logic of this emphasis the 1997–2002 *Loi-Militaire* cancelled funding for the Future Long-range Aircraft [FLA], the new-generation transport aircraft, but said that it would purchase FLA 'off-the-shelf' if the aircraft was developed by a private European consortium.

In terms of hardware the army is scheduled by 2015 to see a 55 per cent reduction in heavy tanks to 420, a 47 per cent reduction in combat helicopters comprising 180 Tigre and NH-90s and the modernisation but numerical stabilisation of light tanks at 350. The airforce is planned to have a 29 per cent cut in operational forces from 89 200 to 63 000 personnel, combat aircraft will fall by 26 per cent to comprise 300 Rafale aircraft, refuellers will be increased by 46 per cent to 16 aircraft, airforce helicopters face a cut of 77 per cent to 84, and – again against the projection logic – a cut in transport aircraft of 40 per cent to 52 aircraft. While the strategic nuclear airforce [*Force Aerienne Stratégique* – FAS] remains organisationally unchanged, the rest of the airforce is being restructured into three elements: a combat airforce, a projection airforce, and a surveillance, intelligence and communications component.

The navy is earmarked for a 34 per cent cut in operational forces from 68 800 to 45 500 personnel, major surface vessels [excluding frigates] will fall by 20 per cent to 81 ships, frigates by 20 per cent to 12, maritime patrol aircraft by 33 per cent to 22 aircraft, and nuclear attack submarines will remain unchanged at 6 vessels. The future with respect to aircraft carriers is less clear. The navy took delivery of the *Charles de Gaulle* nuclear-powered aircraft carrier in February 1997 and it was scheduled to enter service and replace the *Clemenceau* in 1999. After a great deal of speculation about whether a second aircraft carrier would be ordered to replace the *Foch,* the 1997–2002 *Loi-Militaire* fudged the issues by agreeing to order a second carrier in 2002 subject to finances and without specifying that it would necessarily be nuclear-powered.

As with the airforce the navy is to be restructured. While the strategic nuclear submarines [*Force Océanique Stratégique* – FOST] remains unchanged, the rest of the navy is being reorganised into four elements: a naval action force [*Force d'Action Navale* – FAN], a submarine action group [GASM], a mine warfare group [FGM] and a naval air group[122].

In terms of the nuclear arsenal the cuts and downsizing were equally far-reaching. In all six major decisions were taken:

(1)   the 18 S3D IRBMs on the Plateau d'Albion were to be quickly scrapped by 1998 and the fissile material from them recycled. To ease the impact on the local economy and make use of the facilities the Plateau d'Albion sites were to be converted for use as a signals and electronic intelligence centre for the French intelligence service, the DGSE [*Direction Générale de la Sécurité Extérieure*][123];

(2)   the anachronistic Hadès mobile tactical missile, the subject of Franco-German tensions throughout the 1980s and mothballed by Mitterrand in 1992, were to be scrapped. This decision was evidently taken after discussions between Chirac and Kohl in which Chirac appears to have tried to make political capital from the move[124];

(3)   the South Pacific nuclear weapons test facilities at Moruroa and Fangataufa were to be closed in line with the French decision to sign the Raratonga Treaty in order to promote the ongoing CTBT process;

(4)   because of the surfeit of fissile material in the French stockpile following the scrapping of the S3D and Hadès missiles, the decision was taken to close the fissile material production and processing plants at Marcoule and Pierrelatte, close to the Plateau d'Albion, ending the production of fissile material;

(5)   as part of the overall restructuring and following the decision to end all physical testing of French nuclear weapons, two CEA/DAM research facilities were also to be shut: the warhead design facility at Limeil to the south-east of Paris and the high-explosives facility at Valjours to the north-west of Paris; and

(6)   finally, the future 'shape' of the overall nuclear arsenal was announced. The French deterrent would rest on four Triomphant-class SNLEs [down from the six the military had originally wanted], the first of which entered service in 1996 and the last of which was due in service in 2005. The submarines would eventually be equipped with the M-51 missile, a simplified version of the M-5 replacement for the M-45 which was proving to be 'over-sophisticated' in design[125]. The M-51 missile in turn was be fitted with the currently-designated TNN warhead, a simplified and 'more robust' development of the TN-75, which is intended to allow France to maintain a nuclear arsenal over the long term without physical testing.

The second leg of France's nuclear dyad was to be provided by the airforce and navy variants of the Rafale aircraft [the latter deployed aboard the *Charles de Gaulle* carrier and probably on a second carrier if eventually built] using a follow-on ASMP missile designated the ASMP-*amélioré*, expected to have a range of at least 500km [the present ASMP has a range of about 350 km] and due in service in 2007. As with the M-51 warhead, the ASMP-*amélioré* missile is to be fitted with a simpler and more robust warhead, itself a variant of the TNN[126].

Professionalisation, conventional and nuclear downsizing and restructuring were aimed at shaping French forces overall into a 'new model army' orientated towards the new geostrategic context. *Une Défense Nouvelle* is quite specific in arguing that French forces are most likely in the decades ahead to be engaged in multilateral operations within NATO/WEU or in *ad hoc* groups, principally on the European continent in the Mediterranean basin or on Europe's southern flank; contributing to international peace and stability under UN or European mandate; or engaged as a consequence of defence accords, particularly with African 'partners', in primarily national, but perhaps later European, operations[127]. The central theme of the national defence of France itself is combined with these roles to describe four '*grandes fonctions'* of French strategy: *dissuasion, prévention, projection* and *protection.* These ideas appeared, as we have seen, in the 1994 *Livre Blanc*[128], but in *Une Défense Nouvelle* they were expanded, refined and moved centre-stage as the *raison d'être* for the 'new model army'.

Une Défense Nouvelle reaffirms the centrality of nuclear weapons in French strategy, insisting that nuclear weapons remain the ultimate guarantee against any and all threats to France and French vital interests. In addition French nuclear weapons are said to contribute to the 'global' *dissuasion* of the Atlantic Alliance, and to be 'called upon' [without specifying by whom] to play an increasingly European role. Finally nuclear weapons are argued to play a role in France's status as a permanent member of the UN Security Council and as a signatory of the NPT. *Une Défense Nouvelle* however offers no definitive statement about the evolution of French nuclear strategy, although in his speech to the *Ecole Militaire* Chirac stated that France had to 'make the most of the respite offered by the present strategic situation in order to rethink [its] nuclear posture'[129], suggesting that the President at least envisaged an on-going and future-orientated nuclear debate rather than one necessarily confined to traditional concepts.

The replacement of a known and direct threat to France in a relatively stable context by disparate threats to France in a less stable context both within and outside Europe has given France the new strategic priority of *prévention. Prévention* is essentially about seeking to avoid the emergence or reappearance of threats to French security, the outbreak of conflicts or the escalation of situations which 'over time could become major threats'. The principal means of meeting these objectives are stated to be through utilising French intelligence capabilities and by the prepositioning of personnel and equipment. With respect to the former, *Une Défense Nouvelle* proposed enhancing intelligence-gathering and analysis by boosting the DGSE and the *Direction du Renseignement Militaire* [DRM – Military Intelligence Agency] and by promoting the space-based intelligence programmes, in particular the Helios II follow-on to the tri-national Helios I optical observation satellite and the Franco-German Horus radar satellite [which superseded Osiris]. The 1997–2002 *Loi-Militaire* consequently prefigured, subject to German agreement, the deployment of three Franco-German Helios II infra-red satellites from 2001 and the deployment of three Horus satellites from 2005. In addition funding was also provided to complete the Sarigue-NG electronic listening aircraft upgrade by 1999 and the Minrem electronic surveillance ship programme from 2001[130].

With respect to prepositioning, preparation for *prévention* is centred on reviewing forces and organisations to determine the appropriate blend of personnel, equipment and location to support French policy in or close to sensitive areas. The accompanying *Loi-Militaire* provides an additional 9–10 per cent for spending on upgrading existing bases but while garrisoned force levels were to be reduced the overall number of bases, in Africa at least, were not[131]. Based on inter-service units, naval surface vessels and combat and transport aircraft, the prepositioning of personnel and equipment is intended to allow France to respond rapidly to crises with sufficient force to defuse, damp down or stamp out potential problems.

The third strand of strategy – *projection* – involves the projection of force outside France 'in Europe and the World' in what Chirac has termed 'the priority for [French] conventional forces'[132]. The key elements of *projection* were to be the Army's four 15 000 combat groups, Leclerc tanks, VBCI [*Véhicule Blindé de Combat Infanterie*] armoured cars, Tigre helicopters, anti-tank missiles, multiple rocket systems and Horizon heli-borne radar; the Navy's air-naval and amphibious units deployed aboard a carrier-group supported by anti-aircraft frigates [Horizon], anti-submarine frigates and Hawkeye airborne radar; and an Air Force capability to project up to 100 combat, air defence, reconnaissance, ground attack and electronic warfare aircraft into a remote theatre supported by refuellers. Underwriting *projection* were to be two key command and control systems: the Syracuse II satcom network [and its follow-on] to provide inter-service and inter-allied command and field connectivity, and the PCIAT [*Poste de Commandement Interarmées de Théâtre*] system to command inter-service and inter-allied operations within a given theatre[133].

The final element of strategy was that of *protection*. In *Une Défense Nouvelle protection* relates primarily not to proliferation threats as in the *Livre Blanc*, but to low-level threats arising, *inter alia*, from terrorism, drug-trafficking, and internal security problems arising from population movements. As such, the emphasis of *protection* has shifted to issues of essentially domestic security, the purview of the *Gendarmerie* [National Military Police Force]. In line with this the reforms presaged an increase in the *Gendarmerie* from 92 230 to 95 600, bucking the general trend of downsizing, and stated that they would be supported by the airforce in relation to airborne approaches to France and by the Navy in relation to seaboard approaches[134].

The final part of the reforms prescribe the restructuring of French defence industries. The plans call for France's state-owned or state controlled industries to be reorganised around four 'poles': an Aeronautics and Space group intended to be centred around a merged Aérospatiale and Dassault Aviation[135]; an Electronics group premised on the privatisation of Thompson-SA which would in turn provoke 'a wide-ranging reorganisation of the electronics sector in France'[136]; a Mechanical Engineering group built around the GIAT industry's domination of French army weapons and the DCN's [*Direction des Constructions Navales*, part of the DGA] dominance of naval construction[137]; and the nuclear sector already in place to provide independence in nuclear weapons and related technologies[138].

The attraction of consolidating and reorganising the defence conventional industries in this manner were stated to be threefold. Firstly the changes were expected to stimulate domestic rationalisation and restructuring to increase the efficiency and competitiveness of French industries. According to some analysts these changes were expected to be 'crucial in resolving France's means/policy goals equation' and one commentator went so far as to argue that 'if defence industry reform cannot be successfully accomplished, most of the hoped-for benefits from the restructuring of the armed forces would prove illusory'[139]. Secondly it was hoped that the French 'poles' would act to facilitate the emergence of European 'poles' by functioning as 'federators' extending existing, and creating new, partnerships with other European defence industry giants. Thirdly it was hoped that these European 'poles' would then be in a better position to compete with US defence industries for international markets while assuring Europe's technological edge and relative defence autonomy.

The potential fragility of this sequential expectation was not lost on analysts nor were the self-serving national motives of the reforms difficult to perceive. If successful the defence industry reforms could in effect underwrite French leadership in European defence projects and, by extension, in European security. Moreover the changes could also be read as a French attempt to get European funding for projects which the French government itself was no longer willing or able to support[140].

The changes were not however without a downside. The Europeanisation of French defence industries in this way, for example, was an open admission that the Gaullist touchstone of national defence procurement independence was now an illusion. Furthermore tying French projects to European funding sacrificed independence in another manner by making some projects conditional on the support of partners. As German foot-dragging over the Helios II project had vividly illustrated, France's partners were not always as clear about the benefits of expensive military technology as France appeared to be, nor were defence interests always aligned.

While as a package *Une Défense Nouvelle* and the 1997–2002 *Loi-Militaire* were subject to widespread support – the latter being voted through parliament promptly in the *Assemblée Nationale* on 5 June 1996 and the *Sénat* on 18 June 1996 – the reforms also cut across many vested interests. In open statements Chirac committed himself personally to the fortunes of the reforms[141] and while he was praised at the time for taking tough decisions he could not escape later the consequences of having done so.

One of the most insightful critiques of the changes was presented by François Heisbourg in a newspaper commentary piece. In it he identified three obstacles to the successful implementation of the reforms. The first, the issue of phasing, drew attention to the problems arising because France, unlike its principal allies, had eschewed prompt reform as the Cold War ended. While allies had made deep budgetary cuts and taken painful decisions France had pressed on with an over-ambitious defence modernisation. As a consequence, Heisbourg argued, France now faced the dilemma of either a rapid adaptation to 'catch up' which could prove socially unsustainable or a longer-term adjustment which risked delaying 'catch up' and exposing the process of reform to the vagaries of time. The second critique pointed to the cost of change: the raft of closures and cuts, downsizing and shake-outs prefigured faced the government with great costs in meeting the social implications of change from French regional problems caused by the disbanding of regiments or industrial downsizing to the subsidies for French Polynesia to smooth over the closure of the French test facilities, to say nothing of the armed forces cuts themselves. Finally Heisbourg raised the issue of communication, arguing that the scope and speed of reform, particularly when there had been little public consultation or explanation, risked what there was of the national consensus on defence. The pace and nature of reforms were in certain respects driving change in French defence policy, for example with respect to European security and relations with NATO, while it remained uncertain that public opinion had yet come to terms with these new relations[142].

As early as June 1995 Chirac's popularity had begun to tumble from his election high in the wake of first the nuclear weapons tests and then creeping public anxiety about his policy of economic *rigeur* to meet the Maastricht criteria which left unemployment unmoved. By November 1995 opinion polls [IFOP and SOFRES] showed his popularity down to 27 per cent and in May 1996 *Le Monde* noted solemnly that 'never, before him, had a president known such a brutal erosion in the confidence of the French people in him'[143]. By the early spring of 1997 opinion polls put Chirac's personal support below 20 per cent and those dissatisfied with his performance as high as 69 per cent[144]. It is a measure of his political paralysis that in an effort to rally support for his economic policies Chirac chose this moment to trade on the public support for his party, dissolve parliament [a step he had resisted in 1995] and bring forward the 1998 legislative elections to 1997. The subsequent and unexpected election of Lionel Jospin's Socialist government on 25 May and 1 June with a relative majority took France into its third period of political cohabitation. It is pertinent to what followed that Chirac's star continued to wane.

Defence policy and the Jospin Government

In appointing his cabinet Jospin ushered in a new generation excluding such Socialist luminaries as Jacques Delors, Jack Lang and Michel Rocard. At the same time his relative majority required compromise with the Greens and Communists and the appointment of members of both parties into ministerial positions. The appointment of the budgetary specialist Alain Richard as Defence Minister signalled both Jospin's intention to exercise tight financial control of defence spending and his intention to avoid a competitor for his own influence in defence matters. The appointment of the vastly experienced Hubert Védrine as Foreign Minister was in turn intended to ensure fidelity with Socialist foreign policy [Védrine was diplomatic adviser and later Chief of Staff to Mitterrand with particularly good relations with Germany] and to handle the Elysée [Védrine having served as the principal liaison between Chirac and Mitterrand during the 1986–8 cohabitation]. Finally the appointment of Paul Quiles, former defence minister, as the chairman of the parliamentary all-party defence committee suggested a tight government grip on parliamentary defence matters[145].

While accepting the broad framework of the decisions made by Chirac from 1996 – that is professionalisation, the main procurement programmes, budgetary constraint, and the need for defence industry restructuring[146] – the new administration nevertheless moved quickly to adjust policy in line with changed priorities and preferences. At least four main areas of policy impact can be discerned. To begin with Jospin insisted, as his predecessors had, on the co-direction of French foreign and defence policy with Chirac and the exploitation of Prime Ministerial prerogative within the '*domaine partagé*'[147].

In his first year in office Jospin made three high-profile visits to armed forces bases, to the air force at Colmar [a base piloting the professionalisation process] on 28 July 1997[148], to the army at Var on 15 September 1997, and to the navy aboard *La Foudre* from Toulon on 3 April 1998[149]. During these visits and at his IHEDN speech in September 1997, Jospin made clear his intention to exercise a strong role in policy and operational defence matters, particularly with respect to the projection of force outside France.

The second strand of Jospin's influence was exerted on the defence budget and armed forces reforms. Jospin accepted the procurement budget of FF88.7 billion, set for 1997 but cut the budget sharply the following year to FF81 billion, well below the approximately FF86 billion per annum agreed by the 1997–2002 *Loi-Militaire* framework. The deep cut provoked consternation in many quarters. The political right attacked the decision as threatening the process of reform, undermining French defence industries and the armed forces themselves[150]. The trade unions responded critically, anxious about the implications for defence sector employment[151]. Even the usually silent military began to grumble. Senior officers like the CEMA Jean-Philippe Douin expressed his disquiet about the cuts[152], while the heads of the Air Force, Army and Navy spoke openly about the risks to the defence reforms and French security[153]. Behind these senior figures the French press carried portentous stories of widespread and growing unease throughout the military at their loss of influence and status[154].

Against this background the government took a number of measures throughout 1998 to defuse concern about defence. On 24 March at a meeting of the *Conseil de Défense* the President and Prime Minister agreed a 'contract' for the four year period 1999–2002 that defence should not be a 'subject of discord' between the cohabitation partners. Accordingly, four-year procurement budget stabilisation was agreed at not less than FF85 billion per annum and the basic and shared elements of policy were affirmed: strategic autonomy, the priority of projection forces and the resulting shift in relations between France and NATO, and the construction of a European defence[155]. With the exception of the budget however the 'contract' agreed ends but said little about means.

Later in 1998 a wholesale clearout of senior military personnel took place. Jean-Philippe Douin was replaced as CEMA by Jean-Pierre Kelche; the senior military figures at the Matignon and Ministry of Defence were replaced[156]; in July the Defence Minister reorganised his staff[157] and later the same month a new *Secrétaire Général de la Défense Nationale* was appointed to replace the outgoing incumbent who resigned for personal reasons[158].

Both the Defence Minister and the Prime Minister then took time in September to address the military directly to assure them of the enduring value and importance of the armed forces to France[159]. Military disquiet nevertheless remains palpable and is likely to remain an important factor in the success of French reforms[160].

Jospin also took a hand in defence industry restructuring, halting the privatisation of Thompson-CSF in July 1997 in a move which greatly irked Chirac[161]. Socialist distaste for privatisation thus emerged as an important factor in Chirac's plans for European defence industry 'poles' not least because several potential European partners for French firms indicated an unwillingness to merge with state-owned or state-controlled French industries. The deeper politics and economics of this process are beyond the purview of this study[162].

Finally the Jospin administration has led important initiatives – which appear to be ongoing – to improve and update the management of defence. In this their case has been considerably helped by a stinging condemnation of defence management published in June 1997 by the independent *Cour des Comptes* [public accounts office]. The report, withheld during the 1997 election campaign, entitled *La Gestion Budgétaire et la Programmation au Ministère de la Défense* attacked the Ministry of Defence for inadequate preparation for defence procurement projects, accepting unrealistic cost estimates for new armaments, programme drift, a failure to monitor or control costs during development and a lack of transparency and accountability[163].

Ordinarily the pronouncements of the *Cour des Comptes* have only limited impact on the closed defence policy-making process, but the timing of publication coincided with a particularly sensitive and politicised debate about defence management as the new government challenged the Elysée's ideas about how to achieve agreed ends. The report of the *Cour des Comptes* was consequently readily exploited by the new administration which moved quickly to review the defence management process. The Defence Minister Alain Richard launched a general review of defence programmes in the autumn of 1997[164] and began the implementation of changes in July 1998.

An early decision was the July announcement of a shake-out of the lo-gistics and support side of the armed forces. In essence this was about bringing these elements into line with the reduced needs of the smaller professional armed force envisaged by *Une Défense Nouvelle*. As such the shake-out, which was expected to cost 6000 jobs, can be understood as the logical working through of the process begun by Chirac in February 1996. In a second move Richard announced the replacement of SIRPA, the armed forces information and public relations organisation with a new body, the *Délégation à l'In-formation et à la Communication de la Défense* [DICOD] with an enhanced mandate to promote the armed forces, boost recruitment and strengthen the bonds between army and nation. Whereas SIRPA was directed by milit-ary personnel and developed a reputation for defensiveness and circumspec-tion, DICOD was to have a civilian head and to be tasked with opening up the military to scrutiny and criticism in a kind of belated *glasnost*. In appointing Jean-François Bureau directly from the *Cour des Comptes* to head up the DICOD the government signalled a clear intention to be bold. At the same time as a confidant of former defence ministers, Charles Hernu and Pierre Joxe, Bureau was expected to be sufficiently 'onside' to overcome mil-itary apprehension[165].

A third initiative, and one which may prove to be the most important, was the decision in September 1998 to facilitate parliamentary oversight of the defence budget by providing costing details of each procurement pro-gramme rather than simply handing parliament a lump sum figure for each branch of the armed forces[166]. Itemised expenditure will be in place by the summer of 1999 and represents a small if important step towards strength-ening the checks and balances in what remains an authoritarian area of the constitution. It may presage the start of a determined effort by the Jospin administration to extend transparency to facilitate oversight and accountab-ility and to allow parliament to exercise tougher scrutiny in defence matters.

Taken together, the initiatives of the Jospin administration have in certain respects challenged the 'norms' of cohabitation defence policy-making, en-larged the scope of government action, and may have established new pre-cedents in the exercise of government power within the existing constitutional framework[167]. Exactly why it has done so and been able to do so are com-plex questions which will probably only be fully answered in the longer term. Nevertheless at the time of writing several factors seem important. To begin with, the political weakness of Chirac and his reduced role in certain ele-ments of defence policy have undoubtedly created space for government man-oeuvre.

Secondly it is evident that the process of defence reform announced by Chirac in February 1996 has unleashed forces and processes over which no-one is exercising full control. Chirac was perhaps the architect of change but it seems that he is not to be the engineer of its implementation. In the turbulence of change there is considerable scope for the Jospin administration to exert influence in a way that is not possible in more settled times.

Thirdly on the cusp of the millennium and informed by a wider political and social malaise in France to which many commentators have drawn attention[168], a modernising broom is sweeping French institutions and methods of which the defence reforms are part. As one commentator put it: '[France] is entering a new phase where behind the unchanging text [of the constitution] a new spirit is being asserted little by little'[169]. In matters of defence, as elsewhere, change in the name of modernisation is finding a wide acceptance, a dynamic the Jospin administration is skilfully exploiting.

Reflecting on almost a decade of the internal reform of defence since the end of the Cold War, it is evident that France was slow to revise defence budgets downward and slow also to orientate defence policy towards the new context. Adjustments begun in the 1992–4 *Loi-Militaire* were only properly articulated in the 1994 *Livre Blanc* and the 1996 defence reforms, and will not be fully implemented until 2015. France has remained committed to its main end-of-Cold War defence programmes – notably the Leclerc tank, Rafale fighter and the Charles de Gaulle aircraft carrier – and even with nuclear downscaling has consequently found it impractical to commit the necessary resources to 'new context' programmes like space-based military assets and strategic airlift. Wide-ranging objectives and limited means have pushed France towards multinational procurement programmes in new context mission areas but this has created a dependence on others for the realisation of French ambitions.

Three periods of political cohabitation have gradually evolved a smoother *modus vivendi* between the President and government in such circumstances replacing the *domaine réservé* with a *domaine partagé*. *Fin de siècle* malaise in the French body politic and French institutions has prompted reform in defence, as in other areas of government activity, which though presently limited may yet open up and perhaps even rebalance the defence policy-making process. Certainly in the area of defence the Gaullist constitution, now more than 40 years old, appears anachronistic and unaccountable by contemporary standards and the weaknesses of its checks and balances seem unworthy of a state with such fine democratic traditions.

Having brought the discussion of the internal reform of defence up to date it is now appropriate to turn the attention to the evolution and future trajectory of French defence in Europe and in the wider international context.

4
French Defence Policy in the New Europe

Introduction

The previous chapter left the discussion of French security adaptation to the 'New Europe' at NATO's seminal summit at Brussels in January 1994, the formal point at which NATO's evolution and French security ambitions for Europe became intertwined through the CJTF concept and the willingness of the United States to see the European role in NATO develop. By 1994 France had begun a meaningful rapprochement with NATO accepting both NATO's continued dominance of European security and that ESDI could only be taken forward within it. It was by no means clear at the time however how far Franco-NATO rapprochement would go nor whether ESDI could become a functional reality. Five years later France's role in both processes is clearer and the likely trajectory of future French policy evident. This chapter is intended to explore these issues.

To do this, the following discussion focuses first on the Franco-NATO relationship after 1994 and assesses how French policy had shaped and been shaped by NATO and where Franco-NATO relations and the construction of ESDI within NATO are going. In order to more fully understand the French role in building ESDI and the implications for France of doing so, it is necessary to look at the prospects for the emergence of an autonomous ESDI in the longer term. The second part of this chapter undertakes this by exploring the French role in EU CFSP, in the relations between the EU, WEU and NATO and by assessing French efforts to build up credible and functional European military forces. The third part of the chapter completes the analysis by looking at the French role in the emergence of a European nuclear deterrent, seen by many in France as a *sine qua non* for a strategic Europe.

France and NATO after 1994

The 1994 *Livre Blanc sur la Défense*, published just a few months after the January 1994 NATO summit, provided a remarkable summary of the evolution of French policy towards NATO at the time. In looking ahead it stated that, in relation to the evolution of the Alliance, French politics would conform to a number of constant principles [*principes constants*]:

(1) the Atlantic Alliance must become an arena where the European security and defence identity can establish itself;
(2) the adaptation of NATO to the new strategic conditions must be resolutely pursued;
(3) the Alliance must be the main arena for consultation between Europeans and North Americans for the major questions which directly affect the security of member states; and
(4) an increasing opening up to other European countries, in particular those in Central and Eastern Europe, must be organised and intergovernmental co-operation on areas of security and defence must be encouraged[1].

What is remarkable is the extent to which these 'principles' are a reversal of the French policy towards NATO articulated at the end of the Cold War, a point which illustrates just how far France had adjusted by 1994. Until late in 1991 France had been pursuing the objectives that ESDI was to be constructed outside NATO, the adaptation of the NATO alliance was to be resisted and NATO pinned back to traditional roles, NATO was to share transatlantic security roles with the then CSCE [and North Americans were to be excluded from Mitterrand's pan-European confederation], and NATO eastward expansion was to be resisted. In less than three years these objectives had all been abandoned in favour of an inescapably Atlanticist position born of French national weakness, European policy failure and necessary pragmatism.

At the same time tension inside France and the ghost of de Gaulle demanded that neither the degree of French compromise nor the future implications for France could be publicly uttered, not least because, as Heisbourg has pointed out elsewhere, no-one in the run-up to the 1995 presidential elections was sure that the French public had yet caught up with, let alone endorsed, Franco-NATO rapprochement. As if to pour oil on troubled waters and reassure the disciples of orthodoxy, the *Livre Blanc* thus confidently opined that:

> [the] different developments don't modify our particular military situation inside NATO. The principles laid down in 1966 (non-participation in the integrated military organisation, free use of our forces and territory, the independence of our nuclear force, freedom to assess our security in periods of crisis, and freedom to choose our means in case of action) will continue to guide our relations with the [Alliance][2].

The reluctance of Mitterrand in particular to see rapprochement with NATO move too far or too fast – evident for example in his decision to block the participation of the CEMA in the April 1994 meeting of the Military Committee[3]– and the requirement to observe Gaullist norms kept France at arm's length and led to all manner of procedural awkwardness as meetings which included France had to hang a sign of informality and French representatives naturally excluded themselves from those elements of discussions which did not relate to NATO's peacekeeping roles. That said, the return of French representatives to collective discussions within NATO – even informally – enhanced the important dynamic of institutional socialisation underway since French involvement in NATO's early Bosnian initiatives.

A decisive foot on the brake of further French-NATO rapprochement was removed when Mitterrand left office in the spring of 1995 and was replaced by Jacques Chirac. At this point it is important to cast an eye over the political landscape. Chirac, despite his impeccable Gaullist credentials, had in the run-up to the Presidential election made known his intention to narrow the gap between France and NATO. His election thus in certain respects mandated further adjustment. At the same time Chirac, precisely because his Gaullism was unquestioned, was arguably in the strongest position to deliver that adjustment, much as US President Richard Nixon's unquestioned anti-communism had allowed him to deliver US-Chinese rapprochement and SALT I in the 1970s.

At the same time NATO's evolution narrowed the divide from the other side creating an opportunity for Chirac to end Europe's 'other Cold War' between France and NATO[4]. The gap between the two, however, remained too wide for Chirac to leap in one bound, particularly with de Gaulle on his back. France's return to the NATO fold thus required further adjustment on both sides and, to understand much of what followed, it is important to understand how France saw these issues in 1995.

Were Chirac simply seeking to return to the NATO fold, the process would have been relatively smooth and brief. The weight of history and the mindset of estrangement however required that France had to be seen to be true to Gaullist principles and to wring adjustment from NATO so that both the point from which France leapt and the point at which France landed could be seen to have been determined, or at least decisively shaped, by France. To achieve this, as one analyst put it, Chirac's objectives were 'to exploit the ongoing European and transatlantic evolutions in order to restore France as a major player in the Alliance and to change and, most of all, to Europeanise NATO from within'[5]. France would return fully only to a French conception of NATO [or as close as possible to it] and only through a Europeanised NATO could France be restored as a major player within the alliance. This was arguably a quintessentially French way of making a virtue out of a necessity, but Chirac's rapprochement with NATO was still for some 'the most ambitious and extreme adaptation of the country's defence and security policy since de Gaulle'[6].

The issues thus became exactly what changes France required of NATO in order to return to full participation. Chirac's problems with NATO were essentially the same as those of de Gaulle [and every French president between them]: the dominance of the United States and the imbalance of power between the United States and Europe; the automaticity of the integrated military structure; and, the lack of political management of NATO relating in particular to the dominance of SACEUR within NATO. In the context of the Cold War, these historic French criticisms were often seen as absurd given the nature of the Soviet threat, the *de facto* reliance of Western Europe on the United States and the political and military need for automaticity. In the post-Cold War context the criticisms had more bite and indeed NATO had since 1990 been moving to address certain of them amongst the issues shaping adaptation to the post-Cold War context. The January 1994 summit statements on ESDI and NATO's European pillar, non-Article 5 missions, and the CJTF concept softened NATO's position on some of France's structural criticisms but, while welcome to France, these adjustments did not amount to the transformation of NATO Chirac was seeking.

To understand what France wants it is necessary to address each of the French criticisms in turn. The overriding issue is that of the dominance of the United States both politically and in the organisation, command and operation of NATO. As has been seen, the French assertion of great power status was premised on national independence and autonomy both of which, beginning with de Gaulle, were felt to be fatally compromised by full NATO membership which subordinated France to the United States and eroded national autonomy. The price France paid for independence from NATO during the Cold War was a loss of influence in the Alliance. In the post-Cold War era Chirac's restoration of France as a 'major player in the Alliance' was dependent on regaining influence with NATO [a need dictated also by the French position on the ground in Bosnia] and risking, under the Gaullist formula at least, a concomitant loss of independence. In line with European security ambitions the restoration of French influence could thus only flow from a rebalancing of US–European relations within NATO without which France's [and Europe's] voice would remain muted. At the same time in the new context the inverse influence-independence dynamic could itself be questioned as the absence of the Soviet threat weakened the arguments in favour of military integration. The Europeanisation of NATO thus offered the prospect of asserting meaningful French influence inside NATO while the emergence of the CJTF and non-Article 5 missions implied NATO's evolution away from Cold War integration.

The second set of issues for France are those relating to NATO organisation, command and operations. France expects to see US–European rebalancing at the political level reflected in the way NATO is organised and functions at the military level and the interface between the two. The legacy of the Cold War in this respect is an organisation nominally organised for shared decision and action but in practice dominated by the United States. For France the new context requires that US dominance is challenged to make 'a European autonomy within NATO possible and [to adjust] the NATO structure to the requirements of collective security'[7].

With respect to the former, 'European autonomy' means the capacity for European member states of NATO to act together without US participation and to do so eventually without a US veto [otherwise autonomy is a fiction]. At present European autonomy is primarily envisaged in relation to non-Article 5 missions since Article 5 missions are assumed to necessarily implicate the United States. To realise European autonomy NATO needs to be structured and commanded in such a way as to facilitate uniquely European operations. This means Europeans in senior command positions, a chain of command which permits European autonomy, and access to the NATO and relevant US military assets – particularly airlift, command and control and intelligence – required to undertake designated non-Article 5 missions.

With respect to the latter, collective security roles require, in the French estimation, the loosening of the integrated military structure because the new context and new roles require a less centralised command structure and one capable of operating with the participation of a sub-set of NATO-member states, given that not all NATO members will necessarily want to participate in non-Article 5 missions. Between 1993 and 1995 France pushed a twin-track approach to solve the mission command dilemma. Under this conception two military structures would have coexisted: one based on the IMS for Article 5 missions, the other a new – and necessarily flexible and autonomous – structure for non-Article 5 missions[8]. When this approach failed, largely due to European fears about duplication and reticence about the implications of realising European autonomy, the French position became a prisoner of its own logic. Given that two military structures were unrealistic and given that the IMS was, according to France, not suitable for non-Article 5 missions, it followed that the IMS had to be mutated into a much looser and flexible structure.

A second issue relating to non-Article 5 collective security roles was the need for enhanced political control of NATO. Within NATO this meant wresting away from SACEUR and the integrated military commands the politico-military functions designated – at least in NATO's formal conception – for the non-integrated Military Committee[9]. In parallel, with a loosening of the IMS, redistributing power from SACEUR to the Military Committee would go a long way towards rebalancing US-European influence within NATO, further enhancing French influence within the Alliance.

Enhancing political control of NATO also meant ensuring that NATO was responsive to the collective will rather than being simply the instrument of American foreign policy. To this end France sought to legitimise the use of NATO power by embedding it either within the United Nations or the OSCE, on the basis that NATO's claim of a collective security function would be more legitimate if mandated by a genuinely inclusive forum which might reasonably claim to be acting on behalf of the international or regional community [however defined]. Not irrelevantly, a consequent advantage for France of embedding NATO in this way in either the UN or OSCE would be to shift crucial elements of political decision-making to fora in which France had equal [or at least more equal] status with the United States while at the same time weakening the importance of US dominance at the NATO level.

Two observations about the French 'preconditions' for returning to full participation in NATO are appropriate at this point. Firstly, it is clear that there has been in France no fundamental questioning of the Gaullist framework of thinking about Franco-NATO relations in the post-Cold War context. This framework is restated in the 1994 *Livre Blanc* and in its application appears to be an unquestionable axiom. Despite the passage of time, the novelty of the unipolar world, the flux in European security, and the diversity of new threats and processes, each step of Franco-NATO rapprochement continues to be tested against the verities of 1966[10]. No-one at the policy level appears to ask whether Gaullism is an appropriate template or whether adherence to it is damaging France.

Secondly, there is a remarkable convergence between France's view of the conceptual adaptations necessary to ensure the relevance and utility of NATO in the post-Cold War context and the changes which will 'restore France as a major player in the Alliance'. Through French eyes at least, what is good for France in Franco-NATO relations is also good for Europe and NATO itself. This picks up a theme which Paul Gebhard has touched upon that France has a diplomatic gift for crafting national policy in collective terms and making acceptance of the French position a 'litmus test' of the collective good[11].

To move closer to NATO France was looking for clear evidence of NATO's Europeanisation, a Europeanisation which of course would enhance the role of France within NATO. The stronger France would be within NATO, the more likely a French return to the fold would be. On this basis France could even return to NATO short of the full realisation of its objectives, confident that NATO's evolution was moving in its favour and confident also that France's capacity to influence NATO from within would increase in line with those trends. The oft-quoted American critique that French rapprochement with NATO was a 'Trojan horse' – France coming into the alliance in order to change it from within – was thus justified even if the modalities of French policy were somewhat oversimplified.

Through 1995, and against the background of events in Bosnia which were underlining the primacy of the United States and NATO in the region, France began the implementation of Chirac's new approach to NATO. Following a review of Franco-NATO policy co-ordinated by the SGDN[12] – which as an interministerial body [see Chapter 1] was intended to rebalance relations between the Foreign Ministry, a bastion of Gaullist orthodoxy which had traditionally dominated Franco-NATO policy, and the Defence Ministry whose role had been growing in line with France's involvement with NATO forces in Bosnia – Chirac made a statement to his Ambassadors at the end of August which set out the intention to seek a path back to NATO. Little over a month later the Defence Minister Charles Millon attended a second informal meeting of NATO Defence Ministers and Jean-Philippe Douin attended a meeting of the Military Committee, the first CEMA to do so since 1966[13].

The unfolding French position was clarified and given further impetus by the French foreign minister Hervé de Charette in his address to the North Atlantic Council on 5 December 1995[14]. In the speech de Charette announced that France would henceforth participate fully in the Military Committee, ending the awkward and limited participation in its agenda, and that the French defence minister would 'regularly' participate in the North Atlantic Council meetings, moves summed up by Defence Minister Charles Millon as France resuming its place in those military elements of NATO which 'respect sovereignty'[15]. De Charette's statement fell short of fully reinstating the defence minister and pointedly excluded any move back to the IMS or NPG, but it nevertheless represented a significant shift. As further evidence of this 'normalisation', de Charette also noted that France would assume its place in those organisations within the purview of the Military Committee – including NATO's Situation Centre [Sitcen], and the NATO Defence College and SHAPE School – and that France would seek to further enhance its relationship with SHAPE[16]. De Charette also looked forward, effectively setting out the offer of France's full return to NATO subject to NATO's evolution.

At the end of 1995 the omens for those final steps seemed propitious: French and US policy convergence on Bosnia had led to the Dayton Peace Accords and French forces under IFOR were directly subordinate to NATO command for the first time since 1966; France had taken important steps back into the NATO fold; and NATO itself appeared to be moving in the direction of meaningful Europeanisation in the aftermath of the 1994 Brussels summit. As the French government wrestled with the final elements of the defence reforms which were to be announced as *Une Défense Nouvelle* in February 1996, the French degree of confidence that its conception of the Franco-NATO relationship was materialising can to some extent be judged by the degree of 'NATOisation' in those reforms. Professionalising the armed forces and tasking them for multilateral and primarily 'out of area' operations looked remarkably to many analysts like shaping the French forces to slot smoothly into, or at least closely beside, NATO[17]. As one commentator put it:

> the irreversible transformation of the French army to a professional army
> obliges France to co-operate more and more with NATO, which, in time,

makes the return of France to NATO's integrated military structure inevitable[18].

To the extent that this was so, France was nailing its colours to a mast that was not yet fully in place, the risk being that if NATO continued the evident Europeanising trends all would be well, but if not France's defence reorganisation might leave it with little option but to continue rapprochement on less favourable terms given that an economically driven denationalisation and multilateralisation of defence policy left France with few options. Within 18 months France was to be disabused of much of its optimism about the trajectory of NATO.

Shortly before *Une Défense Nouvelle* was published, President Chirac flew to the United States to press both the President and a depleted Congress [by some accounts up to 90 per cent of House and Senate members boycotted or absented themselves from Chirac's speech in response to the testing resumption or as a result of indifference[19]] for a more equal security partnership between the US and Europe[20]. Chirac's uncomfortable American trip, from which he returned essentially empty-handed, and de Charette's December announcement on French steps back to NATO, provoked criticism in France, particularly from the Socialists, that under Chirac France was conceding much to the US with little to show in return[21]. Under pressure Chirac needed a demonstration of NATO's Europeanisation if he was to calm domestic dissent and take Franco-NATO rapprochement forward on more favourable terms. Three processes held out the prospect of such a 'success': the CJTF concept, the practicalities of which were being debated in the run-up to the NATO Berlin summit in June 1996; the reorganisation of NATO and in particular the distribution of senior command positions between the US and Europeans; and the enlargement of NATO. It is important to consider each in some detail.

After the January 1994 NATO summit a whole set of issues emerged to be addressed if the CJTF was to move beyond the conceptual stage, including the precise role of the WEU and its relationship with NATO, authority and leadership, command structures and the relationship to SACEUR and the IMS. In the debates between January 1994 and June 1996 France sought a Europeanist resolution of each of these issues, premised on maximal autonomy and power for Europeans within NATO.

In the event the 3 June 1996 Berlin summit package on the CJTF contained considerable evidence of NATO's accommodation of the French position but it fell short of delivering the genuine European autonomy France sought. With respect to the WEU the CJTF could be WEU-led but not WEU-authorised and European assets within NATO were to be 'separable but not separate' to allow uniquely European operations. To further facilitate European autonomy the summit agreed to NATO/WEU 'dual-hatting' an arrangement under which certain European NATO commanders would simultaneously be designated a WEU command role from the deputy SACEUR [DSACEUR], who would be the senior WEU commander, down the command chain. The possibility of European autonomy in turn eroded the dominance of SACEUR and while no agreement to abolish or adapt the IMS was made the possibility of action by a uniquely European sub-set of NATO members implied its loosening. All of this could be presented as a diplomatic triumph in Paris.

On the other hand the United States and European Atlanticists were unwilling to cede full European autonomy. The role of the WEU was to be conditional on NATO agreement, the crude division of tasks between NATO for Article 5 and the WEU for non-Article 5 missions was rejected as NATO asserted its relevance for non-Article 5 missions, and France [and other Europeanists] was denied the use of NATO military assets merely as a 'tool box'[22] for European security ambitions.

With respect to the latter point the United States' reasoning was simple enough: discussion about NATO assets being available to the WEU often overlooked the point that NATO's own assets [essentially theatre command and control, logistics, and AWACs] were largely confined to the NATO theatre. The assets Europeanists sought for the WEU such as 'out of area' command and control, airlift and intelligence were in fact US national assets. The maximal French position was thus dependent on the United States – in the name of rebalancing – handing over national assets without a veto or control over their use. That this required the United States to accept precisely the erosion of sovereignty and autonomy de Gaulle so completely rejected for France in 1966 is an irony which seems lost on the French, or at least on those in France who genuinely believed this arrangement would be acceptable in Washington.

The Berlin summit thus contained something for France in terms of building ESDI within NATO, albeit subject to US veto, but left the details of the accommodation for the future. It also, of course, left untested the issue of what exactly Europe could or would do if push ever came to shove. Could it act as a cohesive whole and autonomously from the United States within the framework agreed or were NATO's European members in practice unlikely ever to act without the United States, as the lessons of Bosnia seemed to imply? For some analysts America was giving Europe just enough rope to hang itself: Europeans would prove, as in the past, divided and unwilling to act independently of the United States, leaving ESDI within NATO as a convenient fiction amongst whose virtues was the ensnaring of the French[23].

The second arena for Chirac's ambitions in relation to NATO was the question of redistributing NATO commands as part of European-American rebalancing. This process led to the 'AFSOUTH spat' which played an important part in bringing Franco-NATO rapprochement to a halt by the spring of 1997. To understand it is necessary to look back to 1994. NATO's post-Cold War adaptation had by June 1994 led to the simplification of the top strata of commands from three to two Major NATO Commands [MNCs]. Allied

Command Channel [ACCHAN] was disbanded and NATO restructured under the remaining Allied Command Europe [ACE] and Allied Command Atlantic [ACLANT]. In the process the forfeiture of British command of ACCHAN ended any European command at the MNC level as both ACE and ACLANT were in American hands. In relation to the process of NATO's supposed Europeanisation this was clearly a retrograde step and one France sought to address. Given ACLANT's role and its base in the United States, the scope for European command rebalancing lay in the ACE structure. The French initially sought a European SACEUR to command ACE [see below] but this was an isolated view and utterly unacceptable to both the United States and NATO. It followed that the adjustment would have to take place at the next level down, namely within the Major Subordinate Commands [MSC] of ACE.

Within ACE a similar process of rationalisation was also underway at the MSC level with the scrapping of the British-led AF North West leaving just two MSCs: the German-commanded AFCENT and the US-commanded AFSOUTH. For France two senior European-led NATO commands had been lost and under the new structure the US was to retain control of three of the four top military jobs in NATO: SACEUR, SACLANT and CINCSOUTH. It was hard to argue that this was anything other than the United States consolidating its grip on NATO's structure in direct contradiction, at least for France, of the Europeanising dynamic. It followed, in the name of rebalancing, that a European should be given the CINCSOUTH job in command of AFSOUTH at Naples as part of the ongoing NATO restructuring.

The fate of AFSOUTH thus became crucial to France on several levels. A European commander for AFSOUTH would be a practical demonstration of NATO's evolution and of the United States' seriousness about Europeanisation. It would in turn strengthen Europe within NATO and thus enhance the prospects of Franco-NATO rapprochement by providing evidence that NATO was moving in a French-favoured direction. At the same time it would be a clear 'success' for Chirac, staying the hand of domestic detractors and strengthening domestic support for rapprochement. Moreover, a European command of AFSOUTH would neatly align European NATO's security concerns with France's own national security interests in the Mediterranean with considerable synergies for France and potentially important implications for ESDI.

The view from the United States was however very different. Having facilitated the CJTF concept, 'separable but not separate' forces, an enhanced role for the WEU, dual-hatting, the strengthening of the Deputy SACEUR [DSACEUR], the strengthening of political control through the Policy Coordination Group [PCG] and enhancement of the Military Committee, and access – albeit conditional – to NATO assets, the US saw no reason to go any further down the path of Europeanisation. This perception was underlined by events on the ground in Bosnia which suggested that a genuine Europeanisation of NATO would mean a weakening of NATO given the divisions amongst European NATO members. Moreover, in the post-Cold War context of instability and proliferation on Europe's periphery in the Middle East, North Africa and the Balkans, US command of NATO's southern region and the control of the militarily dominant US Sixth Fleet made AFSOUTH more important than ever to the United States. Underpinning the American position was the domestic political debate in the United States which would not allow NATO's Europeanisation to go too far without asking questions about why the United States continued to bear a heavy European defence burden.

Although characterised as a 'spat', the stand-off which emerged between the United States and France over AFSOUTH during 1996 and into 1997 was thus in fact the expression of the much deeper debate about American versus European influence within NATO. While the positions of the two main protagonists were sharply drawn and appeared mutually exclusive, there was still considerable shared interest in reaching some sort of agreement. Had the debate remained low-level and behind the scenes, that agreement might have been possible around one or more of the compromises being floated such as separating the US Sixth Fleet from AFSOUTH so that the US could control the Sixth Fleet directly through SACEUR, having two commanders of AFSOUTH [an American for the Sixth Fleet, Middle East, collective defence and peacekeeping involving the United States and a European commander for European peacekeeping missions], or an agreement to rotate the CINC-SOUTH post between the US and Europe with the US remaining in command in the short term[24]. The projection of the issue into the public domain in the second half of 1996 however led to the entrenchment of respective positions and effectively ended the prospects for compromise. Exactly how and why this happened is the subject of considerable controversy.

Some analysts have pointed to misunderstandings and misperceptions between France and the United States about the whole process for Franco-NATO rapprochement and NATO Europeanisation. According to this view France understood the NATO evolution evident at Brussels in 1994 and Berlin in 1996 as meaning that NATO was indeed evolving to meet French conditions for rapprochement and it followed that the United States would accept the French requirement for an extension of real European authority within NATO such as a European commander for AFSOUTH. For its part the United States understood French rapprochement with NATO [particularly the acceptance of NATO command in IFOR] as an acknowledgement of US leadership in Europe and that, in pursuit of ESDI within NATO, France was looking for a raised profile for Europeans [not least as a cover for reintegration] which the Brussels and Berlin summits had met, but *not* for real authority, the latter both because there was little support for it elsewhere amongst the other European members of NATO and because Europeans had in practice shown themselves unable and unwilling to exercise real authority at least in relation to practical security demands. The French demand for a European AFSOUTH commander was thus seen as a logical [and domestically important] next step in France while it was viewed in the United States as France introducing a new and unacceptable condition for French-NATO rapprochement[25].

These factors were undoubtedly important in colouring each side's handling of the AFSOUTH issue but to understand its public expression and French overreach it is necessary to return to December 1995. De Charette's announcements of French steps back to NATO were an expression of the French perception that NATO's Europeanisation was genuine, a point underscored by the concessions wrung from the United States in the run-up to the Berlin summit by the French, Germans, and British [in the so-called 'three plus one' discussions] on the CJTF and DSACEUR questions in particular. Buoyed by this success and evident 'European' cohesion, France effectively overreached itself with respect to the speed and ambition of the next steps by both overestimating how far German and British support would continue and how much more the US would concede.

Initially two French ideas were floated to take NATO Europeanisation forward: the more radical being that SACEUR would become a European post [balanced by a US SACLANT] and US status would be preserved by the appointment of a NATO military 'supremo' in a command position above MNC level and the less radical being the French-led [and initially German-supported] idea to appoint a European as CINCSOUTH. The German perception was that, given that the United States had happily subordinated its forces to a German CINCENT in AFCENT, it would presumably be content to see its forces subordinated to another European commander in AFSOUTH. While US and NATO opposition to the more radical French idea sank it without trace the less radical option was evidently not torpedoed, leaving the French with the distinct impression initially, based on the apparent absence of opposition, that it was viable and that a European CINCSOUTH would in effect be a consolation for the US and NATO rejection of the more ambitious model of a European SACEUR.

In the aftermath of Berlin France began to articulate the AFSOUTH idea more openly and, seeing it gain some momentum in Europe, the United States responded by quietly stating its opposition to the idea. In an exchange of letters in September and October 1996 with President Clinton, President Chirac cut across the carefully crafted language of his advisers and ministers inserting, reportedly in long-hand, a personal demand for AFSOUTH[26]. The leaking of Chirac's letters thrust the issue into the public domain and, as a consequence of Chirac's language, cast the issue in stark 'AFSOUTH or nothing' terms in relation to French rapprochement with NATO. Whether Chirac had his back to the wall as a result of his well known lack of diplomatic tact or whether he did it deliberately to try to strengthen his hand is of only marginal interest given that the effect was to elicit a firm and highly public American no to the proposal, making clear the relative power of the two protagonists. To compound French frustration France was attacked by the German government at the NATO defence ministers' meeting at Maastricht in October 1997 for isolating itself, and risking damage to NATO and European defence, over AFSOUTH[27].

Chirac's third area of possible Europeanising success was the NATO expansion process, first articulated in 1994 and made feasible after the Dayton agreements stabilised the Balkan conflict at the end of 1995. In line with most European states France had pushed for the initial number of new NATO members to be five to include Romania and Slovenia in addition to the three – the Czech republic, Hungary and Poland – favoured by the United States[28]. To understand Chirac's objectives the French position on NATO expansion requires elaboration at this point.

As Pascal Boniface points out, in one of very few substantial French articles on the issue, the debate within France on NATO enlargement [*élargissement*] was limited, being both a marginal issue in relation to France's central domestic issues and being overshadowed by the more important, as France saw it, security issues of Franco-NATO relations and US-European NATO rebalancing[29]. The lack of debate did not mean that the French position was unclear. From the end of the Cold War until 1994 the French position under Mitterrand was essentially one of outright opposition to a process which was seen to underwrite continued US hegemony in Europe, enhance the relevance of NATO in the new context, and undermine French-led attempts to promote pan-European processes like the CSCE/OSCE and Mitterrand's ill-fated European Confederation.

From the January 1994 summit however the French position shifted. Given the renewed importance of the United States, the fact of deepening NATO engagement with the former Warsaw Pact and the prospect of NATO's Europeanisation, the calculations changed to support *élargissement* on the grounds of boosting the European membership of NATO, aligning NATO and EU expansion, and conditioning Germany's engagement with Central and Eastern Europe by embedding it within NATO's inclusive Partnership for Peace [PfP] process. In Mitterrand's final year of office he nevertheless stuck to the view that the pace of expansion should be determined primarily by progress on NATO's internal reform and conditioned by sensitivity to Russia and Russian interests. The latter sought to trade on the historically good links between France and Russia and was picked up by Chirac as a major theme following the acceleration of the enlargement process from the end of 1995[30].

The widely shared French arguments for supporting Romanian and Slovenian accession were extensive: the states appeared to meet NATO's own admission criteria with respect to democracy and human rights as well as economic and political stability, they would balance the accession of the northern Czechs, Hungarians and Poles, and would provide important additions to

NATO in view of their geographic proximity to the unstable Balkans providing both a 'beachhead' for NATO into the region and a bulwark against the spread of regional instability[31]. [If NATO expansion was indeed 'projecting stability' where better to project it than into the Balkans?] An unstated subtext for the French was that Romania in particular, with its historically close links to France, would provide a supportive ally for France within NATO and a means to enhance French influence in the Balkans, in contrast to the perceptibly pro-US Czechs, Hungarians and Poles.

In the run-up to the NATO Madrid summit in July 1997 at which the first wave of new members were to be formally invited to begin the process of accession the United States made the unilateral announcement that only its three preferred states would be in the initial wave, to the dismay of the French and Romania's and Slovenia's other supporters[32]. Although there were real issues impacting on the American decision – including cost, the better-preparedness of the three preferred candidates for accession, and anxiety about Congressional support for ratification – the manner and evident finality of the American position was portrayed in France as hegemonic unilateralism[33]. Chirac's decision to drop his determined pressure for Romanian accession in the week prior to Madrid, under pressure not to disrupt the summit, served only to underline the miserable position of France in relations with NATO by the summer of 1997[34].

By then Chirac's three opportunities to demonstrate French influence with NATO to an increasingly sceptical strategic community had yielded little. French preferences on the CJTF concept, NATO command reorganisation and enlargement had effectively been overruled by the United States leaving Chirac open to the criticism that he had 'played all France's cards in advance'[35] while few of the promised changes – particularly with respect to NATO's internal adaptation – were in place.

The substance of the Madrid summit itself merely confirmed that the Europeanisation of NATO, presaged at Brussels and taken forward at Berlin, had ground to a virtual halt. Madrid's emphasis on NATO enlargement in a largely American-set agenda left France little to do but squabble about which countries would pay for the accession of the three new members and which would benefit from the sale of the military hardware necessary to bring the new members' armed forces up to NATO standard[36].

Chirac was saved the task of terminating Franco-NATO rapprochement by the election of the Socialist Jospin administration which, critical of the process, was happy to do it for him. The reality by April 1997, however, was that Franco-NATO rapprochement was already dead in the water, leaving the partly returned France in a curious limbo seen as little more than a continuation of the old French-NATO relationship by some but as a novel and uncomfortable position by others. For many the fact that French-NATO rapprochement had stopped did not negate the gains made, particularly since 1994, nor require France either to complete reintegration or reverse the steps taken since 1993[37]. As one Franco-centric analyst put it at the time, as far as France was concerned 'NATO can wait'[38].

France's limbo however was not cost-free. Above all the decision to terminate rapprochement risked some loss of French influence inside NATO presenting France with a problem: by limiting its engagement France was depriving NATO's European members of the most vociferous advocate of Europeanisation, namely itself. The French decision to end rapprochement thus threatened precisely the core objective France herself sought within NATO. One can argue that Europeanisation was becalmed by 1997 due to US intransigence and the failure of France's European partners [not least Germany and the UK] to back France when it mattered, but it was hard to escape the conclusion that the more distant France remained, the less likely NATO was to Europeanise. In sum, if a lack of French influence put a limit on Franco-NATO rapprochement it was equally the case that a lack of Franco-NATO rapprochement put a limit on French influence.

The erosion of French influence mattered for France not only in relation to Europeanisation. To begin with, NATO did not stand still while France wavered, NATO in that sense did not wait. Enlargement moved quickly ahead after Madrid leading to the accession of the three nominated states formally to NATO membership in March 1999. NATO's new command structure was agreed on 2 December 1997 based on two Strategic Commands SCAtlantic and SCEurope, the former with three Regional Commands [RCWest, RCEast and RCSouth-East] and the latter with two Regional Commands [RCNorth and RCSouth] in a restructuring which cut the number of military headquarters in NATO from 65 to 20 and perpetuated the United States dominance of three of the four most senior posts in NATO [SCAtlantic, SCEurope and RCSouth][39]. Finally, the development of a new strategic concept, in which France played a part, moved ahead to complete the adaptation of the Alliance to be unveiled at NATO's fiftieth anniversary summit in Washington on 23–25 April 1999. As in the early 1990s France was swept along by these changes whether or not it was shaping them and thus risked being marginalised by the self-imposed limits of its engagement.

Similarly, at the operational level France remained bound to NATO by events in the Balkans, as a consequence of national military downsizing and restructuring, and by the need to assert the relevance of France and the Europeans in the only organisation capable of responding to Europe's major security concerns. Non-participation in IFOR, for example, was simply not an option as France could hardly reduce Europe's role in the force by dropping out nor could it be seen to be doing less than, for example, Bulgaria or Latvia for European security. With this kind of operational imperative in place the erosion of influence within NATO risked widening the gap between French influence and operational obligations, the very gap which had been instrumental in pushing France back to NATO during the Bosnian conflict.

As in the past while NATO members debated the details of organisational changes, events on the ground in the Balkans contrived to move the agenda forward. An important test of the limits of NATO Europeanisation and the willingness of the Europeans within NATO to exercise the functional autonomy carved out since 1994 arose with the question of the succession to NATO's IFOR which had gone into Bosnia in the wake of the Dayton agreements in December 1995. The mandate for the 60 000 strong multinational force which included both NATO and former Warsaw Pact states under the PfP rubric, expired at the end of 1996 to be succeeded by the smaller NATO stabilisation force, SFOR, of around 35 000 troops [of whom approximately 7400 were American and 3300 French], the latter with an 18-month mandate which ran through to June 1998.

In 1996 and 1997 the participation of United States' ground forces in IFOR and SFOR was seen as the *sine qua non* for the success of the Dayton agreement both by most of Balkan signatories and by the NATO members themselves. In the run-up to the expiry of the SFOR mandate in mid-1998, however, while the need for an SFOR follow-on force was widely accepted, the question arose as to whether that force could comprise exclusively European troops. Potentially for proponents of ESDI, the French included, the prospect of an all-European SFOR II presented a clear opportunity to demonstrate European cohesion, functional autonomy and a willingness to act. If Europe could act, the case for ESDI within NATO would have been demonstrated, the WEU as the presumed command organisation would have been greatly empowered, and the organisational processes for building ESDI, effectively stalled since the Berlin summit, would have been revitalised.

By the spring of 1998 a number of factors had come together which appeared to make the option of a European SFOR follow-on both practical and desirable. Firstly, something approaching peace had taken root in Bosnia as the situation on the ground stabilised, permitting the NATO-led forces to be scaled down. As a result the Europeans [NATO and PfP partners] already had adequate forces on the ground to accommodate an American withdrawal. Secondly domestic pressure in the United States, primarily from Congress, was pushing the Clinton administration towards disengagement particularly as the United States had already extended its original one-year IFOR commitment a further 18 months through SFOR[40]. Thirdly the mechanics of at least a form of ESDI within NATO were theoretically in place requiring only implementation. Fourthly, set against the bold claims made for the CFSP and ESDI in the early 1990s in relation to the Yugoslav conflicts, the assumption of SFOR follow-on seemed like a relatively straightforward and risk-free task. Finally, there were the implications for ESDI if Europeans were seen *not* to assume the responsibility. As one analyst put it: 'if a Bosnia that had been at peace for four years as a result of a US-led military presence proves too much for Europe to take on, ESDI will be exposed as a myth rather than a nascent reality'[41].

In the event the United States remained in the slightly reduced SFOR II force [around 32 000 troops instead of 35 000] which took over from SFOR I on 2 June 1998. The chance to realise an all-European force was lost in the interplay of US hegemonic and structural reasons for continued US engagement. One was the need, given that the US was to remain in the driving seat politically, to maintain a US ground presence in order to avoid the US-European separation which had created so many problems in Bosnia between 1992 and 1995. The French [and the British] made it clear that they would leave if the United States did[42], there being evidently no political will in Europe to assume responsibility for Bosnia nor an American political will to relinquish it.

A second stated problem was Europe's lack of credibility in the region. Impotent without the United States between 1992 and 1995 it was arguably uncertain that an all-European force would command the respect of the parties to the Dayton accords or that it would be capable of deterring or containing renewed fighting[43]. Perhaps so, but arguably the presence of former Warsaw Pact forces in SFOR II could have given an all-European force a different kind of credibility in the region. Arguably too, 'lack of European credibility' arguments were something of a smokescreen both for Europeans who had no wish to meet the challenge of SFOR II and for the United States which had no wish to see its commitment to European security questioned [on the grounds that it expected to lead] nor to risk a pan-European stabilisation force whose complexion could raise questions about NATO's role and pave the way for enhanced Russian influence.

The US decision to stay, however, remained a body-blow to ESDI for at least two reasons. Firstly, because the very 'European autonomy' formula towards which France in particular had been working [i.e., European action supported by US logistics, command and control and so forth] was exposed as unworkable even in relatively benign circumstances so clearly in Europe's backyard. Secondly because the SFOR II decision did not resolve the 'European impotence' issue but merely ensured it would arise indefinitely whichever US forces were on the ground in Bosnia. SFOR II thus became a constant reminder that Europe had flunked yet another test of unity and willingness to act, casting a mocking shadow across whatever organisational steps towards ESDI unfolded within NATO.

Notwithstanding the pressures which bound France to SFOR and which kept the United States on the ground in Bosnia, France continued to look for room for manoeuvre to resolve the question, as a *Le Monde* editorial put it, of: '*comment ne pas être dehors sans être dedans*' [how not to be outside without being inside][44]. France's CEMA Jean-Philippe Douin argued enigmatically in December 1997 that France would not be 'running back to NATO like a schoolboy running home from school'[45] and elements of French behaviour in 1997 and 1998 upheld the view that France would continue to 'cultivate [its] difference and continue its solitary game'[46].

France held one of the deputy SFOR command posts at SFOR headquarters in Sarajevo and commanded one of the three SFOR divisional headquarters at Mostar in SFORs's Sector South-East [the others being headed by the UK and US at Banja Luka and Tuzla respectively][47]. Within Sector South-East, France had a minor triumph in encouraging Germany in the transition from IFOR to SFOR to send 2600 combat rather than support personnel from the Franco-German brigade [headquartered in Bosnia at Rajlovac] to make up the Franco-German contribution to SFOR. Europeans may not have been ready to act without the United States, but while the United States was engaged there was little point in wasting the opportunity to exercise even limited European operational defence co-operation.

Fears that the French 'cultivation of difference' was periodically giving rise to real problems for SFOR were given substance in a series of accusations of maverick French activity [or at least the activity of French mavericks] in dealings with Serbia and Bosnian Serbs. The first of these arose in December 1997 when The Hague's International Warcrimes Tribunal prosecutor Louise Arbour claimed that Bosnian Serb war criminals were being effectively shielded by French forces in the French SFOR sector from SFOR's own efforts to bring war criminals to justice[48]. The second was a related claim, by the American *Washington Post* newspaper in April 1998, that a French commander Hervé Gourmelon, working at the behest of the French government, had thwarted an SFOR operation in Palé to arrest the Bosnian Serb leader Radoran Karadzic on war crimes charges[49]. The third occurred when a French army major Pierre Brunel, the *Chef de Cabinet* to the General who represented France on NATO's Military Committee, reportedly confessed in November 1998 to having passed information to Serbian agents about NATO bombing targets against Serbia in October 1998[50]. In a revealing vignette, sources in Paris and Brussels attempted to play down the significance of the espionage by arguing that Brunel's knowledge was limited because of France's semi-detachment from NATO[51], despite the fact that France had returned fully to the Military Committee in 1995 and had a close working relationship with NATO on peacekeeping matters[52].

All three claims were either denied or ignored in Paris and by the French commanders in the Bosnian theatre but, as with similar accusations of French links with Serbia and Bosnian Serbs which emerged between 1992 and 1995 in Bosnia [see Chapter 3], the accusation served as *Libération* noted to 'reinforce the bad reputation which the French army suffers, notably amongst American and British armed services, in the former Yugoslavia'[53].

Storm clouds gathering over the Serbian southern province of Kosovo in the spring of 1998 presaged an escalating conflict which was to pose deep questions for France particularly in terms of relations with NATO. Omitted from the Dayton agreement, Kosovo, with a population of 2.1 million of whom around 90 per cent were ethnic Albanians, became in 1998 the scene of intensified hostilities between the separatist Kosova Liberation Army

[KLA] fighting for independence and the Serbian army and paramilitary police forces resisting KLA 'terrorism'[54]. Ostensibly in pursuit of the KLA, evidence emerged throughout 1998 of Serbian brutality in Kosovo against the Albanian population, sparking fears of renewed ethnic cleansing.

A UN resolution [1203] passed in August 1998 called for an immediate cease-fire and created space for the Holbrook-Milosevic accords which appeared in October to have achieved a framework peace deal, centred on Kosovan autonomy, though not independence. In support of the deal an OSCE Kosovan Verification Mission [*Mission de Vérification au Kosovo*, MVK] created on 25 October 1998 was earmarked to send 2000 OSCE 'verifiers' into Kosovo to monitor Serbian and Kosovan compliance with the cease-fire terms. Despite the deal, conflict and Serbian atrocities continued in Kosovo prompting NATO to use air-strikes to encourage Serbia to come to the negotiating table with Kosovan representatives at Rambouillet in France, a process which began in early February 1999.

France proposed and had accepted a NATO extraction force to support the OSCE verifiers as part of Operation *Joint Guarantor*, based in the neighbouring republic of Macedonia[55]. US reluctance to contribute ground forces to the extraction force paved the way for an all-European deployment of French, German, British, Dutch and Italian troops initially numbering 1800 [of whom 800 were French][56] commanded from the French base at Kumanovo, north east of the Macedonian capital Skopje. It was not lost on the French that the all-European make-up of the extraction force could be exploited to offset European 'weakness' *vis-à-vis* SFOR II. At the same time the absence on the ground of the United States [though it was present in the air operation *Eagle Eye* to provide theatre surveillance], the size of the extraction force and the lightness of its armour clearly indicated that it was not intended to enter Kosovo except perhaps under the most permissive circumstances.

In the spring of 1999 events spiralled rapidly out of control. The Serbian government of Slobodan Milosovic backed away from the Rambouillet negotiations largely on the grounds of having been bullied into them by the United States and NATO and a clear perception that the United States was not serious about enforcing compliance [a perception shaped by the use of OSCE observers in Kosovo, the small all-European extraction force in Macedonia, and repeated statements that NATO would not use ground forces][57]. As violence escalated inside Kosovo the OSCE force, of whom only around half the intended number had actually been deployed, was hastily withdrawn against the background of up to 40 000 Serbian forces massing ominously on the Kosovan–Serbian border while NATO threatened imminent and substantive air bombardment of Serbian targets. In the event Serbia did not sign the Rambouillet accord.

The commencement of NATO cruise missile and aircraft attacks on 24 March 1999 which quickly involved up to 20 French aircraft based at Istrana in Italy as well as the aircraft carrier *Foch* in the Adriatic[58], precipitated not the prompt capitulation of the Serbian leadership but a ferocious escalation of ethnic cleansing inside Kosovo. In all, amid reports of widespread and systematic Serbian massacres and rape, within a month more than 700 000 Kosovars, around 30 per cent of the population, were displaced into neighbouring territories in addition to more than 750 000 others displaced or murdered within Kosovo itself.

NATO stepped up the bombing and committed itself publicly to a set of war objectives aimed initially at restoring the situation *ante bellum* and subsequently at returning and fully ensuring the safety of Kosovan refugees. The Yugoslavian capitulation on 3 June and the open-ended commitment of KFOR ground forces on 12 June have enormous implications for NATO and the future of European security which cannot be considered here. What does matter here is the nature of French participation in NATO operations and the insight this offers about the trajectory of French policy.

One of the most striking features of the French participation in the NATO bombing of Serbia and Serb targets in Kosovo was the early and explicit commitment of the French government [Elysée, Matignon, Quai d'Orsay and rue St Dominique] to the *Alliance* war aims. This degree of alignment is important for several reasons. To begin with, it limited French diplomatic room for manoeuvre, confining France to attempts to meet aims set primarily by the United States and Britain rather than brokering a French deal [as for example Mitterrand attempted in the run-up to the 1990/1 Gulf War] or seeking to exploit Franco-Serbian or Franco-Russian relations in the name of French diplomatic distinctiveness. France thus fell wholly into line behind the United States.

A second feature was that, although the French government argued that French participation in the NATO bombing was consistent with its role in IFOR, SFOR and operation Deliberate Force and represented no change of policy, this is doubtful for at least two reasons. The first is the absence of an explicit UN mandate for the bombing. In the post-Cold War era France has been careful to position participation in NATO operations within a UN framework, not least to avoid the charge that it was acting as a tool of US foreign policy. In this case however, though arguments can be made about humanitarian intervention as the basis of legality, there was no unequivocal UN mandate for NATO action. In addition doubts have been expressed about NATO authority and war aims by a spectrum of critics within France including the socialist government's own interior minister Jean-Pierre Chevènement[59], the former RPR leader Philippe Seguin[60], the communist party[61], French intellectuals[62], trade unions[63], and even a former senior French commander in Bosnia[64]. That France fell into line behind the United States in the absence of a clear mandate is a novelty. Another reason to doubt the continuation of French policy is because for the first time France struck targets in Serbia itself, novel both because Serbia is a sovereign state and because of the previous 'distinctiveness' of Franco-Serbian relations.

Continuity of policy in changed circumstances amounts to an evolution of policy.

At the military level French forces remain under NATO operational command and in this respect there is no change from the French IFOR and SFOR deployments, and incidentally no distinction between French subordination to NATO and that of, for example, the British or Germans. As at the political level, however, the context introduces novelty because French forces attacking Serbian targets were under NATO command without an express UN mandate and in operations against a sovereign state.

Exactly why France should have thrown its lot so openly and comprehensively in with the Americans takes us to arguably the core issue France now faces in European security [and has arguably faced since the end of the Cold War]: in the new context the United States is more important to Europe [and to France] than the reverse, and NATO is so vital to European security and the United States so vital to NATO that there is quite literally no choice for France other than to fall in line behind the United States.

France cannot stand aside from NATO without its relevance and any expectation of European relevance within NATO being eroded. It is consequently obliged to do all it can to achieve NATO objectives at the cost of diplomatic and operational independence. Given that France has, at least since Dayton, been committed by structural factors to NATO support and operational participation, the decision not to fully reintegrate with NATO makes little sense as it limits French influence without providing France with any tangible benefits *vis-à-vis* NATO's other members states. France cannot escape the obligations of NATO membership, yet is not fully at the heart of decision-making. The irony is that the Gaullist mindset which keeps France from fully reintegrating with NATO is now placing France in the situation of having obligations with only limited influence, exactly the situation which led de Gaulle to partially withdraw from NATO in 1966. Given that France is unlikely to be able to cast off the obligations in NATO which have arisen in the post-Cold War context, it follows that those obligations should be matched by influence, a logic which, if accepted, points decisively towards France's prompt reintegration with NATO.

At its fiftieth anniversary summit in Washington DC from 23 to 25 April 1999, NATO unveiled its new strategic concept. It confirmed NATO as the principal guarantor of an expanded European regional security beyond the Article 6 theatre, allowed that NATO could act without a UN mandate [a point France disputed but obviously accepts] while recognising the UN's 'primary responsibility' for maintaining international peace and security, and recognised a European Union role in defence and the latter's right to approve military action where a European sub-set of NATO members alone were involved[65]. NATO's dominance of the European regional security agenda and the explicit linkage of NATO and the European Union point the way forward for ESDI in the medium term. To understand the role of France in this process and the implications for France, it is necessary to turn attention to the ESDI process itself.

ESDI: The Long Card

From the vantage point of the year 2000 and in relation to French expectations in the immediate aftermath of the Cold War, ESDI remains objectively a fragile and etiolated concept. Sown in the stony ground of geostrategic change, European indifference and institutional complexity and quickly choked by the vigorous growth of NATO, ESDI has remained stunted and undeveloped. For critics it is presently little more than a 'myth' and a 'convenient fiction', while even for a more generous analyst ESDI at present 'does not exist in any practical terms apart from the possibility of WEU-led CJTFs, presumably using Eurocorps or other Euro-designated NATO forces'[66].

But for some ESDI is in the eye of the beholder. In France it is seen not primarily in terms of its present weakness [though this is fully recognised] but in terms of the oak tree into which it can eventually grow, and in this there is more than an echo of de Gaulle's fidelity to a vision of France rather than a France of fact. In France the discourse on ESDI is focused largely on the future, on the unquestioned need to construct a European security and defence competence and on the processes and trajectory of that construction. Viewed in this way, Franco-NATO rapprochement and the coming together of ESDI and NATO's evolution after January 1994, for example, is seen as a *temporary* accommodation with NATO on the road to a distinct and genuinely autonomous strategic Europe. This underlying assumption surfaces repeatedly in French official statements about, and analyses of, Franco-NATO relations as, for example, following France's decisive steps back to NATO at the end of 1995 when it was pointed out that France was being 'more Atlanticist today in order to be more European tomorrow'[67] or when a leading analyst noted that the path to 'European defence passes *through* NATO [emphasis added]'[68].

The notion of a temporary accommodation with NATO means that for France Franco-NATO rapprochement is not about being permanently reconciled to the US-dominated *status quo* but rather is about building up ESDI within NATO to the point at which US-European balance is achieved and European strategic autonomy becomes a functional reality. At that point NATO will either continue nominally as the umbrella for a completely new transatlantic partnership between the US and Europe or it will become an irrelevance.

Recognising the dominance and importance of NATO in the medium term, however, means that the challenge for France and other Europeans, as Alain Juppé himself put it, is 'to make a European defence without unmaking the Atlantic Alliance'[69]. Making European defence is a dualistic process requiring both a European willingness to do what is necessary to realise it and an American willingness to allow it to happen. The suggestion that the United States may eventually give up some relative influence in Europe implies that the United States may in return demand something of Europeans.

This trade-off has led David Gompert and Stephen Larrabee to posit in the longer term a new transatlantic bargain: European willingness to acknowledge and support American dominance at the global level in exchange for European responsibility and American support for Europe on the European continent[70]. Under this kind of conception France has a decisive contribution to make on both counts, having shown itself – despite Gaullist antagonism to US hegemony – both to be one of the United States' most reliable partners in post-Cold War world order roles [*cf* the 1990/1 Gulf War, Somalia, Bosnia and Kosovo] and a leading architect, if not the leading architect, of ESDI.

The prospect of this kind of new transatlantic bargain, however far down the line, would tend to confirm the argument made in the previous section that Gaullist objections to French reintegration in NATO should be dropped as counter-productive and France should assert itself fully at the heart of NATO decision-making. As a reliable long-term partner for US global ambitions, France could reassure both the United States and European Atlanticists that ESDI is a complement to, and not a threat to, US objectives. Conversely French opposition to the fact of US global hegemony may serve only to undermine US and European confidence in ESDI which will appear, particularly if French-led, to be a challenge and a threat to the United States.

The present study is no place to pursue the wider questions of how the future of European security will pan out and whether a new transatlantic bargain will be struck. But issues like these will inevitably impact on the key questions here, namely how exactly does France intend to realise its vision of ESDI and what in turn are the implications for French defence and security policy of pursuing ESDI.

To begin with, we need to understand how France sees ESDI. The maximal vision of ESDI towards which France aspires is reasonably clear. It is centred on a cohesive and unitary European entity, Europe as *l'Europe-puissance*, functioning as a distinct strategic actor in a multipolar world co-operating and competing with other centres of power – the United States, Russia, China and perhaps others – in what is seen as a 'realist' global context[71]. Certain features of this European strategic entity are discernible. As a unitary actor, Europe will have achieved the common foreign and security policy presaged at Maastricht and thus will largely have set aside national approaches to foreign and security policy or will at least have found a way to realise national interests through collective processes. To function as an autonomous strategic actor, Europe will have developed a common defence and thus will be nuclear-armed [unless the rest of the world has denuclearised] and will have conventional military forces capable of meeting the entire range of Europe's defence needs, not least *vis-à-vis* other centres of power. Moreover, given French political and intellectual leadership on ESDI and

French 'comparative advantage' over most other European states in defence and security matters – based on the possession of nuclear weapons, a space-launch capability, and a global but limited military reach – Europe as a strategic actor will be cast largely in the French image and thereby will preserve French status within and outside Europe and will be able to act as a 'force multiplier' in pursuit of French-led objectives. Through Europe France will thus realise many, if not all, of the historic Gaullist ambitions it has been unable to realise nationally: a permanent resolution of the German issue, continental stability, continental leadership, strategic autonomy from the United States and the collective expression of French *grandeur* and *rang*[72].

The boldness of this conception of a strategic Europe is premised on few delusions either about the distance between the present reality and the future vision or the obstacles to moving from the former to the latter. There is, on the contrary, a realism underlying French ambitions which recognises that France is not in control of the events and processes which will ultimately determine whether or not a strategic Europe is realised but which also recognises that France need not simply drift or react to unfolding events and process dynamics. In pursuing a French conception of ESDI, with the emphasis on a security and defence *capability* rather than a mere *identity*, France is using its vision of a strategic Europe as a star to steer by, guiding each incremental step of French policy. As such the maximal view of French conceptions of ESDI is seldom fully articulated and most French statements about ESDI focus on intermediate or interim positions. The underlying vision is however implicit in the wording of the Maastricht treaty and annex, in the establishment of the Eurocorps, the revitalisation of the WEU, the Europeanisation of NATO, in French national defence reforms, and in French efforts to take ESDI forward since 1994.

The clarity of vision in France about the destiny of European defence and security has several important implications. Firstly it casts France as an *avant-garde* state pushing at the limits of what the United States will allow and often running ahead of the European 'centre of gravity' on ESDI, witnessed for example in the AFSOUTH problems. Secondly, it means that France is agenda-setting by virtue of its forward momentum while many other European states prefer to avoid change, to avoid hard choices, or seem unable to reconcile competing pressures. But thirdly it also means that France's clarity of vision is, like all zealotry, accompanied by a narrowing of focus. French blinkers on ESDI assume that what is good for France is good for Europe and NATO, often overstate and overplay French influence, often fail to perceive or accept alternatives and, by seeking compliance on French terms, generate tensions and obstacles and, not infrequently, result in French isolation.

Although many factors will impact upon them, there appear to be three main determinants of the eventual realisation of an ESDI of some sort. The first is the trajectory of the European project which has accelerated since the end of the Cold War via Maastricht and the introduction of the European single currency in January 1999. The logic of Maastricht in particular is that states which come together economically and politically will in due course evolve a common foreign and security policy [CFSP]. While evolving the latter, European states will need and will increasingly be able to develop a common defence. The eventual materialisation of ESDI is consequently the simple working through of this logic: common economics and politics in due course equals common foreign and security policy which in due course equals common defence.

The second is the move towards the realisation of ESDI through the interplay of relations between NATO, the EU and the WEU. Whatever one's views about the present reality or otherwise of ESDI, it is difficult to discount the conceptual and organisational advances towards ESDI made between the early 1990s and the explicit roles accorded to ESDI, the EU, and WEU in NATO's new strategic concept announced at the April 1999 Washington summit[73]. ESDI is firmly established in Europe's security architecture and is being given substance within NATO and outside NATO in the deliberations and activities of the EU and WEU. Even allowing that the pace on ESDI has slowed since 1996, there is clearly a process unfolding with some form of ESDI as its end point.

The third is the functional realisation of ESDI through practical and exclusively European defence co-operation in terms of organisational and operational developments and in terms of defence industry research, development and procurement [where US–European distinctiveness and competition is real and pregnant with implications]. France has an important hand in all three of these mutually-informing strands and it is consequently important here to examine these to assess some of the implications for French defence and security policy of focusing on a European defence future.

To deal first with the European Union, the obstacles to the EU developing a defence competence seem immense and to understand these obstacles it is necessary to say a little about the EU itself. The EU, at the time the EEC, has had mechanisms for the co-ordination of economic and political aspects of foreign policy at least since the Davignon Report of 1970 and the creation in 1974 of the European Council. This role was taken forward by the 1986 Single European Act which transformed the EEC into a broader political community and, *inter alia*, formally agreed to 'consistency' between political co-ordination inside the EC and the EC's external relations. As a 'civilian power' the EU [and its forerunners] has gradually become an important and usually cohesive actor in the international system. It exercises political influence through solidarity as a group both generally and within other fora [for example within the UN or OSCE], through moral suasion as an exemplar of how regional or inter-state relations can be conducted, and through bloc diplomacy by which other states and groups of states have sought diplomatic links directly with the EU [and its forerunners]. The EU also exercises important economic influence, often in parallel with political influence, through aid, trade and sanctions[74].

The EC's failure to find a collective response to the Gulf War of 1990/1 informed the Maastricht decision to build a CFSP to deepen existing foreign policy co-ordination and to extend EU competence into the military-security sphere, the limits and complexities of which were quickly exposed by the wars in Yugoslavia. For two of France's leading defence analysts, François Heisbourg and Pierre Lellouche, the wars in Yugoslavia rendered 'null and bypassed by History . . . the treaty of Maastricht, and in particular . . . the sections on "foreign policy and common security"'[75]. This proved to be an exaggeration and the announcement of the death of CFSP premature, but the criticisms did bring the difficulties of developing CFSP sharply into focus.

One set of obstacles to the development of CFSP, and in due course a common defence, relate to the nature of the EU itself. Despite the erosion of sovereignty arising from political and economic harmonisation, the 'high politics' agenda of the military-security of member states has remained resolutely national and intergovernmental within the EU. The EU requirement for unanimity of decision-making strikes at the most sovereign element of policy-making [decision-making autonomy] and subjects a given state to the potential veto of other members. Attempts to move away from unanimity, for example by creating a 'security council' of Europe's big five – France, Germany, Italy, Spain and the UK – accorded special rights and/or by moving to some meaningful form of majority voting, have been resisted by states which do not want the EU to be 'hijacked' by a sub-group of states and by those who prefer to hide behind unanimity to avoid hard choices. The nature of EU decision-making thus emerges as a fundamental obstacle to the evolution of an EU military-security competence as few states will willingly choose the quagmire of CFSP in the pursuit of 'high politics' objectives[76].

Another problem with the EU is the democratic deficit arising from the weakness of the European parliament and the limits of the accountability of the European Commission. The idea that the EU could one day order the military forces of member states into action is premised on a degree of EU legitimacy and democratic accountability which today seems fanciful. The problems of building towards that legitimacy and strengthening EU-level accountability, to the point at which this most sovereign function of states could be shared with or even transferred to the EU, is a second fundamental obstacle to an EU defence competence[77].

A third problem and one which appears to contradict, at least in the short term, the Maastricht logic is the paradox that the economic and financial rigour demanded of states in order to meet the Maastricht criteria for EMU has forced cuts across Europe in government spending and thus in defence budgets, thereby undermining the capacity of states to build up the very military capability required to realise a functional ESDI. While this situation may not persist, particularly if the single currency is a success and the European economy grows, it is pushing into the future the point at which European states can begin to seriously address their collective military weaknesses.

The scale of the mountain to climb to meet these requirements can to some extent be judged by the development of CFSP since Maastricht. Between 1992 and the Amsterdam IGC which began in 1996, CFSP stagnated due, as Jacques Santer the EU President at the time himself noted, to 'the lack of political will, the absence of a common definition of our essential joint interests, the difficulty of activating the unanimous decision-making system, the crippling budgetary procedures, the ambiguity of the roles of the Presidency and the Commission, the European Union's lack of legal identity, and the problem of its external representation'[78].

The 1996–7 Amsterdam IGC, widely expected to take the CFSP agenda significantly forward, proved to be a disappointment to those anticipating great advances for CSFP. Although it agreed, amongst other changes, a strengthened Secretary General of the Council of Ministers to act as a High Representative, a policy-planning and early warning unit, crisis management mechanisms and some important innovations on voting around the idea of 'constructive abstention', Amsterdam did not substantively address the unwillingness of member states to establish supranational authority within the CFSP field[79]. As a result the EU is little nearer developing the kind of powerful institutional framework on CFSP which obtain in other areas of EU activities such as economic and social policy. This weakness is important because it means there is little institutional conditioning of the national defence and security policies of member states[80], another reason for questioning the inevitability of the Maastricht logic.

Since Amsterdam, and notwithstanding the appointment of Javier Solana, the NATO Secretary General as the EU CFSP supremo, the lack of European cohesion on major security issues has continued to disappoint advocates of CFSP witnessed in disagreements about the Albanian crisis in 1997, the air-strikes against Iraq in 1998, and the NATO bombing of Serbia in 1999.

A second set of obstacles to CFSP arise from the eastward expansion of the EU. Set originally to begin in the year 2000, the EU now looks unlikely to take in any new members from Central and Eastern Europe before 2005 and the process seems certain to be painful and protracted given the divisiveness of budgetary, Common Agricultural Policy and structural funding issues. On the basis that they cannot be permanently excluded from the CFSP pillar, whatever the interim arrangements, new members will inevitably introduce new security issues into the EU and complicate further the search for consensus and the capacity for collective action. Perhaps even more importantly, the enhancement of CFSP within the EU, which will amount to the creeping militarisation of the EU, will make an eastwardly expanding EU appear as an additional threat to Russia already anxious about an expanding NATO. For some the need to reassure Russia is a powerful argument against CFSP and this argument can be expected to play amongst some states within the EU.

What then are the implications for France of pursuing ESDI through EU CFSP? The answer in short is that the EU will likely prove to be the graveyard of many French defence and security ambitions because of the gap between the present reality and those ambitions and because of the gap between France and many of the other EU member states on ESDI.

At present the EU has no defence character and binds members into no formal defence relationship or obligations, nor does it have any meaningful institutional power to begin to generate a defence character and condition the national defence policies of member states. Realising ESDI through CFSP, assuming the immense obstacles outlined above can be addressed, will necessarily mean empowering the EU on defence and security matters and thus will inevitably mean the erosion of national sovereignty in these areas. This erosion of national sovereignty, combined with the requirement for agreement [even amongst a sub-set of EU members] and the willingness to act, will lure France on to common ground determined not primarily by French interests but by the ability to agree a relevant consensus. Paradoxically the French Gaullist preference for intergovernmentalism may reinforce this state of affairs because activity will be confined to what can be agreed amongst member states rather than to a potentially more inclusive integrationist agenda of what is for the collective good.

Being lured on to common ground would matter less if there was general alignment between France and the other EU member states on defence and security matters, but there is not. To begin with, many European states do not share the French preference for *l'Europe-puissance* and see Europe instead as *l'Europe-espace* which, depending on the point of view, means Europe as a region organised around concepts of common security, a free market area, a zone of economic and political stability with shared norms and values, essentially introspective and relatively quiescent, at least in French terms, on the international stage. Secondly, France is noticeably out of step with many EU member states on key issues of defence and security including, in particular, relations with NATO, nuclear weapons policy and foreign 'adventurism'.

The search for a way out of the lowest common denominator conundrum has led France towards the 'two circles' or 'two-speed' solution already established in other areas of EU policy. The idea, however, that EU CFSP problems can be addressed by having an outer circle of states committed to a minimum framework and an inner circle of states with an 'enhanced solidarity' [*un cercle des solidarités renforcées*] is itself fraught with complexities flowing from the fact that there is no clear 'inner circle' within the EU which shares and can help realise French defence and security ambitions in Europe.

EU member states are riven by national interest in relation to defence and security matters and a number of important cleavages are evident amongst

member states including Europeanists/Atlanticists, federalists/integrational-
ists, neutrals/non-neutrals, northern/southern Europeans, those with extra-
European obligations and those without, and nuclear weapons states and
non-nuclear states. At least two of Europe's 'big five', the states which might
most credibly form the basis of *l'Europe-puissance* – Britain and Germany[81] –
are opposed to the idea. At least three of the 'big five' – Britain, Germany
and Italy – are strongly pro-NATO and will resist any developments which
threaten the transatlantic partnership. Britain, which has shown itself willing to co-operate militarily with France 'out of area', opposes merging the EU
and WEU, while Germany, France's key partner in the European project, is
antagonistic to foreign 'adventurism' and cautious, perhaps now even hostile, to European nuclear weapons[82]. New member states of the EU will only
add to the centrifugal forces which mean that the larger the EU grows, the
less unified the aspirations of its membership will be[83].

Without a stable 'inner circle' France will, as in the recent past, continue
to be forced down the path of *ad hoc*-ism, participating in bilateral and multilateral ventures with any group of willing EU states with which it has
shared interest. Whether this results in France leading the EU towards a
French conception of European security or whether it results in the entangling of France and the erosion of French ambitions in the cross-winds of
compromise, will be determined by whether these *ad hoc* groups can be
forged into a cohesive whole and how willing the common members of
these groups are to act in relation to Europe's 'real world' security problems.
There is nothing logically imperative about the former and to date, despite
the organisational advances, precious little evidence of the latter in the military-security sphere.

Reflecting on the foregoing, it seems that building CFSP and ESDI within
the EU will require the sacrifice of at least some French ambitions, the more
so because the process of strengthening the defence and security character
of the EU will undermine the sovereign capacity of France [and all other
member states] to pursue a national agenda. Arguably France has few illusions about this and, while acting wherever possible, and to a degree consistent with its own interests, to build EU competence and cohesion on
defence and security, it is looking to the interplay of relations between the
EU, WEU and NATO to take ESDI forward. Given that we have said something about the prospects of CFSP/ESDI within the EU and have discussed
the development of ESDI within NATO, it is appropriate to concentrate here
on the WEU, the often-termed 'bridge' between the EU and NATO.

The agreement at NATO's January 1994 summit that the WEU could have
a formal role in the conduct of CJTF operations was clearly predicated on
the WEU developing into a more substantive organisation. Despite a decade
of WEU 'revitalisation' after 1984, there were few illusions in 1994, even
amongst WEU advocates, about the limitations of the organisation[84]. The
WEU consequently moved to enhance its competence and role, setting out

at its Madrid summit on 14 November 1995 a Common Concept in which all ten full WEU members and all 17 associate and observer members agreed a joint analysis of European security issues and a shared set of approaches to address them[85]. Raising the WEU's profile and enhancing its nominal presence was accompanied by a series of practical steps which included the creation of a new WEU headquarters in Brussels to facilitate the deepening of relations with the EU and NATO, the establishment of a round-the-clock crisis management Situation Centre, a defence planning staff, and a satellite interpretation centre to exploit uniquely European space-based intelligence [see below][86].

Exactly how the WEU fits into the EU–WEU–NATO relationship is a complex question, the answer to which is determined largely by the viewpoint of the commentator. There is a clear consensus amongst all members on the need to strengthen the organisation and general agreement that the WEU is the European framework organisation for the evolution of ESDI within NATO, but the consensus breaks down on the issues of the role of the WEU and the relationship between the EU and WEU. All are however agreed that the reconciliation of different views is complicated by the membership issue which arises because four European members of NATO are not in the WEU [Denmark, Iceland, Norway and Turkey], three members of NATO are not in the EU [Iceland, Norway and Turkey] and four members of the EU are not in NATO [Austria, Finland, Ireland and Sweden]. The prospect of harmonising these memberships in the medium term seems remote given the problems of bringing 'neutral' states into defence pacts and accommodating states like Turkey into the EU. As such, EU–WEU–NATO relations will develop in the context of these membership complications.

The question of exactly what roles the WEU should seek to fulfil is the source of some tension within the organisation. There are disagreements or differences of emphasis about the WEU's role in the defence of Europe [i.e., in relation to WEU Article 5], about Article 5/non-Article 5 overlap with NATO, and about WEU engagement with Central and East European states. In terms of Article 5 missions, the WEU remains redundant because not even the French are presently suggesting Article 5 defence without the United States[87]. The persistent French theme of uncertainty about long-term US commitment to Europe does however indicate that France looks in the longer term to the WEU to accrete influence to the point of bearing uniquely European Article 5 missions, whereas Atlanticists do not. In terms of the WEU's St Petersberg tasks, some Europeanists continue to press for the NATO Article 5/WEU non-Article 5 division of labour which has been rejected by NATO since the early 1990s and which makes little sense as long as Europeans are unwilling to act when it matters and as long as NATO is expected to stand behind the WEU in case a given situation escalates and NATO is drawn in. In relation to engagement with the Central and East European states, Atlanticists in the WEU see risks in developing a pan-European arrangement in relation to defence and security matters which excludes the North Americans[88].

  These questions are, however, less problematic that those relating to relations between the WEU and EU. The cleavage on this issue remains that of Maastricht between the Atlanticists, for whom the WEU is NATO's European pillar, and the Europeanists for whom the WEU is unequivocally also the defence arm of the EU. NATO's CJTF concept did not reconcile these differences, but merely allowed them, Janus-like, to be welded together. Atlanticist states like Britain argue that the WEU should be strengthened but that institutional relationships should remain essentially unchanged. Underpinning this view are a set of criticisms which make a powerful case against merging the EU and WEU. These are briefly that the EU has no defence competence and that any development of a defence competence will open national defence and military security policies to integrationalist interference from EU bodies; that merging the EU and WEU compromises EU eastward expansion by 'militarising' the EU; and that European vacillation and the patent inability of the EU to decide on important, and often even minor, security questions will undermine the ability of the WEU to act[89].

  For France and other Europeanist states the merger of the EU and WEU is essential, a view made clear in the Franco-German communiqué, issued after a meeting of French and German foreign and Europe affairs ministers at Freiburg in March 1996, that argued for a 'medium term' merger[90]. A few months later the French defence minister Charles Millon reiterated the point stating that: 'the rapprochement of the WEU and EU must proceed with the clear aim of long-term integration, by making the WEU the undisputed military arm of the Union'[91].

  Attempts to close the WEU–EU gap at the Amsterdam IGC summit in April 1997 confirmed that the Atlanticist/Europeanist approaches would be difficult to reconcile. The British, in particular, resisted attempts to formally move towards merger of the EU and WEU – insisting that Article J4 of the Maastricht Treaty [Annex] be left essentially unchanged – and strengthened the treaty references to NATO, in the words of two commentators 'codifying the subordination of the European defence identity to the Atlantic Alliance'[92]. As the British Prime Minister Tony Blair put it himself at the time: 'getting Europe's voice heard more clearly in the world will not be achieved through merging the EU and WEU or developing an unrealistic common defence policy. Instead [Britain] argued for, and won, the explicit recognition, written for the first time [into the Treaty], that NATO is the foundation of our and other allies' common defence'[93]. The French took a different view and in a speech to the IHEDN in February 1998 the French Defence Minister Alain Richard argued that the Amsterdam Treaty *had* taken the EU-WEU relationship forward by including the WEU St Petersburg tasks in the treaty, by facilitating closer co-operation between the EU and WEU and by allowing, at least in theory and subject to unanimity, the merger of the EU and WEU[94].

The important issue here is to understand the French preference for merger and the implications for France of such a merger. On the face of things the French position seems contradictory. On the one hand France is working towards merger primarily on the grounds that it is essential for the European project, for the stability of the Franco-German 'deal' [German acceptance of economic unity for French acceptance of political unity], and for the eventual realisation of a strategic Europe [*l'Europe-puissance*], that the WEU becomes *de facto* the defence competence of the EU. On the other hand France prefers an intergovernmental approach and is antagonistic to creeping integrationalism in relation to defence and security matters [and is thus for example opposed to an enhanced role for the Commission and European Parliament in these areas]. France, moreover, well understands the risk that strengthening the hand of the EU may lead to the further paralysis of the WEU.

France expects to square the circle by proceeding with merger but using the '*coopérations renforcées*' formulation and '*ad hoc*-ery' to both hold back integrationalist tendencies and ensure that EU/WEU membership complexities and EU decision-making difficulties do not stymie the activities and evolution of the WEU. It is a tall order and, as discussed above in relation to CFSP, may not work because of the different perceptions and ambitions of France and other Europeans, even some of those within the 'inner circle'. There is, for example, disagreement on this issue even between France and Germany, with the latter favouring a more integrationalist approach to EU–WEU merger and being more reluctant about the use of the WEU[95].

A second set of problems for France arise from viewing the EU–WEU–NATO relationship from the other end and seeing it in terms not of the Europeanisation of NATO but in terms of the NATOisation of the EU. NATO's dominance of the European security agenda and the evolution of ESDI to all practical purposes wholly inside NATO contrasts sharply with the EU's lack of competence in the defence and security field and the immense difficulties of building that competence. The effect of this is to situate the centre of gravity of European security in the WEU–NATO relationship with the clear implication that for the indefinite future European *ad hoc*-ism, defence and security variable geometry, and '*coopérations renforcées*' will be centred on NATO–WEU at least in relation to any meaningful security questions Europeans have to address. The attempt to weld the EU to the WEU may only shift this centre of gravity further towards NATO because the EU will inevitably cramp the capacity of the WEU to act. NATO's decisive action in Kosovo and the 'civilian power' role of the EU in the Balkans context would appear to confirm a seemingly natural division of labour between the two.

It seems that any prospect for a meaningful ESDI emerging lies in the NATO–WEU link in relation to which the EU is a complicating factor if not a positive hindrance, a point almost wholly obscured by the French long-term desire for a strategic Europe. It seems clear also that France could advance the vision of a meaningful ESDI by dropping its self-created obstacles to reintegration with NATO. A fully reintegrated France and a France less bent on forging an autonomous strategic Europe through the EU would find the British and Germans more willing partners in the creation of an enhanced European military capability.

The third pathway to ESDI is that of forging uniquely European functional military forces from the bottom up to put in place the operational capability which could wean Europe away from its pattern of military dependence on the United States. At first glance this seems quickly realisable. Between them NATO's 17 European member states have almost 2.9 million men and women under arms [and at least 3.8 million in reserve], deploy at least 19 100 main battle tanks, 6500 heavy weapons, 7 aircraft carriers, 3600 fighter aircraft, and 1200 transport aircraft, and have a collective defence budget of around \$195 billion, roughly 70 per cent of the US defence budget[96]. In relation to the threat spectrum [with the possible exception of Russian nuclear weapons] this is in principle more than enough firepower to fashion a coherent and autonomous European military capability.

This kind of bean-counting is of course misleading. Duplication, lack of interoperability, low skills, conscription, static operational concepts, and hardware weaknesses mean that uniquely European military means are far weaker than numerical indicators alone would suggest. In relation to the projection of military force in particular, only a tiny proportion of these forces could be used. Some estimates suggest that Europe's present capacity to project forces is not much better than it was at the time of the 1990/1 Gulf War and that Europeans would find it difficult to project more than 60 000 or 70 000 troops and would be unable to support or sustain them out of theatre for any length of time[97]. The comparatively high levels of overall European defence spending do not appear to be able to quickly address these problems because the money is not being spent on forging a single European force but instead is supporting the national armies of 17 individual states with consequent duplication and redundancy. Underpinning these problems is the lack of any political will to begin to move substantively away from national to collective concepts of defence.

Attempts to forge something substantial from Europe's standing military forces such as Alain Juppé and Pierre Lellouche's 1995 suggestion that the 'big five' – Britain, France, Germany, Italy and Spain – should each contribute 50–60 000 troops to a European army of approximately 300 000 under the WEU[98], or the less ambitious proposal for a WEU *Force Armée Européenne de Métier* [FAEM] of 75–125 000 troops[99], have amounted to nothing because of uncertainty about the relation of such a force to NATO, a lack of political will and the limitations of existing European military forces.

The obstacles to achieving the kind of grand transformation required to make a large European army even a medium-term reality have steered France towards the incrementalist approach of building numerous smaller-scale defence relationships and gradually developing European defence capabilities in the expectation that at some point in the future these elements can either be brought together or may naturally coalesce into a cohesive and potent European military force.

A useful place to start an overview of these uniquely European 'Forces Answerable to the WEU' [FAWEU] is with the Eurocorps [*le Corps européen*]. Originally a Franco-German force agreed at la Rochelle on 21 May 1992, the Eurocorps was intended to lay the foundations of a European defence capability building on the limited success of the earlier Franco-German brigade. Joined by Belgian forces in 1993, Spanish forces in 1995 and forces from Luxembourg in 1996, the Eurocorps eventually comprised around 35 000 troops. It was agreed on 20 September 1993 that the Eurocorps would be at the disposal of the WEU for non-Article 5 Petersburg tasks as well as for WEU Article 5 tasks in the 'setting' [*cadre*] of the European Union[100]. In the WEU context the Eurocorps could be used subject to the agreement of the participant states and the WEU council under the direction of the WEU military planning staff. Within the NATO context Eurocorps could be the basis of a uniquely European CJTF.

In December 1993 shortly after the Eurocorps–WEU linkage was formalised, Italy, together with France and Spain, proposed the creation of a *Euroforce Opérationelle Rapide* [Eurofor]. Premised on the contribution of a brigade from each state, Eurofor, based in Florence, was to have division strength [roughly 10 000 troops] centred on Petersburg tasks and was intended for Mediterranean theatre operations at the behest of the WEU. Within NATO Eurofor could constitute the basis of a southern CJTF[101]. The projection of Eurofor immediately butted against the requirement for an amphibious [i.e., sea-land] operational capability and this in turn led to the French-Italian-Spanish decision on 15 May 1995 to create a *Force Maritime Européenne* [Euromarfor]. Euromarfor was inaugurated on 2 October 1995 at Rota in Spain and developed thereafter in tandem with Eurofor based on the participation of the original three states together with Portugal[102].

Euromarfor was a substantial enhancement of existing European naval co-operation which, outside NATO, until 1995 comprised primarily the *Force Navale Franco-Allemande* [FNFA]. The FNFA had its origins in April 1992 in a temporary arrangement between the French navy and the German Bundesmarine. Formal co-operation began in May 1992, since when the bilateral relationship has continued under alternating commands [Toulon in 1992, Wilhelmshaven in 1993, Toulon and Brest in 1994, Warnemünde in 1995 and so on][103].

In May 1996 France was able to take European naval co-operation a step further by signing, at British initiative, a bilateral 'letter of intent' on Franco-British naval co-operation[104] designed to formalise joint maritime operations in support of trouble-shooting and peacekeeping roles. While for the British at least this move was 'not seen as a first step towards a Franco-British or European naval force' it nevertheless fitted, as one analyst commented 'into a pattern of increasingly formal links between European armed forces which could ultimately mesh with political moves towards a European defence arm'[105].

The third element of growing European defence co-operation is that of air forces. The key step forward in this respect was the creation on 30 October 1995 of the *Group Aérien Européen Franco-Britannique* [GAEFB], also known as the European Air Group [EAG]. Based at High Wycombe in the United Kingdom, initially under French command, the GAEFB is intended to allow France and Britain to co-ordinate air operations in support primarily of Petersburg missions. As such the GAEFB is not to be confused with the national air defence co-operation of the two countries but is centred on providing air-transport, combat aircraft and air-defence capabilities for 'out of theatre' missions[106]. Notwithstanding British perceptions, it was clear from the outset that France saw this as the start of deepening European airlift co-operation. As Charles Millon, the French Defence Minister, put it at the time: 'the EAG has resources available to it in all of our respective airforces and . . . once expanded to include other European partners it may constitute a very significant force'[107].

A few months after the creation of the GAEFB, France set up an exchange system between the French air force's *Force Aérienne de Projection* [FAP], responsible for military air transport, and the German Luftranskommando as a further step towards strengthening European airlift. Under the exchange the two countries have been able to utilise each other's aircraft for military airlift and within a year of the agreement being signed French aircraft had carried German forces between Germany and Canada for training and the Luftwaffe had rotated French forces between France and French bases in Africa, the Antilles and Guyane[108].

In addition to European forces led by or involving the French, the WEU also has four other NATO FAWEU at its disposal: the British-Belgian-Dutch-German Multinational Division [Central], the British-Dutch Amphibious Force, the Spanish-Italian Amphibious Force and the 1st German/Dutch Corps headquarters. All together these eight ground, sea and air forces [and the numerous related bilateral co-operations] which are formally linked to the WEU and centred on Petersburg missions, involve all of Europe's 'big five' states together with a clutch of smaller states. While this does not necessarily imply a willingness to act, it does imply a growing competence to do so, particularly if the underlying French logic that these arrangements will expand and deepen proves to be correct.

In parallel with these organisational relationships, France has been pushing European co-operation in operational contexts, the result of which has been to advance the capacity for functional co-operation. France played a major part in leading a number of European states in multinational deployments in Bosnia as part of UNPROFOR between 1992 and 1995, worked with the British and Dutch to deploy a small Rapid Reaction Force into Bosnia in June 1995 to beef up the UN presence, persuaded the Germans to contribute combat troops from the Franco-German brigade to SFOR deployment in January 1997, took the lead with Italy in the all-European 'Operation Alba' in Albania in 1997, and played a key role in the initially all-European Macedonian Extraction Force supporting the OSCE in Kosovo at the end of 1998.

The emergence, however tentatively, of a European air-land-sea projection capability and the growing joint operational experience of some European militaries highlights the shortfall in what for many is the Achilles' heel of European military capability: space-based assets for intelligence and command and control. In these areas the level of European dependence on the United States continues to heavily qualify any claims that Europeans are moving towards genuine operational autonomy. It is this dependence, after all, which is the basis of European reliance on US assets within NATO and President Chirac himself has identified European space-based military assets as a 'necessary precondition for achieving [European] strategic autonomy'[109].

French and European weakness in these areas, exposed so clearly by the 1990/1 Gulf War, prompted an ambitious set of French-led programmes to develop a range of military intelligence satellites, and the successful launch of the first of these, Helios I, on 7 July 1995 was widely heralded in France as a defining moment in Europe's birth as a strategic actor[110]. With a resolution reportedly down to a metre[111] and coverage of the earth from geosynchronous orbit of south-east Europe, the Mediterranean and the Middle East, Helios I delivered significant intelligence capability. Moreover, it quickly exposed the implications of intelligence dependence on the United States when in September 1996 US claims of large-scale Iraqi troops and tank movements against Iraqi Kurds, which prompted US airstrikes against Iraq, were disputed by France on the grounds that Helios I had seen no such large-scale movements[112]. Discrepancies of this kind were subsequently to play a role in the French distancing themselves from 'Anglo-Saxon' policy in Iraq.

In the wake of the success of Helios I, however, the pace and ambition of the French-led space programmes have nose-dived. Helios II and Horus were taken through turbulent waters by budgetary constraints, industrial wrangling about the relationship between the lead French and German companies and about the links between space programmes and other Franco-German military projects [helicopters, missiles, and aircraft], and political differences about the implications of strengthening European intelligence autonomy [with Germany cautious about the implications for the US-European relationship and French-led 'out of theatre' adventurism]. The schedule for Helios II [initially due in 2002] was stretched and Horus was abandoned[113]. Of the remaining programmes, Cerise and Zenon went forward only in low-key system experiments while the Alerte early warning system fell victim to the French decision to pull out of the US-French-Italian-German MEADS [Medium-Extended Air Defense System] programme, the latter because of costs and the perception that BMD systems weaken *dissuasion* [why develop BMD unless you doubt the capacity of your nuclear weapons and other forces to deter?] and may provoke BMD proliferation which could in the longer-term undermine the French deterrent[114].

In parallel with the development and deployment of military satellites France has worked to strengthen the European capacity to utilise the resultant intelligence. The centrepiece of this effort is the WEU Space Centre [WEUSC] at Torrejon de Ardoz in Spain. Operational since March 1993 and funded by France, Germany, Italy, the UK [17 per cent each], Spain [13 per cent], and Belgium and Holland [8–9 per cent each], the centre initially began to operate using French SPOT satellite images to give the WEU a limited interpretation capability[115]. Since 7 May 1996 the WEUSC has been able to take Helios I images on the basis of a measure of priority to the three Helios funding states, France, Italy and Spain [France controls Helios I from Toulouse, and the three funding states have direct downlinks at Colmar, Lecce and Maspalomas respectively] and subsequent availability to the WEU. This, however, is only the most sensitive part of a flood of data which the WEUSC also uses from SPOT, the European Space Agency's ERS satellite, the US Landsat, Canadian Radarsat, Russian Ressource-F, and Indian IRS-IC[116]. While it should not be exaggerated, the WEUSC does, as a result, have considerable utility in support of Petersburg-type missions, crisis management, arms control verification and non-proliferation, a utility which has allowed it to number NATO, the OSCE and UN amongst its clients.

With respect to space-based command, control and communication, France is developing a Franco-German successor to the Syracuse II network, which presently provides French forces with virtually global connectivity and is due out of service around 2005, having failed to reach any agreements with the British and Americans after several years of discussions on a joint successor to Syracuse, Skynet and DSCS[117].

For the sake of completeness one final area of European defence co-operation, defence industries, needs to be briefly considered. French ambitions to forge a more cohesive and competitive European defence industrial base were discussed in the previous chapter in the context of the 1996 defence reforms but, a few years on, the obstacles remain largely unchanged. Notwithstanding the roles of the *Organisme Conjoint de Coopération en matière d'Armament* [OCCAR] and Western European Armaments Group [WEAG][118], national priorities, not least in terms of employment and technical edge, continue to act against co-operative synergies. In France the problems are compounded because high levels of unemployment and instability in the defence industry sector, presently restructuring, continue to act against the possibility of European mergers and rationalisation[119].

As if these problems were not enough, a new challenge is now visible over the horizon, namely the implications of the American-dominated 'Revolution in Military Affairs' [RMA]. According to some analysts, Europe is in danger of falling qualitatively behind the United States in military technologies, sliding into the second of three tiers of technological competence. This is obviously a complex question which cannot be fully explored here, but several observations are important. Firstly, the prospect of constructing an autonomous ESDI is in no small part going to be conditioned by the European uptake of RMA technologies. Secondly European failure to assimilate these technologies fast enough will reduce European military effectiveness in a context in which other centres of power are adopting these technologies. Thirdly, failure to keep pace with the United States will both increase European dependence on the US and reduce the relative effectiveness of Europeans in US-led 'world order' operations. The latter is particularly important if one accepts the idea that the ability and willingness of Europeans to work with the United States in 'out of area' roles is a key part of the cement which binds the transatlantic alliance in the post-Cold war era. As one analyst put it: 'the ability of the European countries to be engaged in high technology is going to be a central part of maintaining the strength of the Alliance'[120]. If Europe is to assert greater autonomy it can ill afford any degree of technological marginalisation.

While there is an ongoing debate about the long-term importance of the RMA in France[121] as elsewhere, France is in a strong position within Europe to exploit its own technological edge to lead Europeans in the development of RMA technologies and to accrue political leadership as a result. In this regard the complexities and obstacles to European defence cohesion, set out in the foregoing in relation to ESDI more generally, are sharply relevant to the capacity of France to lead Europe towards a collective RMA.

The final piece of the ESDI jigsaw is the French attempt to exploit the political and security utility of its nuclear weapons to forge a European deterrent and realise a strategic Europe.

Finessing the European nuclear future

A useful way into French nuclear weapons policy in Europe in the post-Cold War era is through the concept of *'dissuasion par constat'*. This argues that there are specific and unavoidable implications arising from the fact that France [and Britain] possess nuclear weapons and are likely to do so for the foreseeable future. In essence the mechanism of *dissuasion par constat* is as follows. France and Britain are nuclear weapons states and also members of the EU, WEU and NATO. As such, French and British nuclear weapons exercise a *de facto* dissuasive [or deterrent] effect on behalf of the members of those organisations – irrespective of formal obligation or whether other members acknowledge or desire that dissuasive effect – because there is an inescapable distinction in the mind of aggressors between political or security organisations which have nuclear weapons states as members and those that do not.

For France a number of important implications flow from the fact of *dissuasion par constat*. The first is that French nuclear weapons enhance the security of the EU, WEU and – in a different manner – NATO. Any state or group of states threatening the EU or WEU has to factor-in French nuclear forces and similarly any state or group of states threatening NATO has to factor-in French nuclear weapons as a second [or possibly third] centre of decision-making. This point had been explicitly recognised in the wording of NATO Ottawa declaration [Article 6] in June 1974 and the WEU Platform on European Security Interests in October 1987 [when Europeans were worried about the INF and US decoupling] and was repeated in very similar language at the WEU's meeting in Noorwijk in November 1994 when the members agreed that 'the independent nuclear forces of Britain and France . . . contribute to the *dissuasion* and overall security of their allies'[122].

Secondly, Europeans 'benefiting' from this enhanced security must, sooner or later, come to terms with the French contribution to their security and with the resultant implications for them as 'consumers' of the French dissuasive effect. Thirdly the United States [and thus NATO] must also, in due course, come to terms with the implications of *dissuasion par constat* operating in the context of growing European political and economic cohesion, US nuclear downsizing in Europe, and the changed threat context.

At the same time France is under few illusions about the altered role and perception of nuclear weapons in the post-Cold War context. Nuclear weapons have lost their place at the centre of security discourse and now seem either irrelevant or peripheral to many of Europe's most pressing security issues. Nuclear weapons have ceded operational primacy in national arsenals to the capability to project conventional power and have lost some legitimacy as a guarantor of international security and stability. In addition nuclear weapons have come under meaningful pressure, as Michael Mazarr put it, from 'arms control devotees [and] hard-headed realists not satisfied with vague calls for disarmament [and now demanding] detailed outlines of potential nuclear end-states'[123].

This does not mean that nuclear weapons are no longer important, but it does means that cost-benefit calculations about nuclear weapons by publics and elites have changed and that nuclear weapons are at risk, as one French analyst commented, of sliding into a no-man's land of security relevance *'entre marginalisation et banalisation'*[124]. The altered image of nuclear weapons introduces an additional complexity into the French efforts to utilise nuclear weapons in pursuit of wider security ambitions. In the new context nuclear weapons remind many states of past iniquities and failures, highlight divisions, and reinforce inequalities. Heisbourg's observation, made in a domestic context, that *'trop de nucléaire tue le nucléaire'*[125], has wider relevance: France cannot play the nuclear card too strongly for fear of isolating itself from those it is seeking to constructively engage.

While French sensitivity has not always been evident, as for example during the 1995–6 nuclear weapons testing furore, on the whole France has handled nuclear policy with considerable subtlety in pursuit of an ambitious post-Cold War agenda. This agenda somewhat simplified comprises four elements: the traditional role of assuring French security, contributing to European security and to international stability, and the 'new context' roles of building ESDI, rebalancing transatlantic relations, and addressing the threat of the proliferation of weapons of mass destruction [WMD]. In relation to the latter, the projection of French nuclear forces outside Europe will be considered in Chapter 5, but the remaining issues will be assessed here.

Implicit in the February 1996 nuclear reforms were changes in French nuclear strategy. To understand this a certain degree of terminological precision is required. France's core nuclear doctrine, that is the fundamental tenet at the heart of French thinking about nuclear weapons, has remained unchanged since the end of the Cold War. This states that France must have 'the will and capability to make any adversary, whoever he may be and whatever means he may possess, fear unacceptable costs that are out of proportion to the stakes of the conflict should he attack our vital interests'[126]. In the Cold War this was assumed to refer primarily to the Soviet Union but to be adaptable at least in principle to any threat to French vital interests. In the post-Cold War context the official qualifications and explanations of the doctrine make it clear that it now relates to a wider range of threats, of which Russia is but one, to states less powerful as well as more powerful than France, and to both nuclear and non-nuclear [chemical, biological and perhaps conventional] threats to vital interests. As a consequence what has changed is the scope of doctrine and exactly how France would ensure its validity.

Part of this is ensuring the credibility of the nuclear forces themselves. France remains publicly committed to the 'non-use' of nuclear weapons and rejects the development of war-fighting but has nevertheless consistently believed that for *dissuasion* to function France needs credible nuclear forces capable of operationally implementing the threats explicit in doctrine.

For at least the next 15 to 20 years, French dissuasive credibility will rest on a nuclear dyad which will eventually comprise four *Triomphant*-class submarines and airforce and naval versions of the ASMP-*Ameliore* stand-off missile – expected to be in service by 2010. The French perception that its nuclear weapons are a dissuasive minimum means that France will stay out of START III and will thereafter resist force cuts which threaten operational credibility.

A second issue in the credibility of French nuclear weapons is maintaining the reliability, safety and security of French nuclear weapons in the context

of a complete physical test ban. This has drawn France towards closer technical co-operation with the United States [and thus with the UK]. Building
on the 1961 and 1985 Franco-American nuclear agreements, France signed a
new 'secret accord' in June 1996 to allow specialists from both countries to
work together to maintain nuclear arsenals in a 'no-test' context[127]. Aside
from sharing test data and simulation modelling with the United States,
France is working with US assistance to duplicate US facilities in a move
which presages long-term Franco-US co-operation. The French *Laser Méga-
joule* [LMJ] facility under construction near Bordeaux and expected to be
operational by 2006 is comparable to the US laser facility at Lawrence Liver-
more; the Airix radiography machine shortly to be constructed at Moronvil-
liers near Reims parallels a similar US facility at Los Alamos and France is
purchasing new Cray computers from the United States to upgrade those it
bought in 1985[128].

Exactly how the streamlined nuclear arsenal is to be used to assure that
France could meet the requirements of its nuclear doctrine is the role of
nuclear strategy. The decisive break with the past in this regard occurred
between 1994 and 1996. The 1994 *Livre Blanc* argued that French nuclear
forces had to be capable of fulfilling two functions: inflicting unacceptable
costs on an adversary and conducting a limited strike as an *'ultime avertissement'*[final warning][129], a statement entirely consistent with the Cold War
formulation of the prestrategic function based on an explicit and permanent linkage between the prestrategic and strategic use of nuclear weapons.
By February 1996, and reflecting the scrapping of the land-based weapons
[Pluton and Hadès] most closely associated with the prestrategic role, the
notion of *ultime avertissement* and the sequential relationship between prestrategic and strategic nuclear weapons had been revised. In its place was a
new formulation which argued that all remaining French nuclear weapons
were strategic systems[130] and that nuclear strategy was to be based on 'two
complementary and modernised components capable of responding together
to all political and military situations requiring *dissuasion*'[131].

The new strategy does not rule out the use of ASMP in a prestrategic-type
role, not least because in the event of Russian revanchism the missiles could
be required for such a role, but it suggests that there is now flexibility in the
operational planning to make *dissuasion* credible. At the very least the
change allows the gap between French and NATO nuclear strategy to be narrowed by taking out of the French formulation the automatic prestrategic-
strategic linkage which has been such a problem for NATO in the past.
Arguably it has also opened the door to the selective use of French nuclear
weapons to the extent that any nuclear weapon or combination of nuclear
weapons in the French arsenal can now, at least in theory, provide the
operational credibility underpinning 'non-use'. This should not be read as
France moving towards the acceptance of war-fighting nuclear strategies:
there is no evidence that France has bought into concepts of graduated use

or nuclear exchanges. However, the decoupling of submarine-launched nuclear weapons and the air-launched ASMP does give France the flexibility to exploit the latter in particular in relation to the Europeanisation of nuclear weapons and perhaps more importantly the projection of French nuclear power into regional crises, including those outside Europe.

France has made no secret of its desire to Europeanise its nuclear forces, by which is meant finding a way for French nuclear weapons to play a role in the construction of ESDI and the defence of the EU and involving European partners in a gradually deepening process of nuclear engagement. Since 1992 when Mitterrand posed the decisive question 'is it possible to conceive a European [nuclear] doctrine?', French thinking about nuclear forces has been gradually denationalised as the emphasis on French security and independence has given way to a growing emphasis on the European context within which French security is embedded.

One of the keys to understanding the trajectory of nuclear Europeanisation, at least from the French point of view, is the issue of converging vital interests. Official French statements have repeatedly stressed the converging European vital interests arising from the Maastricht logic. In March 1995 Jacques Chirac, for example, asked 'how can we imagine the pursuit of European construction if France does not recognise that it is increasingly difficult to separate her own vital interests from those of her closest partners?'[132]. For France the link is clear: French nuclear weapons are the indispensable guarantor of French vital interests; as French and European vital interests converge, it is axiomatic that French nuclear weapons will come to be the indispensable guarantor of European vital interests.

The notion of converging vital interests introduces an important complication for France. French doctrine has never fully spelled out what French vital interests are and to date they remain, as Bruno Tertrais points out, what the President says they are[133]. France though has spelled out what *some* of its vital interests are and these include threats to national survival and sovereignty, about which there can be little argument from other Europeans. But French vital interests also include threats arising in regional crises inside and outside Europe and threats to the DOM-TOMs. The notion of converging vital interests and indeed statements like the Franco-British assertion in October 1995 that 'we do not see situations arising in which the vital interests of either France or the UK could be threatened without the vital interests of the other also being threatened'[134] thus need to be squared with the evident diversity of French vital interests.

There appear to be three possibilities. Converging European vital interests relate to only a sub-set of those of France and thus the French position will be that it has some vital interests which overlap with those of Europeans and some that do not, though this would appear to be at odds with the Franco-British statement. Alternatively convergence could mean that France will curb the scope of its vital interests to align them more closely with those of its European partners. Finally, it may mean that, on the contrary, convergence will require Europeans to enlarge their own vital interests to encompass not only France's wider interests but presumably also those of other Europeans too. There would appear to be adequate flexibility in the wording of the EU Amsterdam Treaty [Article 11] to bear any of these interpretations, but the key point here is that the forging of common vital interests will require France to be more explicit about those interests. In a European context vital interests cannot simply be what the French President [or anyone else] says they are.

The French attempt to take the Europeanisation of nuclear weapons forward in many respects parallels, and is of course deeply intertwined with, the process of promoting ESDI. France has taken a gradualist approach in which the Europeanisation of nuclear forces is conceived as taking place towards the latter stages of the process of European defence harmonisation. This is implicit in Lionel Jospin's remark that 'although national today, French nuclear forces will one day serve a Europe which has acquired a more precise identity in the area of defence'[135]. Moreover, European reticence demands that the process is advanced in a low-key manner. As one analyst noted:

> the legitimacy of nuclear deterrence is very much called into question in Europe and the path to be trodden [by France] is narrow...European deterrence is like a mole which has to progress underground before emerging into the daylight[136].

As the basis of a nuclear relationship between France and other Europeans, *dissuasion par constat* is a usefully discrete concept but it is essentially static with no dynamic or process for building European nuclear cohesion. The two most obvious conceptual approaches to Europeanisation – extended deterrence [*dissuasion élargie*] and shared deterrence [*dissuasion partagée*] – both raise immediate complexities and problems for non-nuclear European states. *Dissuasion élargie*, that is offering to explicitly extend the French nuclear deterrent to cover other states, is highly problematic. It is premised on an inequality of means and thus could appear paternalistic to recipient states who would be dependent on France; it creates divisions between those under the French nuclear umbrella and those outside it; it raises questions about the credibility of the French guarantee [the more so because an historic theme of French policy has been doubt about the American nuclear guarantee]; and it raises the question of whether France would be seeking to displace or replace the American guarantee[137]. The alternative, *dissuasion partagée*, seems equally problematic. It raises a raft of problems about exactly what sharing means at the political and military level; it creates divisions around the question of which states participate; and it raises issues about the relationship between the shared European deterrent and the US nuclear guarantee.

The immediate solution lay in the concept of *dissuasion concertée* [concerted deterrence], first mooted by the Socialist defence minister Jacques Mellick in 1992 but brought into the centre of the discourse in 1995 by Alain Juppé who asked 'must our generation shy away from contemplating, not shared deterrence, but at a minimum a concerted deterrence with our main partners?'[138]. Exactly what *dissuasion concertée* meant and what it did not mean was clarified by Chirac himself in June 1996 when, in a speech before the IHEDN, he stated that concerted deterrence:

> is not about unilaterally extending our deterrence or imposing a new contract on our partners. [It] is about drawing all the consequences from a community of destiny, of a growing and intertwining of our vital interests. Taking into account the difference in sensitivity that exists in Europe about nuclear weapons, we do not propose a ready-made concept, but a gradual process, open to those partners who wish to join[139].

Evidently concerted deterrence can be understood as an open, gradual and potentially inclusive process in which France seeks to engage its European partners in a deepening dialogue on the whole range of issues pertaining to *dissuasion*[140], something David Yost has helpfully referred to as 'deterrence supported by continuing consultations and substantive dialogue'[141]. The diplomatic genius of *dissuasion concertée* is that it is a dynamic process which takes nuclear Europeanisation beyond *dissuasion par constat*, yet by postponing all of the hard questions posed by extended and shared deterrence it allows European states to come into a process of dialogue commensurate with their respective sensitivities, while legitimising French nuclear weapons in the new context and providing them with the new rationale of playing a role in the construction of ESDI.

This process is undoubtedly underway. Since 1992 Franco-British co-operation on nuclear issues has been formalised in the Joint Commission on Nuclear Policy and Doctrine [JCNPD]. The JCNPD comprises a small number of senior figures on both sides and its work has been particularly secretive to facilitate frankness and a wide-ranging agenda. The clearest insight into its work has been provided in a remarkable article by Stuart Croft[142]. According to Croft, the discussions have centred on three main areas: (1) concepts and practices of deterrence, which have evidently built on earlier co-operation, have encompassed threat perception, strategy, doctrine and may have included operational matters such as submarine patrols; (2) co-ordination of policy on disarmament procedures and processes in the former Soviet Union as part of the West's efforts to manage and control nuclear downsizing in the former Soviet states; and (3) co-ordination of arms control positions particularly with respect to the NPT and START processes. Little has emerged about the JCNPD since Croft's 1996 article. The general perception however seems to be that the commission made rapid progress in its early years reaching unexpected levels of agreement and co-operation but that it has now butted against the limits of what is practical and progress has consequently slowed.

The second element of the French-led European nuclear dialogue has been Franco-German discussions on the issue. Perhaps surprisingly there is a brief history of Franco-German nuclear weapons co-operation dating back to the 1950s. In 1957 France and [West] Germany drafted and initialled a secret bilateral agreement entitled 'Common Research and Utilisation of Nuclear Energy for Military Purposes' which involved German contributions to the production of French warheads and ballistic missiles in exchange for German access to French nuclear weapons 'under French control in periods of crisis'[143]. The deal was promptly killed by de Gaulle in 1958 and Germany subsequently signed the NPT in 1969 since when, according to former foreign minister Klaus Kinkel, Germany's non-nuclear status has been 'unambiguous and forever'[144].

In relation to German nuclear weapons this may be so, but Germany has been sharing control of US nuclear weapons in NATO and participating in NATO nuclear consultations since the 1950s. There is thus considerable latitude for Germany to participate in *dissuasion concertée* without abrogating its NPT or domestic constraints. Germany agreed to consultation on the use of French nuclear weapons in 1986, and the path towards closer co-operation appears to have been cleared by the mothballing of Hadès in 1992 which freed the Germans from any residual perception that German territory was a French nuclear battlefield[145].

The clearest statement of a German willingness to participate in further nuclear discussions with France can be found in the 9 December 1996 Nuremburg 'Common Concept' which stated that 'the independent nuclear forces of the UK and France, which have a deterrent role of their own, contribute to the overall deterrence and security of the Allies. Our countries [i.e., France and Germany] are ready to engage in a dialogue concerning the function of nuclear deterrence in the context of European defence policy'[146]. To date this has not translated into a formal arrangement to parallel the JCNPD and indeed the recently elected German SPD-Green coalition has indicated that it would prefer any strictly European nuclear discussions to be postponed to a much later stage of the construction of a European defence[147].

Outside this Franco-British-German triumvirate[148] it is unclear how much wider participation in *dissuasion concertée* can be extended for the foreseeable future. Chirac himself has argued that due to their sensitivity nuclear discussions would necessarily be confined, at least in the medium term, to 'the Germans, the English [sic.] and the Spanish'[149], our closest neighbours'[150]. For Philippe Séguin a European nuclear 'inner core' would be the Eurocorps states – France, Germany, Spain and Belgium [and from 1996 Luxembourg][151] – while for some analysts the group could be slightly wider to include Italy and the Netherlands[152]. Attempts to extend dialogue beyond this group to the WEU and EU would quickly impact on attempts to bring the EU and WEU into a closer relationship and on EU expansion.

What complicates the task of building a European deterrent for France is that all of these 'inner core' states, including France, are NATO members and all are committed to the view, stated in the Franco-German Common Concept, that 'the supreme guarantee of the security of the Allies is provided by the strategic forces of the Alliance, particularly those of the United States'[153]. As with ESDI it seems that the framework for a European deterrent lies within NATO.

Before turning to the question of the implications of NATO and the US nuclear guarantee for a European deterrent, it is worthwhile asking what a European nuclear deterrent might look like since the answer will necessarily condition the relationship between a European and NATO nuclear deterrent. While both Jacques Mellick[154] and André Dumoulin[155] have offered models of deterrent concepts, the French analyst and defence ministry official Bruno Tertrais has gone a stage further to offer four models of functional European deterrent forces: a mutualised deterrent; a common deterrent; a joint deterrent and a single unified deterrent[156].

Tertrais places these models on a conceptual ladder on the basis that a move from one rung to the next would mean crossing an important political rubicon. The simplest model, a mutualised deterrent, is premised on an explicit mutual defence and nuclear commitment, an inclusive nuclear policy co-ordination committee in the WEU, and closer Franco-British operational co-operation which, while decision-making authority would remain in national hands, would include joint targeting. This is considered by Tertrais to be 'not unrealistic' in the short to medium term.

The second rung, a common deterrent, is similar to NATO nuclear-sharing arrangements. Its elements include national decision-making authority, a formal European nuclear planning and consultation arrangement, the operational pooling of French and British nuclear weapons, European sharing of costs and operational responsibilities [including European states accepting ASMP missiles for European aircraft], premised on a shared view of risks and vital interests.

The third rung, a joint deterrent, would comprise a fused Franco-British deterrent in joint national decision-making hands subject to the negative veto of other EU members and would thus be the end of effective national control. European nuclear weapons would be operated by multinational personnel with unified targeting. It would mean a degree of nuclear-sharing well beyond NATO arrangements.

The apogee, a genuine *dissuasion commune*, would be a single unified deterrent in which a European federal body would have sole authority over a fully integrated European deterrent force.

The road to a *dissuasion commune* would seem to be long and hard. If one accepts the widely shared view that a European nuclear deterrent would only become possible towards the latter stages of the construction of a European defence, then the obstacles to ESDI also stand as obstacles to the possibility of a European deterrent. That said, progress on the lower rungs of Tertrais's ladder could begin well before a common EU defence was realised. Exactly how far such co-operation could go would be conditioned by many factors in the linkages between nuclear and wider ESDI issues. It would be conditioned also by issues internal to the process of building a European deterrent including the difficulties of closing the gap between French and British nuclear forces in the context of close British-American nuclear relations, the complexities of sharing sovereignty in relation to the production, decision-making, operation and use of nuclear weapons, the vexed question of overcoming structural obstacles to nuclear-sharing such as the status of EU member states in the NPT, and public resistance particularly amongst the non-nuclear and 'neutral' states.

Tertrais reports that in the context of the NPT the United States assured Germany that a European federal state would not be bound by the NPT[157], presumably on the grounds that a federal Europe would inherit the international nuclear legitimacy of its nuclear-armed member states. The anomaly would appear to be that while the nuclear 'end-state' of *dissuasion commune* would be permissible, if the United States is right [though Russia, China and others may have different views], the steps in its construction would seem to violate NPT principles against nuclear transfer between signatory states.

The entire debate about the obstacles and complexities of constructing a European nuclear deterrent as well as the implications of doing so for France are further complicated because the debates are taking place in the context of NATO and the US nuclear guarantee. President Chirac's 8 June 1996 statement that France did not intend 'to substitute a French or Franco-British nuclear guarantee for the American deterrent' but rather intended European nuclear co-operation to be 'a reinforcement of the overall deterrent'[158] appeared unambiguously to shelve the idea of a genuinely autonomous European nuclear deterrent in favour of a nuclear accommodation with NATO in which strengthening the European deterrent pillar would enhance the transatlantic deterrent and contribute to rebalancing US-European relations. This would appear to be entirely consistent with the wording of the Franco-British summit statement in October 1995 and the Franco-German Nuremberg 'Common Concept' in December 1996.

The way forward for France is thus substantially determined by the limits of what the states identified by Chirac as the 'inner core' of any enhanced European nuclear co-operation – the British, Germans, and Spanish – are prepared to do. The 'Euroenthusiasm' of the British since St Malo has not been accompanied by a substantive adjustment to the basic British position that 'co-operation between the UK and France must be seen as strengthening the European contribution to the Alliance [by] adding to deterrence'[159]. For Germany the election of the SPD-Green government in October 1998 seems only to have underlined the view that 'relations with France must not be allowed to jeopardise the nuclear protection offered by the United States'[160]. Spain is similarly committed to the primacy of the US nuclear guarantee[161].

The prospects for French-led nuclear co-operation within this group, and even more so in a wider European group, is consequently conditioned by the willingness of France to act towards strengthening the European pillar while carefully not undermining the US position. As with the other elements of ESDI, France is being 'more Atlanticist today in order to be more European tomorrow' in the expectation that a European deterrent pillar can coalesce and accrete power to the point of shouldering an autonomous European role. Paradoxically, if there was European willingness, autonomy could be more realistic in the nuclear field than in the conventional because the shrinking Russian arsenal, the START framework, and the 'proportionality' of nuclear weapons – that is, one does not need parity to deter merely the capability to inflict disproportionate costs under any circumstance – could lead to French or Franco-British nuclear weapons bearing a genuine European deterrent role rather earlier than European conventional power could be built up to give Europe conventional military autonomy in relation to regional threats and world order roles.

The surest way France could build up the European deterrent pillar within NATO [and it can be built nowhere else] would be to assure its European partners of its good intent by joining the NATO NPG at the earliest opportunity and returning fully to the heart of NATO decision-making. Outside the NPG any European co-operation with France is a problem for NATO and thus European engagement with France will necessarily be cautious, limited and gradual. Inside the NPG any co-operation with France is unambiguously strengthening the European pillar and thus progress could be expected to be smoother, more rapid and more ambitious and creative. It is for these reasons that Germany and the United Kingdom have been working to bring France back into NATO's nuclear fold.

The many obstacles to France taking this step boil down to the 30-year-old *'non'* of Charles de Gaulle: the widely accepted and still largely unchallenged view in France that to return to the NPG would subordinate France to the United States and thus compromise French autonomy[162]. After a decade of change since the end of the Cold War this view is anachronistic and needs to be questioned.

The first flaw in the Gaullist argument is the issue of subordination to the United States. Unlike Britain which is dependent in many respects on the United States for its nuclear weapons, France meets its own nuclear hardware requirements. France is though in an increasingly close nuclear relationship with the United States in the wake of the CTBT. This co-operation, which has a long-term future, is helping France to maintain its own systems rather than creating a creeping dependence on the United States. There would consequently be no technical subordination of the French to the United States in a return to the NPG.

The suggestion that France would become operationally subordinate to the United States within the NPG is also questionable. At the height of the Cold War at Nassau in 1962 even the dependent British managed to secure an agreement from the United States that British and NATO interests need not be the same and that Britain needed a degree of operational autonomy in the event of 'supreme national interests' being at stake[163]. Participation in the NPG is not, and never has been, predicated on the surrender of operational national autonomy. In the post-Cold War era when European vital interests are converging and transatlantic vital interests are very close in relation to the threat spectrum [from Russian revanchism, through regional instability to proliferation], anxieties of this kind make even less sense. French and NATO vital interests overlap and even where they diverge there is demonstrable latitude within the NPG for national interests, for example in France's case the DOM-TOMs, to be defended.

Nor does joining the NPG mean institutional subordination to NATO. Spain, which like France is also outside NATO's IMS after a referendum on 12 March 1986, has nevertheless remained inside the NPG and sees no contradiction in its position. Furthermore strategy convergence between France and NATO around the idea of nuclear weapons as weapons of last resort has removed the prospect of France being forced into adopting nuclear war-fighting strategies. This does not mean that there are no nuclear issue difficulties remaining between France and NATO or France and the United States [there are many, not least on 'no-first use'[164]] but there are also overwhelming shared interests. The key point is that there is no reason why difficulties cannot be addressed *within* the NPG, not least because there the cleavages are much more diverse that simply Franco-American, and inside the NPG France is in a much stronger position to take the European nuclear debate forward.

The second flaw in the Gaullist position is the defence of an overstated French autonomy. One argument made in this chapter is that French autonomy, understood in a maximal Gaullist sense, is an illusion based on a failure to accept the degree of overlap between France and Europe as a result of the single currency and intertwining vital interests, a failure to understand that being tightly bound to states like Germany, Britain and Spain which are themselves tightly bound to NATO *de facto* binds France to NATO, and a failure to accept how dependent Europeans presently are on the United States. Events in the early 1990s demonstrated that standing aside from NATO cut France off from its European allies yet did not spare it the consequences of NATO evolution. Events in Bosnia and Kosovo have demonstrated that in relation to real world security problems it is virtually impossible for France to stand aside from its operational obligations in NATO. Gaullist defence of French autonomy in Europe *vis-à-vis* NATO is consequently fatally flawed because the obligations and constraints already exist irrespective of France's formal position, and the only remaining issue is the degree of influence France wishes to assert.

Reflecting on the French position in European security more widely, it has been argued here that trends and pressures in Franco-NATO relations and the obstacles to constructing ESDI outside NATO point decisively to the need for France to return fully to the NATO fold. The continued adherence to Gaullist verities on this issue, particularly when in so many other areas of French defence policy in the post-Cold War era the Gaullist paradigm is either crumbling or no longer relevant, makes little sense and seems to be contrary to French interests.

Fully within NATO France could maximise its influence in the key processes shaping European security – NATO's evolution, the construction of ESDI within NATO and the transatlantic response to Europe's 'real world security concerns' – and maximise its capacity to forge a cohesive European response to these issues. The force of these arguments seem likely to be decisive in France's return to NATO sooner or later.

5
French Defence Policy in a Global Context

Introduction

The emphasis placed on *prévention* and *projection* in *Une Défense Nouvelle* requires that a study of French defence policy gives attention to the French capacity to exercise military power outside Europe. At the same time French interests and French involvement around the world raise so many diverse and complex issues that a single chapter cannot but simplify analysis. To adapt to this limitation this chapter views French military reach through the lens of the armed forces. It describes the French military presence around the world and their roles in the 1990s, examines the implications of the 1996 defence reforms, and assesses the future trajectory of French policy by looking at trends in the *projection* of French conventional and nuclear power outside Europe.

A useful place to start the discussion is in 1994 when France had a post-Cold War peak of 40 389 military personnel deployed outside Western Europe[1], approximately 10 per cent of its total forces[2]. Of these 8621 were deployed across sub-Saharan African by virtue of formal accords or agreements between France and host states in *'missions de présence'*, 21 540 were stationed on or operated from French sovereign territory in *'missions de souveraineté'*, while the remaining 10 228 were deployed on *'missions de paix'* with the United Nations[3]. Figure 5.1 shows the standing French military commitments around the world in the 1990s [excluding *'missions de paix'*]. These consist of garrisoned forces in sub-Saharan Africa and military forces deployed around the world in France's *Départements d'Outre Mer* and *Territoires d'Outre Mer* [DOM-TOMs], the territorial *'confettis de l'Empire'*.

In each of the types of *mission* deployments the basis of French participation, the objectives of French policy, and the role of French forces is different and it is consequently important to consider each in some detail before the wider issues can be explored.

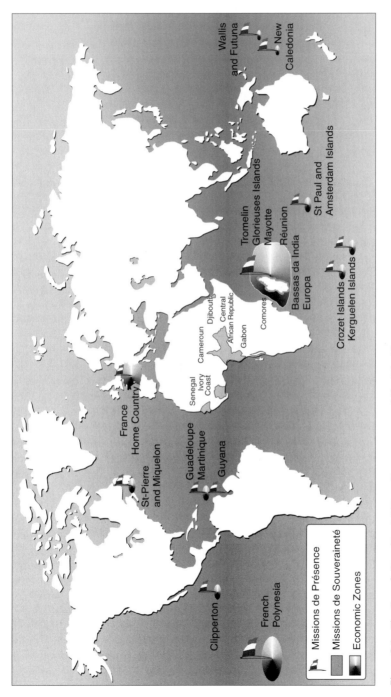

Figure 5.1 The French Standing Military Presence around the World.
Source Defense and the Armed Forces of France (SIPRA, June 1998) pp 4–5.

France and *missions de présence*

Until very recently the most important remnants of the French colonial empire were the *Francophonie* states in Africa. In engineering their nominal independence in the late 1950s and early 1960s, Charles de Gaulle ensured a transition from colonial to neo-colonial dependence[4], the latter defined as 'the survival of the colonial system in spite of formal recognition of political independence in emerging countries which [thereafter became] the victims of an indirect and subtle form of domination by political, economic, social, military or technical means'[5].

Over the following 35 years or so, and despite the Cold War, France exercised a 'virtual empire' in sub-Saharan Africa, premised on the cultural, economic, linguistic and personal ties forged during the colonial period and, somewhat less plausibly, on 'geographic proximity'[6]. While accepting these elements shaped French policy they did not fully explain it. At its heart French policy flowed from national interest and national advantage. In post-colonial Africa France found an exclusive sphere of influence, a *'pré carré' africain*, on which to base its claims of *grandeur* and great power status and on which also to base claims of third world leadership[7]. In Africa France found strategic resources – particularly oil and uranium – and a ready market for French goods, French culture and French ideas. With Francophone Africa France was always more than a middle-sized European state.

One of the most remarked upon features of the French presence in Africa has been the continuity and stability of policy[8]. In understanding this we can discern something of the character of French involvement in Africa. One reason for the French dominance of Francophone Africa [though one which is deeply unpalatable in France] was American indulgence. Had the French hand faltered in maintaining the stability and pro-Western orientation of Francophone Africa during the Cold War, it is unlikely that the exclusivity of French influence would have remained. Similarly the French diplomatic distance from Washington kept the Soviet Union from serious engagement in Francophone Africa for fear of drawing in the United States.

Policy continuity through the presidencies of de Gaulle, Pompidou, Giscard d'Estaing, and Mitterand, was underwritten by a complex interplay of factors. Firstly, the terms under which the transition to independence were made ensured a pro-French orientation in the emergent states. Secondly structural arrangements – currency links, the deep penetration of French companies, Franco-African political and economic fora, close personal links between French and African elites, and a dedicated French Ministry for Co-operation – served both to bind Francophone states to France and to preserve the *status quo*. Finally, one cannot pass over the enduring influence of Jacques Foccart. Foccart served in the Elysée staff as Secretary General for African and Malagasy Affairs under de Gaulle. His closeness to de Gaulle and the subsequent persistence of Gaullist ideas ensured a continuity of influence for Foccart who, first from the Elysée and later from his home, exercised enormous influence on French-African policy often in competition with both the Ministry of Co-operation and the Foreign Ministry. Foccart's networks of personal contacts with African leaders and other key figures [his *réseaux Foccart*] and his intimate links with French intelligence services and France's diplomatic representatives in Africa allowed him to exercise a near fiefdom in Francophone Africa underpinning his importance to each of de Gaulle's successors[9].

The final element in both the continuity and stability in Francophone Africa has been the role of the French military[10]. Following independence France established a variety of formal military accords and agreements with Francophone states in Africa, adding the former Belgian colonies of Burundi, Rwanda and Zaire to this group in the mid-1970s. These arrangements have allowed France to maintain hegemony and regional stability by force if necessary, a power France has not shied from exercising. In all, between 1962 and 1995 France intervened militarily 19 times in African states, not including French participation in UN operations in Angola as part of UNAVEM and in Somalia as part of UNITAF/ONUSAM II. These interventions involved Senegal [1962], Gabon [1964 and 1990], Chad [1968–72, 1978, 1983 and 1986], Mauritania [1977], Zaire [1978 and 1991], Central African Republic [1979], Togo [1986], Comores [1989 and 1995], Rwanda [1990–3, 1994], Djibouti [1991], Benin [1991], and Sierra Leone [1992][11]. In all but a few of these interventions, French action was to protect French nationals, subdue rebellion [irrespective of its legitimacy], and prop-up pro-French rulers including some of the most despotic individuals in post-Colonial African history[12].

At the time Chirac came into the Elysée, France had active military agreements with 23 Francophone African states: Benin, Burkina Faso, Burundi, Cameroon, Central African Republic, Chad, Comores, Congo, Djibouti, Ivory Coast, Gabon, Equatorial Guinea, Guinea-Conarky, Madagascar, Mali, Mauritius, Mauritania, Niger, Rwanda, Senegal, Seychelles, Togo and Zaire. In six of these France deployed standing *mission de présence* forces: Cameroon [10], Djibouti [3500], Gabon [610], Ivory Coast [580], the Central African Republic [1500] and Senegal [1300] and France also had 850 troops in Chad under the terms of a temporary bilateral military assistance agreement[13].

Standing forces and a propensity to intervene were only the visible dimension of French military influence. Less obvious, but no less important, was the influence of the French military on the national defence policies and national armies of the Francophone African states. As Guy Martin has argued: 'the size, location and mobility of French forces... in effect results in France's strict control over the defence of *Francophonie* states – one of the foremost attributes of their sovereignty –... further exacerbat[ing] their already acute dependence on Paris'[14]. Since their creation at the time of independence African *Francophonie* armies have been trained and largely equipped by France for close co-operation with French forces and as a result are often functionally dependent on France. It is a model France continued to use as new states joined the Francophone military group. As Robin Luckham has trenchantly observed: '[T]he presence of French troops and military advisers, the consolidation and reproduction of national military structures through external support, and the transmission through military training of metropolitan [i.e., French] skills, tactics and ideologies *constitute a permanent intervention*' [emphasis added][15].

It is a measure of the insularity of the *Francophonie* African region that the end of the Cold War had relatively little immediate impact on French policy. France continued to support the *status quo*, maintained existing patterns of relations, intervened where it felt necessary and maintained stable standing force levels. Between 1994 and 1997 however France's place in Africa changed significantly. The decisive fracture from the patterns of the past began with the human catastrophe in Rwanda.

French involvement in Rwanda started with a July 1975 bilateral accord on military technical assistance. The nature of French military influence, discussed above, provides evidence that, while this agreement was nominally a limited relationship, it obscured far-reaching French influence on Rwandan national defence. The operational involvement of French troops in Rwanda began in October 1990 when in Operation Noroît France introduced troops to protect French nationals working in Rwanda and to protect the Hutu-dominated and corrupt government of Juvénal Habyarimana from guerrillas of the Tutsi-controlled Rwandan Patriotic Front [FPR]. French troop levels were boosted to around 550 during 1992 and 1993 while at the same time France stepped up the supply of military hardware to government forces.

Habyarimana's death in a plane crash in April 1994 sparked a genocidal 'final solution' led by Hutu militia which systematically slaughtered between 500 000 and 1 000 000 mainly civilian Tutsis. In the chaos the FPR advanced across Rwanda defeating the French-supplied and trained government army and in the FPR's wake at least two million Hutus, including thousands implicated in the killings, fled across the borders into Zaire and Tanzania[16].

In April 1994 France sent around 500 troops into Rwanda in Operation Amaryllis to evacuate French nationals and '*personalités rwandaises*', the latter including key Rwandan government figures some of whom were complicit in the unfolding slaughter. As the scale of the killings became clear to the rest of the world the UN mandated action in Resolution 925 but circumscribed intervention to humanitarian objectives, a limited duration [two months], and neutrality in the security and protection of displaced persons, refugees and civilians[17]. While the UN debated troop deployments, France led a multinational force of mainly Francophone African states [Chad, Congo, Guinea-Bissau, Mauritania, Niger and Egypt] in Operation Turquoise, which was hastily mandated by the UN under Resolution 929 the day before French troops arrived in the theatre on 23 June 1994. France contributed 2550 troops [including 1500 from its standing forces in Africa] to the total force of 3000[18].

The objectives and role of Operation Turquoise have become the subject of much controversy[19]. Defenders of the French intervention argue that it was intended to halt the killing, stabilise the situation, advance the principle of humanitarian intervention and uphold the role of the United Nations. It was not irrelevant either that it demonstrated the continued relevance of France in Africa premised on a French willingness and ability to act while other vacillated[20]. To others, Operation Turquoise was a cynical operation to first attempt to prop up the crumbling regime as the FPR swept across Rwanda and thereafter to create 'safe havens' for fleeing Hutus to protect *'les génocidaires'* and *'les amis de France'* from the FPR and wider accountability[21].

With hindsight the 'shockwaves' from Rwanda played out as they did in France because of the past patterns of French behaviour in Francophone Africa. In Rwanda France supported, armed, and trained a corrupt regime and sent troops to defend it from rebellion. In April 1994 it sent troops in to pull out its own nationals and some allies and promptly withdrew leaving Rwanda to its fate. The deployment of Operation Turquoise was too little too late either to prevent an FPR victory [of the 3000 troops deployed on the Zaire/Rwanda border France had around 1000 troops inside Rwanda but faced 20 000 FPR troops] or to prevent the killing which was largely over by June but which continued nevertheless at a lower level in the French controlled zones 'under the noses of French troops'[22].

The events in Rwanda were arguably the point at which the past patterns of French policy in Francophone Africa ceased to serve even narrowly defined French interests. France had sleepwalked into a disaster on the assumption that past policy would continue to work and could not subsequently escape the consequences of either its association with a murderous regime or of its failure to provide security for that regime or the Rwandan people.

It was not immediately evident however that these lessons were assimilated by French political elites. Analyses of Mitterrand's and Chirac's African policies between 1994 and 1997 point to the continuity of ideas, evidenced by continued intervention [in Comores in October 1995, in the Central African Republic in 1996, and in the Congo from March 1997], static standing force levels, the scrapping of an early Chirac government idea to subsume the Ministry for Co-operation within the Foreign Ministry and even the reappearance at the Elysée of an aged Jacques Foccart[23].

The 'double death' of Foccart in March 1997 [i.e., the death of the man and his influence on Francophone African policy] seemed nevertheless to symbolise the passing of an era[24]. This notion was confirmed within a few months by events in Zaire. The collapse of the regime of Mobutu Sese Seko in May 1997 and the US-brokered transfer of power to the forces of Laurent Kabila crystallised the erosion of the French position. France had clung to the support of Mobutu until the last, prolonging the agony and forfeiting legitimacy with the new regime, at the same time French forces were unable to save Mobutu and indeed at the end, so irrelevant had France become, Mobutu did not even inform the French of his decision to go. As the former defence minister François Léotard commented 'events in Zaire have produced a triple failure for France: tactically because Mr Kabila was backed by the United States and Anglophone African countries, morally because France had given the impression [sic] of supporting the discredited Mobutu to the end, and geopolitically because Zaire was an essential element to the French presence on the continent'[25].

In the wake of Zaire the new administration of Lionel Jospin announced substantial cuts in France's standing forces in Africa. The existing forces of more than 8000 troops were to be cut to around 5500. Moreover France was to close its bases in Cameroon and the Central African Republic leaving bases operational in just five African countries [including Chad]. The troop level reductions would be implemented as follows: in Chad from 840 to 550; in Gabon from 600 to 550; in Djibouti from 3250 to 2800; those on the Ivory Coast to remain the same at 550; and in Senegal from 1300 to 1100, in total a 40 per cent reduction[26].

These cuts have been largely determined by budgetary constraints and are consistent with the overall force level cuts anticipated by *Une Défense Nouvelle* and the 1997–2002 *Loi-Militaire*. The cuts are nevertheless contextualised by the changing French position in Africa and it is important to briefly summarise this. A loss of French influence, displacement from the Great Lakes region, and the encroachment of the United States, Britain and Anglophone African states [particularly Nigeria and South Africa] has ended the exclusivity of French influence in Francophone Africa. France is no longer dealing with Francophone Africa as a bloc and seems increasingly likely to 'cherry pick' close links with the few Francophone states which best serve its interests[27]. In the aftermath of Rwanda the key novelty has been multilateralism – an 'Africanisation' of regional security[28] and a 'multinationalisation' of Western interest exemplified by Franco-British and American-led initiatives to create regional peacekeeping forces[29]. André Dumoulin's argument that French interests in Africa will also 'Europeanise' seems unlikely if understood as meaning that France will co-opt the EU or WEU in pursuit of its interests in Francophone Africa[30], but to be highly prescient in suggesting that existing European interests across Africa are likely to be harmonised as France, for example, looks to Anglophone states like South Africa to extend its influence and to partners like the United Kingdom to co-ordinate responses to events on continental Africa[31].

At the same time the pace of these changes should not be overstated. France has yet to divest itself of any of its existing military agreements with Francophone African states and there are many in France still arguing for policy continuity on the grounds that the French presence acts for continental stability[32]. The catastrophe in the Great Lakes region and instability from Sierra Leone to Somalia and from Chad to Angola does not appear to have intruded on this viewpoint. Perhaps the safest conclusion to draw at the start of the new Millennium is that French security policy in Francophone Africa is on a new trajectory which seems certain to take it further and further from the norms and patterns of the past.

Before leaving the discussion of the French military presence in Africa, it is important to say something about France's largest and arguably most important overseas base at Djibouti. Slightly smaller than Kuwait, Djibouti sits on the horn of the Africa coast on the strategically important Bab-el-Mandeb strait between the Red Sea and Gulf of Aden, the maritime gateway to the Indian Ocean and Asia-Pacific. Occupied by France from 1862 [first as the French Somali Coast and then the French Territory of Afars and Issas] to challenge British regional dominance, Djibouti gained its independence on 27 June 1977 having already agreed protocols for the perpetuation of the French military presence[33].

France explains its role in Djibouti in terms of the defence of Djibouti against Ethiopia, Eritrea and Somalia and against regional instability[34], but the base is of far greater geopolitical and geostrategic importance than that. From Djibouti France has a role in the Horn of Africa [and in Africa more generally], the Middle East and the Indian Ocean, as well as influence in one of the World's key strategic waterways. Djibouti gives France an *entrée* into the Arab League and regional co-operative fora, a further influence in the Organisation of African Unity [OAU] and an ally in the UN. In Djibouti France also finds itself at an intersection of revolutionary Islamic influence between Sudan, Yemen, Somalia and Iran[35].

From Djibouti France is able to project force into three theatres: firstly into Africa where, for example, troops from Djibouti took part in Operations Oryx and ONUSAM II in Somalia in 1992–3, in Operation Turquoise in Rwanda in 1994, and in Operation Azalée in Comores in 1995[36]; into the Middle East as, for example, in the 1991 Gulf War and subsequent operations relating to UN sanctions against Iraq[37]; and finally as a vital link between France and its territories in the Indian Ocean and beyond.

The size and range of forces at Djibouti reflect these diverse interests. For most of the 1990s force levels have been relatively stable at around 3500 personnel comprising 2200 army, 900 air force, and 300 naval forces plus support. The key elements of this force are the 5th *Régiment Inter-Armées Outre-Mer* [RIAOM] and the 13th *Demi-Brigade de Légion Etrangère* [DBLE]. The 1997 force level cuts for Djibouti [from 3250 to 2800] will not substantially alter either the range of forces at the base or the role and importance of Djibouti for France.

France and *missions de souveraineté*

Somewhat overstating the case Jacques Chirac is quoted as once remarking that 'without the DOMs and TOMs, France would only be a little country'[38]. The substantive basis of his remark is that the littoral territories and numerous small islands scattered around the globe which comprise the DOM-TOMs give France a considerable global reach, if not necessarily a global grasp, and without them French claims to great power status or even distinction would be much diminished. The DOM-TOMs, comprising territory of more than 120 000 square kilometres and a population in excess of 1.5 million French citizens, litter the world's three great oceans – the Indian, Pacific and Atlantic – and the Economic Exclusion Zones [ZEE: *Zone Economique Exclusive*] which surround them give France control of almost 11 million square kilometres of ocean, the world's third largest zone of maritime control[39]. From this vantage point France has a role not only in relation to strategic waterways but also in relation to the politics, economics and security of vast regions of the world. Elaborating these points requires that we consider each of the three regions of DOM-TOM influence in turn, but before doing so it is useful to make some general comments about the DOM-TOMs themselves.

The DOMs and TOMs are distinguished from each other primarily by the fact that the former comprise territories seized by France in the 1630s and the latter territories taken by France from the 1840s, after Napoleon. As a result of this division, there are slight but important differences in the nature of relations with France. The DOMs – Guyane, Guadeloupe [and dependencies], Martinque, and Réunion – have the same political and legal position in France as the *départements* within France itself, hence the name. The TOMs however – comprising most of the rest of French sovereign possessions – have a less formally assimilated relationship with France and exercise, at least nominally, a degree of autonomy. Between the two are the *Collectivités Territoriales* – Saint-Pierre et Miquelon and Mayotte – which were seized in 1763 and 1843 respectively and which, for a variety of reasons, sit constitutionally somewhere between the DOMs and TOMs.

The complex historical, political, legal, economic, linguistic, cultural and personal ties that bind France to the DOM-TOMs need not concern us here[40], but it is important to note that the DOM-TOMs are a net cost to France [variously put at 0.2 to 2.3 per cent of GDP[41]], that most if independent would be economically unviable, and that amongst the indigenous peoples of the DOM-TOMs there are many who oppose the French presence, often violently so, particularly in Guyane, Nouvelle-Calédonie and Tahiti[42].

Indian Ocean

The DOM-TOMs in the Indian Ocean are all in the southern hemisphere. They comprise the Iles Eparses in the Mozambique Channel between Mozambique and Madagascar [the Iles Eparses themselves comprise the Iles Glorieuses, Bassas da India, Europa, and Juan de Nova], Mayotte the easternmost of the Comores Islands, the disputed reef of Tromelin, Réunion the linchpin of the French Indian Ocean presence, and the Terres Australes et Antarctiques Françaises [TAAF] comprising the Terre-Adélie, France's Antarctic territory, and the Iles Crozet, Kerguelen, Saint-Paul and Amsterdam strung along the southern Indian Ocean between South Africa and Australia.

The islands [and Terre-Adélie] constitute the military *Zone Sud de l'Océan Indien* [ZSOI], headquartered at Saint-Denis on Réunion, and provide France with control over 2.7 million square kilometres of sea. Amongst these assets Réunion is pivotal[43]. With a population of around 650 000, Réunion administers the rest of France's Indian Ocean possessions and is host to most of the 4500 to 5000 French military personnel of whom around 1000 are 'projectable' forces[44]. These forces have a considerable range of military functions centred on securing the French position and projecting French influence. With respect to the former, ZSOI forces garrison or visit French DOM-TOMs to assert French sovereignty, protect sensitive sites on the relevant islands [for example airports, French administrative offices, etc.], maintain internal security, and conduct surveillance within their zones. With respect to the latter, French ZSOI forces are on stand-by to protect French nationals in neighbouring states, they act to implement bilateral military agreements with neighbouring states, are available to reinforce French forces outside the zone and participate in humanitarian operations[45]. In practice this has meant primarily low-level and localised roles in Comores[46] and Madagascar.

Analyses of the wider importance of the French military in the Indian Ocean centre on control of strategic space ['*espace stratégique*'], the control of strategic waterways for oil flows, strategic minerals, and France's arms trade, and on contributing to regional stability and security[47]. The forces of primary importance in relation to these roles however are those serviced and resupplied or those based at Djibouti.

The arguments for a wider role for ZSOI forces in the southern Indian Ocean are more tenuous. Analyses highlight the control of local strategic waterways [most notably oil flows through the Madagascar Channel in the event the Suez canal is closed or cannot handle the traffic[48]] and the French contribution to stability and security, and argue that the French islands constitute '*un réseau de points d'appui*'[49] [a network of points of support] for French naval vessels as well as for the projection of military power more generally. Objectively the Indian Ocean French territories seem geographically remote, have little history of involvement in significant military operations, and do not seem to have the same strategic importance that, for example, the more northerly Diego Garcia has for the United States. This may change if France strengthens its links with Anglophone states in Southern Africa. For their part the TAAF DOM-TOMs are demilitarised, unpopulated [aside from around 60 French military personnel stationed there], extremely remote geographically or some combination of these. Their limited relevance – other than in terms of maritime control – may increase if Antarctica opens up. Notwithstanding these limitations, a 1998 IHEDN report which gave attention to the issue concluded that France's long-term military presence in the Indian Ocean seemed stable and assured[50].

Pacific Ocean

The French military presence in the South Pacific is divided into two sectors: the *Zone de la Nouvelle-Calédonie* [ZNC], headquartered at Nouméa in Nouvelle-Calédonie [New Caledonia] which takes in the territory of Wallis and Futuna and the *Zone de la Polynésie Française* [ZPF] headquartered at Papeete on Tahiti. The many disparate islands which comprise these zones – and which include the remote island of Clipperton off the Mexican coast, administered from Papeete – provide France with control of 7.6 million square kilometres of the South Pacific, a Francophone 'liquid continent'.

Around 3700 French military personnel are deployed in the *Zone de la Nouvelle-Calédonie*, almost all of them on Nouvelle-Calédonie itself. As with ZSOI forces, the primary function of these forces appears to be the maintenance of the French presence and the protection of French interests centred on internal security, protection of French nationals, the defence of key sites [airports, French administration, etc.], and surveillance. On Nouvelle-Calédonie the issue of internal security has been at the forefront for much of the past 15 years as the result of political unrest, primarily from the indigenous Melanesian [Kanak] population, which flared up from 1984 and culminated in 1988 in violence, hostage abductions, and a military siege at Ouvea[51]. In the wake of the violence, France sought a political accommodation through the Matignon Accords, agreed in 1988, which strengthened economic development, led to a political restructuring which brought the Kanaks into the political process, and paved the way for greater self-determination in Nouvelle-Calédonie which came to some fruition a decade later in April 1998 in the Nouméa Accord which agreed a shared sovereignty deal[52].

Geostrategically the French presence in the ZNC provides an important liaison between the French presence in the Indian Ocean and France's main Pacific interests in French Polynesia, as well as providing France with an influence on the strategic waterways which link the Indian and Pacific Oceans[53].

The focus of the French Pacific presence, French Polynesia, comprises five island chains: the Society Islands [of which Tahiti is one], the Marquesas Islands, the Austral Islands, the Gambier Islands and the Tuamotu Islands, the latter including Moruroa and Fangataufa, the French nuclear weapons test facilities. In all, until 1998, around 3800 French military personnel were deployed across French Polynesia, around 1800 dedicated to the French nuclear weapons test facilities [*Centre d'Expérimentation du Pacifique* – CEP] and most of the rest on Tahiti from where French forces across French Polynesia are commanded. With respect to the CEP, French military forces were based primarily on Moruroa and on the '*base arrière*' at Hao Island, geographically situated between Tahiti and Moruroa, which served as the sub-regional command facility for the nuclear weapons tests[54].

The French test facilities were moved to the Moruroa and Fangataufa atolls in July 1962, following Algerian independence, and the first of 193

South Pacific nuclear weapons tests was conducted at Fangataufa on 2 July 1966[55]. Since 1966 the nuclear weapons tests have dominated both the French presence in the region and relations between France and regional states and territories. While it is impractical here to rehearse this history in any detail, certain features of the French presence are important for the wider discussion. France has maintained its right to test nuclear weapons in the South Pacific by virtue of its sovereignty over French Polynesia and has pointedly rejected arguments about either French neo-colonialism or environmental harm resulting from the tests. It has made much of the geographic isolation of Moruroa and Fangataufa and pointed to political support for French administration on the South Pacific islands[56]. Each of these arguments is highly contentious but even those willing to give France the benefit of doubt have found it hard to defend French acts such as the sinking of the Rainbow Warrior, the use of military force in Nouvelle-Calédonie, and the resumption of nuclear weapons testing, from charges of French 'hooliganism' and indifference to regional concerns[57].

Much of the tension in relations between France and others in the region was eased in 1996 by Chirac's decision to end nuclear weapons testing, close the test facilities at Moruroa and Fangataufa, and sign the Raratonga South Pacific Nuclear Free Zone [NFZ] treaty. As these decisions have been implemented they have led to a considerable downsizing and recasting of the French regional presence. The closure of the CEP and support facilities has meant the phased withdrawal of almost all related French military personnel, while elsewhere small adjustments will leave the French presence in New-Caledonia and non-CEP French Polynesia largely unchanged[58]. Geopolitically, with the testing issue out of the way, France is benefiting from what it sees as a cautious regional re-evaluation of its role. Amongst regional states and territorial elites – even those openly hostile to testing – there has been a creeping recognition of the contribution France makes to regional economic and political stability and security[59].

The anomaly amongst French territories in the Pacific Ocean is the tiny uninhabited atoll of Clipperton Island. Situated 1300 kilometres off the Mexican coast and annexed by France in 1858, the 7 square kilometre atoll was utilised by the United States for a short period during the Second World War [1942–5] but has since been bereft of infrastructure or personnel. The French possession of Clipperton nevertheless yields two main benefits. Firstly the atoll provides France with a maritime ZEE of 425 220 square kilometres [larger than that around Metropolitan France at 340 000]; secondly Clipperton gives France a nominal presence at least in a region of some strategic importance given its proximity to the Panama Canal and important shipping lanes[60].

Atlantic Ocean / Caribbean

The remainder of the French DOM-TOMs are those located in the Atlantic/ Caribbean region. These comprise the Latin-American territory of Guyane

[French Guyana] bordered by Brazil and Surinam; the French Caribbean territories Guadeloupe and its dependencies and Martinique; and the anomalous islands of Saint-Pierre et Miquelon situated in the northern Atlantic just 25 kilometres off the Canadian Newfoundland coast.

Guyane is the largest French overseas possession and arguably the most important because of the presence at Kourou of the *Centre Spatial Guyanais* [CSG], the French-led European space vehicle launch facility for Ariane. Despite its size [roughly one sixth the size of France] Guyane has a population of only around 150 000, 65 per cent of whom live in the capitol Cayenne, 27 per cent along the Guyane coast and only 8 per cent in the densely forested interior. As a French *département* Guyane has no independent political or legal identity but returns two *deputés* to the French *Assemblée Nationale* and a *sénateur* to the *Sénat*. Through Guyane France has membership of the Organisation of American States [OAS] and involvement in other less important regional fora.

France's early space programme was based in Algeria [the first French satellite was launched from there in 1965] but following Algerian independence the CSG was established in 1964, launch programmes transferred there in 1967, and the first rocket was launched in 1970. Organisationally the CSG is shared between the multinational European Space Agency [ASE – *Agence Spatiale Européenne* – in which France has a 46 per cent stake] which directs launch programmes, finances Ariane and contributes to the CSG budget and the French national space agency CNES [*Centre National d'Etudes Spatiales*] which is responsible for the infrastructure at Kourou and operations at the base[61]. The importance of Kourou for France is difficult to overestimate. Through the CSG France has the lead role in Europe's only independent space programme, a demonstration of French technical prowess, modernity and distinctiveness and of Europe's capacity to work together and compete internationally with the United States and others. The European space programme is hugely important economically, generating billions of dollars of annual business activity, and hugely important politically not least in terms of Western Europe's autonomy in a pivotal area of technical activity. The CSG is also of great strategic value to Western Europe in facilitating European military space programmes including the French Syracuse communications network and the autonomous strategic intelligence Helios programme.

Given that there are 'no realistic military threats to either Guyane or Kourou'[62], the standing French military presence in Guyane is small, at least in relation to territory. Around 3600 personnel[63] are stationed in the *Zone de la Guyane* [ZG] headquartered at Cayenne. The formal role of these forces is to affirm and defend French sovereignty, monitor the Maroni and Oyapok rivers which constitute most of the borders with Surinam and Brazil respectively, and to maintain a climate of security within the *département*. The latter relates both to the protection of the CSG at Kourou and of CSG-related activities elsewhere in Guyane and to the maintenance of internal security against 'clandestine immigration, social unrest, and internal destabilization'[64].

In addition to standing forces, France has in place the means to rapidly reinforce the French military presence in Guyane within 48 hours using forces redeployed from the French Antilles or directly from France. The main elements of these reinforcements are the capacity to boost the standing army/ *gendarmerie* presence and the capacity to boost the air defence of the CSG through the prompt introduction of AWACS, Crotale air-defence missile systems and Mirage 2000 fighters[65]. These measures testify above all to the value of the CSG, but also to the significance of the small but volatile independence movement in Guyane.

The French military presence in the Caribbean on the islands of Martinique and Guadeloupe and its dependencies [including the islands of Le Désirade, les Saintes, Marie-Galante, Saint-Barthélémy and the French part of Saint-Martin] is organised as the *Zone des Antilles* [ZA] headquartered at Fort-de-France on Martinique. In all, around 4700 military personnel are based on the islands, around 2800 on Martinique and 1900 on Guadeloupe[66], in addition to around 220 personnel aboard French naval vessels assigned to the *Zone Maritime Antilles-Guyane* [ZMAG][67].

The role of these standing forces reflect those elsewhere in the DOM-TOMs: the affirmation and defence of the French presence, internal security [including in relation to autonomist or independence movements], the protection of French nationals [26 000 *métropolitains* on Martinique and 8000 on Gaudeloupe[68]], and the surveillance and control of the French Caribbean ZEE which together with Guyane coastal waters adds up to around 355 000 square kilometres of sea[69].

The regional influence of these forces is more difficult to assess. Aside from the reinforcement of Guyane, discussed above, the presence would appear to be important in terms of a French influence on strategic waterways through the Caribbean and as a liaison between France's principal DOM-TOM assets around the world [Réunion – French Polynesia – Guyane]. As regional players, French room for manoeuvre in the Caribbean is limited by the dominance of the United States and by an economic and political insularity which keeps the islands dependent on France and outside the main regional economic fora. French forces have played a role in responding to some regional natural disasters, they participated in a multilateral intervention to defend the government of Eugenia Charles in Dominica in 1982[70], and they took part in the UN Mission in Haiti [UNMIH] established in September 1994, but generally the French military profile in the Caribbean has been lower than elsewhere in the DOM-TOMs[71].

France's final outpost is the small island group of St-Pierre et Miquelon off the Canadian coast. Originally a *département*, the islands became a *Collectivité Territoriale* in June 1985 ceding a degree of political autonomy to the roughly 6500 inhabitants who nevertheless still return a *député* and *sénateur* to

France. Because of the close proximity of the islands to Canada, French claims to a ZEE around them are contested by Canada and considerable tensions have arisen particularly in relation to fishing rights[72].

France has a standing presence of around 30 military personnel on St-Pierre et Miquelon whose principal internal roles match those of French forces elsewhere in the DOM-TOMs. Regionally St-Pierre et Miquelon has served at times as an irritant to Canadians concerned about French involvement in Quebec separatism, while the islands' proximity to northern transatlantic shipping lanes has persuaded some of the French presence's wider strategic relevance.

Having said something about the French military presence across the DOM-TOMs it is important to look ahead and assess the impact of the 1996 defence reforms on the French presence, and to do this it is necessary to say a little more about the nature of French deployments. The force levels discussed above should not, with the exception of the majority of naval forces, be understood as standing combat presences. The figures in fact include *gendarmerie*, those serving on *Service Militaire Adaptée* [SMA] – that is those fulfilling their conscription obligations in non-military related roles[73] – and conscripted local non-metropolitan French citizens [for example a Guyanaise French citizen serving his conscription duties in Guyane]. The legal and political limitations on the deployment of conscripted French citizens outside France in turn explains why only a percentage – typically around one third or less – of the standing forces are 'projectable' into combat theatres outside the territory on which they are deployed.

The 1996 defence reforms impact on this presence in two main ways. Firstly the force level cuts in the defence reforms are reducing the overall numbers of forces deployed across the DOM-TOMs [down from around 21 540 in 1994 to around 17 500 by 2002]. However, this reduction is almost wholly accounted for by the closure of the South Pacific CEP [and related facilities including Hao] and aside from this force levels are remaining essentially static[74].

Secondly the ending of conscription means that professional forces will eventually replace most conscripts across the DOM-TOMs. The implications of this change are quite complex. On the one hand, with professionals replacing conscripts and the force levels [except CEP and related] remaining roughly the same, the cost to France of garrisoning the DOM-TOMs as a proportion of the overall defence budget is increasing.

However, there are substantial benefits from professionalisation too. Firstly, it allows for greater flexibility in the rotation of units between France and the full range of military deployments overseas. Finding adequate numbers to garrison the DOM-TOMs is unlikely therefore to be a major problem because forces can be rotated from elsewhere and because some local non-metropolitan French citizens will still be recruited by the attractions of a volunteer or professional career, often in the context of relatively few alternatives.

More importantly, professionalisation is boosting the 'projectability' of standing DOM-TOM forces as the constraints on conscripts become irrelevant. If France was merely wanting to maintain pre-1996 levels of projectable forces across the DOM-TOMs, it could afford to allow garrison levels to fall and thereby reduce costs. Despite the downward pressure on the defence budget, however, this is not happening. In practice the total number of theoretically projectable standing forces on the DOM-TOMs is in the middle of doubling from around 8400 in 1994 to a theoretical maximum of 17 500 by 2002, representing an increase from around 3 per cent of the total professional armed forces in 1994 to perhaps 5.5 per cent in 2002[75]. This trend is in line with the shift in the balance of French interests from *métropolitain* defence to the protection of interests and the shouldering of responsibilities on the international stage[76].

France and *missions de paix*

Since the end of the Cold War, France has been arguably the most active of the Security Council members, the most active European state and with one or two exceptions – perhaps India and Pakistan – the most active state anywhere in United Nations military operations [*missions de paix*]. Since 1990 France has participated in 13 UN operations in Europe, Africa, Asia and Latin-America and in September 1998, the most recent date for which official figures are available, had around 5000 troops [down from a 1994–5 high of over 10 000] in Albania, the former-Yugoslavia, Angola, Central African Republic, Western Sahara, Haiti, Georgia, Lebanon, Kuwait, Sinai, and the UNTSO presence [Israel/Egypt/Syria/Jordan][77].

In the context of this chapter, understanding why France maintains this high UN profile and understanding something of French thinking about *missions de paix* is important because it provides further insight into the projection of military force outside France and because it allows the relations between *missions de présence, missions de souveraineté,* and *missions de paix* to be explored.

France has made much of the alignment between the nature of UN military operations and France's historic role as a global player and *nation phare*, the motherland of democracy and human rights, and the steadfast friend of the downtrodden, the oppressed and the weak. In *missions de paix* France has found a contemporary context for French exceptionalism and a modern expression of the historic *mission civilisatrice*[78]. France's high profile in UN operations [like the high profile of the French NGO's *Médecins du Monde* and *Médecins sans Frontières* in humanitarian operations] is consequently in no small part about the promotion of a positive humanitarian image for France and the support of French norms and values in the international system.

Peel away this presentational veneer, however, and French military participation in *missions de paix* is revealed to be premised on far more pragmatic and self-serving motivations. One of these undoubtedly is an interest in the promotion of international stability and order and France, at least in the post-Cold War era, has championed the United Nations and legalistic pre-scriptions for international problem-solving albeit in defence of the *status quo*. France's participation in UN operations does play a part in strengthen-ing the United Nations, in upholding the rule of international law, and in promoting concepts of collective security, however loosely the latter is defined and however selectively applied. But this explanatory layer in turn is under-pinned by French national interest. As Marie-Claude Smouts has pointed out:

> France's participation in peacekeeping operations is not part of some abstract logic of collective security. It serves material and immaterial aims of national interest such as security imperatives, self-image and interna-tional prestige[79].

The strength of this argument – that high rhetoric obscures base motives – becomes clear if one reflects on the history of relations between France and the United Nations. While it is inappropriate here to address this history in any great detail, it is important to sketch out those elements of past French-UN relations which directly inform contemporary policy[80].

French thinking about the United Nations is dominated by France's role as one of the five permanent members of the Security Council [P5]. Since the UN's inception France has worked to strengthen the prestige and capab-ilities of the Security Council, and through it France's own status and influ-ence, and has also done all it reasonably could to demonstrate that it merited its place as a Security Council member. In the few years between the end of the Second World War and onset of the Cold War, France conse-quently, and despite its dire economic and political condition, contributed military observers to UN missions in the Balkans and Indonesia in 1947 and to the United Nations Trust Supervision Organisation [UNTSO] in the Middle East in 1948[81].

Franco-UN relations began to sour after 1950 due to the outbreak of the Korean War. United States determination to co-opt the UN in the struggle with communism and the temporary absence of the Soviet delegation from the UN at the time paved the way for US Secretary of State Dean Acheson to use the 'Uniting for Peace' resolution [377(V)] to transfer power from the Security Council to the UN General Assembly in relation to 'maintenance of peace' issues in an attempt to circumvent the troublesome Soviet Security Council veto. Empowering the General Assembly was deeply unwelcome in Paris because it diminished the status of the Security Council and thus of France, it strengthened US hegemony by *de facto* disabling the veto, and per-haps worst of all, it opened the door to unrestrained UN criticism of French colonial policy. France was virtually forced to go along with the change because of its dependence on United States aid, but at least found the embroiling of the UN in the struggle against communism a useful cover for its own colonial war in Vietnam.

French relations with the UN slumped further as a result of the Suez crisis in September–October 1956. It is important here to pull out the key points of this crisis for France. The first was US-Soviet collusion in using the Acheson resolution to circumvent the French and British Security Council vetoes which confirmed the French perception that the empowerment of the General Assembly undercut French status and influence. The second was the humiliating withdrawal of French and British forces without even a face-saving role in the subsequent control of the Suez Canal which created [arguably unfairly] political and public fury against the UN in France. The third was the UN decision to deploy a peacekeeping force UNEF 1 [United Nations Emergency Force] without the participation of any of the 'big five' in what became known as the first 'Chapter VI-and-a-half' mission which had the effect of eroding the primacy of the Security Council in Chapter VI operations[82].

If Suez confirmed the deteriorating relations between France and the UN, the nature of the tensions between the two only fully became evident when de Gaulle returned to power in 1958. De Gaulle's emphasis on French independence, the primacy of the nation-state and his belief in great power relations as the main determinants of international order clashed with the 'idealist' and supranational agenda of the UN Secretary General Dag Hammarskjöld[83]. Seven years of confrontation [including non-payment of contributions to UN operations, the 'empty chair' arms control policy, and the denial of French assets such as airbases in Africa to the UN] followed, punctuated by the Algerian transition to independence which de Gaulle worked assiduously to keep out of the UN purview, and the 1960 UN operation in the Congo [UNOC] which encroached on Francophone Africa and which, by intervening in the internal affairs of an African state, was necessarily pregnant with implications for French African policy.

Although relations improved from the mid-1960s as a result of the completion of the decolonialisation of France's African possessions, the death of Hammarskjöld and his replacement by the more prudent and less institutionally ambitious U Thant, France's relations with the UN did not normalise until the primacy of the Security Council had been fully reasserted.

The decisive break with established patterns of French-UN relations occurred with the United Nations Interim Force in Lebanon [UNIFIL 1] in March 1978. For a variety of reasons including traditional French interests in Lebanon, President Giscard d'Estaing's determination to enhance the French role in the UN, and a situation on the ground which effectively precluded the US and USSR, France was presented with an opportunity to act. It subsequently deployed 1380 troops, almost one quarter of the initial UNIFIL deployment of 6000 drawn from 14 countries [the French presence was cut to 530 troops in 1986 and has remained at that level since][84].

UNIFIL was a watershed in other respects too. Firstly, it broke the established, though unwritten, rule of non-participation of the 'big five' in UN military operations after the Korean War [arguably that rule had in fact been broken by British logistical support of UN operations in Cyprus in 1964]. More importantly, however, it involved the United Nations in a new kind of operation. Before 1978 the UN had been involved only in situations in which the warring parties had already agreed some form of cease-fire and only at the invitation of the warring parties. The UN role in these so-called 'first generation' deployments had been based on the use of unarmed observers and monitors introduced to oversee and apply cease-fire agreements. UNIFIL, the first of the so-called 'second generation' deployments was different. There was no cease-fire – temporary or otherwise – in place, and the mandate of the UN was ambitious and complex and included UN imposition between warring parties, persuading Israel to withdraw from the strip of Southern Lebanon it occupied in March 1978, and the restoration of Lebanese government control in South Lebanon. The demands on the UNIFIL forces thus went well beyond observation and monitoring, yet the lightly-armed UN personnel found themselves confined to passive roles and under direct threat from some of the warring groups.

For French forces the experience of impotence in a dangerous context [in all, UNIFIL was eventually to cost the lives of 15 French personnel and injure 42[85]] – particularly when UN forces, who could scarcely even defend themselves, were powerless to stop the reinvasion of South Lebanon by Israel in 1982 – was a key point in the evolution of national doctrine for *missions de paix* which was to be so important in the post-Cold War context[86].

After the UNIFIL deployment, France refused new UN commitments for more than a decade. Part of the explanation for this was undoubtedly the 'Second Cold War' which reparalysed the UN Security Council and focused French attention on European security, but part of the explanation also can be found in France's reluctance to be embroiled in the military operations of an organisation which seemed unable or unwilling to adequately empower the forces deployed in its name.

The ending of the Cold War, and with it the unfreezing of the Security Council, ushered in a new era of potential both for the United Nations and for France. In his address to the UN General Assembly on 24 September 1990, Mitterrand made his assessment and hopes clear:

> Long paralysed, the United Nations now, forty-five years after its birth, at last appears [to be able] to fulfil the mission determined for it by the San Francisco Charter...Let us through the UN ensure that law, solidarity and peace reign in these new times[87].

In making specific reference to the founding San Francisco Charter, Mitterrand was placing centre stage the 'idealist' agenda of the UN as the instrument of collective security and the framework organisation for international stability premised on the rule of international law. The unstated subtext of this rhetoric, however, was that France, by virtue of its Security Council place, its global military presence and its internationalism, had in the new context a chance to empower the Security Council [and thus its own global status and influence], to multilateralise US hegemony by situating US actions within the UN, and to directly assert its own influence and status on the international stage.

France consequently expressed itself vigorously after 1990 in UN operations. Most French activity was in small and low-key deployments in 'first generation' missions. These included the United Nations Iraq-Kuwait Observation Mission [UNIKOM] where France had 15 personnel from April 1991; the UN Observation Mission in El Salvador [ONUSAL] established in May 1991; the UN Mission for the Referendum in Western Sahara [MINURSO] to which France contributed 30 troops in October 1991; a French representative at UNCONSMIL in Cambodia for a year between May 1994 and May 1995; and a handful of military personnel at the UN Mission in Haiti [UNMIH] from September 1994, the UN Observer Mission in Georgia [UNOMIG] from October 1994, and the UN Angola Verification Mission [UNAVEM III] from March 1995[88].

The crucial missions though, in so many respects for the French, were the 'second generation' UN deployments in Cambodia, Somalia, the former-Yugoslavia and Rwanda and it is important to give some attention to each. As co-president [with Indonesia] of the international conference to realise a post-Khymer Rouge settlement in Cambodia and as a former colonial ruler, France had a particular *entrée* into the Cambodian situation[89]. France participated in the two main UN deployments contributing 114 personnel to the UN Advanced Mission in Cambodia [UNAMIC] between 12 November 1991 and 15 March 1992, and then sending almost 1500 military personnel to form part of the follow-on UN Transitional Authority in Cambodia [UNTAC] between March 1992 and September 1993.

In many respects UN operations in Cambodia were a success, stabilising the situation on the ground to allow elections, rebuilding part of the communications infrastructure, disarming and demobilising some of the warring parties, reversing some of the internal displacement of peoples and repatriating some of the hundreds of thousands of refugees from across neighbouring borders, mainly in Thailand. That said, the bulk of UN forces pulled out in September 1993 before any long-term solution to the central struggle between the Khymer Rouge [with strongholds in the country's north and west] and government forces was in place and indeed fighting between the two promptly broke out in January 1994. Outgoing French forces [who had suffered two dead and 33 injured] left Cambodia with few illusions about the ability of the UN to impose peace in an unstable political context or to engineer the reconciliation of those who had suffered or been adversaries during the conflict[90].

The introduction of 2050 French forces into Somalia under the French appellation of Operation Oryx, as part of US-led UNITAF/Operation Restore Hope in December 1992, to beef-up the initial UN Operation in Somalia [UNOSOM 1] which began in April 1992, was potent with implications for French *missions de paix* and UN operations more widely. Operation Restore Hope, mandated by the UN in resolution 794 on 3 December 1992, was undertaken in direct response to the worsening famine in Somalia and as such it was the first UN deployment to be carried out both without the invitation of the warring parties and for an overtly humanitarian objective.

The humanitarian motivation for action in Somalia, and for similar action in Bosnia, steered the UN into uncharted, or perhaps more accurately un-chartered, waters and precipitated within France a wide-ranging debate about the *devoir d'ingérence* [the duty to intervene] and the *droit d'ingérence* [the right to intervene]. The Mitterrand government quickly took the moral high ground with Prime Minister Pierre Bérégovoy explaining on 8 December 1992, just a few days after the UN mandate, that 'France intends to be present, always under UN auspices, wherever the law must be respected or human lives preserved'[91]. This apparently open-ended commitment was consistent with humanitarian intervention already being an established theme of French policy[92]. France had been the lead nation in the adoption by the UN on 8 December 1988 of Resolution 43/131 which laid the foundation for UN *droit/devoir d'ingérence* by accepting the case for humanitarian assistance in cases of natural disasters and 'emergency situations of the same order' [*situations d'urgence du même ordre*][93]. France also took the lead in April 1991 in creating the safe zones inside Iraq for Iraqi Kurds in the aftermath of the Gulf War, and in July 1991 Mitterrand himself both claimed credit for France and set some of the parameters of the debate when he opined that:

> It is France which has taken the initiative in this new right, quite extraordinary in the history of the world, which is a sort of right to intervene in the internal affairs of a country when one part of the population is the victim of persecution[94].

His view however was just one of many in a sharp and often polarised debate within France informed, *inter alia*, by issues of state sovereignty and international law, the potential conflict between national interest and moral foreign policy, and the rights of self-determination and risks of neo-colonialism[95].

The French experience within Somalia itself did little to resolve this debate. The chaotic and lawless situation particularly in the capital Mogadishu gave the UN little framework for its activities. France took control of a northern sector of the country, headquartered at Hoddur, and provided logistic support for wider UN operations which addressed the immediate problem of the Somali famine by facilitating the access of aid and humanitarian assistance[96]. But, as in Cambodia, the UN could find no means within the limits of its mandate to impose peace or to construct a stable political framework[97]. The death of 18 US servicemen on 3 October 1993 in a botched attempt to kidnap senior Somali warlords in Mogadishu precipitated an ignominious US withdrawal accompanied by most of the remaining Operation Restore Hope forces, including the French.

If the experience of the French military in Somalia fed into the evolution of French thinking about *missions de paix*, it was in the former-Yugoslavia that the dilemmas and constraints in French UN policy were posed in their starkest form. France was the lead nation in UNPROFOR between February 1992 and December 1995 with more than 6000 troops deployed at its peak and suffered the heaviest UN casualties [53 killed and almost 600 injured]. Whilst the UN deployments were essential from a humanitarian point of view and while the UN succeeded [with the EC and other parties] in preventing the conflict from spreading beyond the territory of the former Yugoslavia, the UN was powerless to prevent ethnic cleansing and widespread killing, torture and rape inside the breakaway republics themselves. Hamstrung by the operational limits imposed by their political masters in New York, French and other UN forces were forced to stand aside while the worst war crimes in Europe since the Second World War were perpetrated[98].

Arguably, the main issues for the French military can be illustrated by a single event, the attempt by General Philippe Morillon to save the Muslim enclave of Srebrenica in March 1993. Surrounded by Bosnian Serb forces, Srebrenica was cut off from supplies and subject to sporadic artillery and mortar bombardment. Morillon entered the town on 11 March 1993 in an attempt to broker a deal and found himself and his few forces held hostage in a less than spontaneous demonstration by a terrified populous anxious for UN protection. In a unilateral move, and without the consent of his political masters, Morillon raised the UN flag and declared that Srebrenica was under UN protection. His move was decisive in bumping the UN into declaring in Resolution 819 that all the besieged Muslim enclaves were, nominally at least, 'safe areas'.

The importance of Morillon's courageous action lay in the fact that he had single-handedly set aside UN neutrality and offered UN protection to one side in the conflict. To Morillon's disgust, this noble gesture and the UN's acceptance of the idea of safe areas did not translate into any substantive change in UN operating procedures or mandate[99]. The safe areas were besieged for more than two years, were then overrun by Bosnian Serbs in July 1995, and the male populations were massacred while the UN stood impotently aside.

The UN did not come off the fence until a confluence of events – the public revelation of the Serbian massacres in the enclaves, the Serbian stranglehold on Sarajevo which was undermining UNPROFOR, the hostage-taking of UN forces in flagrant violation of existing agreements, and another bloody mortar attack on a Sarajevo market on 28 August 1995 – and French pressure in particular persuaded the United States and thus NATO under the UN mandate to move towards a more robust response. The French-British-Dutch Rapid Reaction Force deployed on Mount Igman above Sarajevo took part in attacks on Serbian positions from 30 August while French aircraft participated in NATO airstrikes on Serbian positions as part of Operation Deliberate Force which was instrumental – along with Croatian and Bosnian federation military gains on the ground – in bringing the Serbs to Dayton. The robust use of military power to effectively impose a peace ushered in a novel 'third generation' *mission de paix*.

Reflecting later, Morillon drew the lesson that the use of force by NATO against the Bosnian Serbs in August 1995 had been 'a watershed for Bosnian history [and] also, in my opinion, *for all future interventions*' [emphasis added]. He added that:

> Abandoning the unworldly attitude which had condemned its troops to powerlessness . . . the UN has understood that [others] had been strong only to the extent that the UN was weak. By finally giving its soldiers the authority and means to retaliate not only when their own lives were in danger, but also whenever their freedom of movement was obstructed, the UN has understood that to limit violence its military forces must be able to implement their mandate whilst throwing down the challenge 'shoot at us, if your dare'[100].

The French deployments as the mainstay of Operation Turquoise in Rwanda between 23 June 1994 and 30 September 1994 showed that some of the lessons of Morillon's experience in Srebrenica had already been assimilated by the French. The 'semi-detachment' of the French forces in Rwanda [see above] ensured that France was able to exercise adequate force in defence of the safe areas. Paradoxically perhaps, the *génocidaires* amongst the multitudes gathered in French-controlled areas thus benefited from a French determination not to see a repeat of UN impotence in Srebrenica[101].

France withdrew its forces from Rwanda in September 1994, leaving Rwanda to the strengthened UNAMIR II, with the integrity of French *missions de paix* and its Africa policy in question. The refusal by the UN to mandate a force for intervention in Burundi in 1995–6, and the unwillingness of France to act independently, appeared to signal that the high water mark of French UN military operations had passed, a perception underlined by events in Bosnia in 1995.

This history of French-UN relations with respect to UN military operations and the defining experiences of the French military in Cambodia, Somalia, the former-Yugoslavia and Rwanda have together forged French doctrine and thinking about UN *missions de paix*. It is important here to bring these elements together in order to assess the current state of the French debate and thereby the likely future trajectory of French policy.

One way to chart the development of French thinking is by following the debates and themes in official French reports and documents dealing with *missions de paix*. France was one of the first states to respond positively to UN Secretary General Boutros Boutros-Ghali's *Agenda for Peace* published in June 1992[102]. In its wake, France made a series of offers and proposals designed to strengthen the capacity of the United Nations to act, including organisational changes to enhance the military competence of the UN, the suggestion that the UN should have a 5000 strong standing military force [a pale echo of the UN's founding expectation that it would have a permanent army for genuine collective security], and the offer of up to 1000 French troops at the UN's disposal on 48 hours notice[103].

Arguably, the first important attempt within France itself to refine French thinking was the February 1994 report of Sénateur François Trucy entitled *Participation de la France aux Opérations de Maintien de la Paix*[104]. In his report to the Prime Minister, Trucy critiqued the lack of clarity in French thinking to date and called for a wide-ranging and deeper debate about *missions de paix*. In seeking to take that debate forward, and reflecting something of the French military experience on the ground, Trucy called for the determining of a typology of missions in which French forces could be involved, and for precision in mandating particular operations under particular parts of the UN Charter.

Some of the answers to the questions posed by Trucy were almost immediately forthcoming in the 1994 *Livre Blanc sur la Défense*. The *Livre Blanc* set out three elements guiding French policy on the issue: firstly, the assurance of the primacy of the Security Council and the promotion of better co-ordination between politicians directing missions and the military forces implementing UN mandates on the ground; secondly, French commitment to work to strengthen the military capability placed at the disposal of the UN; and, thirdly, the requirement that French forces subordinated to the UN be given precise political and organisational conditions for UN operations[105]. The first point in asserting the primacy of the Security Council was arguing, in effect, that with its veto no UN *missions de paix* should take place without French consent. With respect to the second point, the *Livre Blanc* went on to elaborate the need to facilitate and strengthen the links between the UN and NATO or the WEU. This is an important strand in French policy because it reflects the desire to embed NATO in the UN to constrain US hegemony, to embed the WEU within the UN to enhance European defence co-operation, and the requirement to beef-up the capability of the UN by *de facto* positing NATO/WEU as its military arm[106].

The third point spoke to the need, above all else, to avoid the impotence of French and other UN soldiers on the ground by arguing the need for a clear mandate, clear objectives and robust rules of engagement which allowed for the use of force. The French experience in Bosnia in particular had already led within the French military to the adoption of a simple maxim informing *mission de paix* doctrine: '*on tire ou en se retire*' [either we fire or we retire], meaning that French forces either had the capacity and freedom to defend themselves and pursue their mandate or they had no business being in a conflict theatre.

The experience in Somalia and growing frustration in Bosnia shifted the emphasis within France towards the issue of humanitarian intervention and this theme was taken up by the *Assemblée Nationale deputé* Jean-Bernard Raimont in his report *La Politique d'Intervention dans les Conflits: Eléments de Doctrine pour la France* published on 23 February 1995[107]. Raimont explored the issues and argued strongly in support of humanitarian intervention, called for the refining of French doctrine to incorporate humanitarian intervention, set out the operational requirements for intervention, and argued the need to enhance the interface between strictly humanitarian roles and other forms of *missions de paix*[108].

Since Raimont, the pace of the French debate has slowed as enthusiasm for *mission de paix* in general has waned and the UN has to some extent been marginalised – at least in the former-Yugoslavia – as the mandating organisation for the deployment of NATO-led forces [IFOR and SFOR in Bosnia and KFOR in Kosovo]. Nevertheless the key elements of French doctrine are now clear. Underpinning all is the French emphasis on the primacy of the UN Security Council as the authorising body for UN *missions de paix*. France is consequently opposed in principle to a role for the UN General Assembly in this respect and thus against, in all but the lowest level commitments, 'Chapter VI-and-a half' missions initiated without UN Security Council member participation.

With respect to *missions de paix* themselves, France rejects what it sees as an artificial separation between UN actions taking place with the consent of warring parties and those where there is no consent. In his defining statement in 1995, the French CEMA Admiral Jacques Lanxade set out the idea of a continuum between three forms of *missions de paix* broadly reflecting the first, second and third generation actions set out above:

(1) *le maintien de la Paix*: traditional peacekeeping operations conducted under Chapter VI premised on the consent of the parties and the existence of an observed cease-fire;
(2) *la restauration de la paix*: peace restoration operations conducted under Chapter VII premised on the absence of a cease-fire and the absence of consent where the UN is intent on restoring peace without identifying a particular aggressor; and
(3) *l'imposition de la paix*: peace imposition operations conducted under Chapter VII where UN forces re-establish or impose peace by the threat or use of force against an identified aggressor[109].

The notion of a continuum and the explicit reference by Lanxade to impartiality as opposed to neutrality ['impartiality' meaning no preconceptions about warring parties but the option to take sides if necessary, 'neutrality' meaning the requirement not to take sides in any circumstances] inform French thinking about what exactly these different types of *mission de paix* and the liaison between them mean for political masters and soldiers on the ground.

At the political level, the French requirement is for clarity about the purpose of the mission, acceptance of the potential for escalation, and flexibility for the political position to evolve in relation to events on the ground. At the military level, the emphasis is on clear and robust rules of engagement for both self-defence and the pursuit of mandate objectives, delegated operational command as opposed to the attempt of the UN to micro-manage military operations, adequate armaments to cope with limited escalation, and the freedom to take sides if necessary.

Because France is inevitably involved in multilateral UN operations, French doctrine also requires that France pushes for the acceptance of its position at the political level in the UN and other fora relevant to *missions de paix* [such as NATO or the WEU] and that it works for doctrinal harmonisation between its own military and the militaries of the other nations with which it co-operates.

The projection of conventional military power

In order to reconcile its ambitious international commitments with its objectively limited means, France has focused the post-Cold War adaptation of its conventional forces [*les forces classiques*] on the exercise of influence outside Europe and the maximisation of the utility of the armed forces at its disposal[110]. To understand this more fully and to assess the trends in the exercise of French conventional power outside Europe, it is necessary to bring together three strands of French policy: the adoption of the new roles of *prévention* and *projection*; the interplay of French forces deployed overseas on *mission de présence, mission de souveraineté* and *missions de paix*; and the restructuring of the armed forces to equip France with four 15 000-strong projectable combat groups with the requisite air, naval, logistic and command and control support.

The central theme underpinning the emerging concepts defining the exercise of French conventional power is the emphasis on dealing with threats to French interests and to international stability by the use of the lowest level of military force possible. As the notion of *prévention* makes clear, France prefers to tackle situations or conflicts before they escalate on the basis that a modicum of involvement today may save considerably riskier commitments tomorrow[111]. If this seems a simplistic observation it is not. The disposition to act before escalation is premised on knowledge and understanding of the potentially escalating situation and on the capability and willingness to act promptly, if necessary in advance of broader international consensus.

These requirements are reflected in the twin pillars of *prévention*: intelligence [*renseignement*] and the prepositioning of forces [*prépositionnement de forces*]. France's efforts to strengthen intelligence capabilities have centred on national responses for theatre intelligence and European-level responses for space-based assets. The former exploits France's global presence, intelligence-gathering facilities on French-controlled territory around the world, and dedicated theatre military intelligence systems, the most important of which are the Army's *Brigade de Renseignement et de Guerre Electronique* [BRGE], the Navy's MINREM system [transferred from the *Berry* to the *Bourganville* in May 1999], and the Airforce's Sarigue-NG, AWACS and reconnaissance aircraft[112].

The idea of *prépositionnement* is centred on exploiting French standing forces around the world. The professionalisation of the armed forces is allowing the troops on the different French missions overseas to be rotated smoothly between operations and is eroding the barriers between deployments paving the way for increased effectiveness. The trends discussed above in French military activity in Francophone Africa, the boosting of projectable forces in the DOM-TOMs and increasingly robust rules of engagement for French forces on UN deployments point towards increasing synergies between the three and the further extension of the kinds of cross-deployments which have recently been possible such as Operation Oryx and Operation Turquoise.

Prépositionnement though is about more than standing forces, it is also about enhancing the permanent infrastructure of France's global presence [airbases, ports, logistics and communications] to facilitate interactions between French forces deployed overseas and to facilitate the projection of more substantive forces if necessary from France itself[113]. To this end considerable resources have been directed by the 1997–2002 *Loi-Militaire* to enhance this infrastructure. The pattern of this spending reveals the prioritisation of France's remaining bases in Africa [including Djibouti] from which one can deduce an emphasis on *prévention* within the African continent, North Africa and the Middle East, an assessment in line with France's stated anxiety about threats to French interests arising from these quarters.

If one accepts the arguments made above about the trends in France's Africa policy, there may appear to be a contradiction between the *prépositionnement* emphasis on the African bases and the evident 'denationalisation' of French policy in Africa. This contradiction is explained in part by the appearance of *Une Défense Nouvelle* and the 1997 *Loi-Militaire* in early 1996, a year before the collapse of the Mobutu regime dealt traditional French African policy a final blow. It is explained in part also by France seeking to renew its relevance in Africa by using its standing presence and infrastructure to facilitate multinational operations. In this way France is either assured a leading role in international deployments in the region or, if other players prove unwilling to act, France may yet have the chance to reassert a unilateral relevance.

The upgrading of infrastructure is also being undertaken, or has been completed, at a number of the key locations across the DOM-TOMs in Guyane, Nouvelle-Calédonie and Réunion[114]. This points to the long-term determination of France to retain its global possessions by strengthening its capacity to reinforce and project forces in the respective regions of interest.

Prévention carries both advantages and risks for France. The advantages flow from the better match between French capabilities and the situational requirements if France acts early in an evolving crisis. In the 1990s France has repeatedly exercised force in deployments below 5000 troops and this willingness and ability to act in relation to low-level threats has enhanced French influence at relatively little cost. The risks in *prévention* arise from the same French predisposition to act. France may be drawn into conflicts unnecessarily because of the way its forces and doctrine are structured and the propensity to intervene risks the perception of French neo-colonialism and adverturism. *Prévention* may also undermine the French capacity to find multinational responses to crises because potential partners may balk at being bounced into action.

The exercise of more substantial force, as the 1991 Gulf War showed, is much more problematic for France, hence the post-Cold War emphasis on restructuring conventional forces for *projection*. Under the terms of *Une Défense Nouvelle* France is putting in place four projectable combat groups, fashioned largely from its previous manoeuvre forces the *Force d'Action Rapide* and the 3rd Army Corps. The new formations will be wholly separate from the Eurocorps and Franco-German brigade and will thus not be subject to German decision.

The fullest official elaboration of the role of these four groups is that the pool of 60 000 personnel, of whom 50 000 are combat troops, [intended to be distinct from standing forces deployed overseas] are expected, when fully in place and resourced, to allow France to project up to 50 000 troops for a short, though unspecified period, into a major combat theatre 'as part of a North Atlantic Alliance engagement' or to project simultaneously 30 000 troops into one theatre and up to 5000 troops into a second theatre, sustainable for up to a year through the rotation of the residual 15 000 projectable troops[115].

In composition the mainstay of the heavy armoured force will comprise 290 heavy tanks, 240 armoured infantry carriers, more than 400 other armoured vehicles and 66 major artillery pieces; the mechanised force will have the same composition but only 130 heavy tanks; the armoured rapid intervention force will have at its heart 192 light tanks, 400 armoured infantry carriers, more than 400 other armoured vehicles and 33 major artillery and 33 light artillery pieces; and the infantry assault force will be structured around 36 light tanks, 700 armoured infantry carriers, 360 other armoured vehicles and 66 light artillery pieces[116]. The heavy armoured force, by way of comparison, has more than seven times the number of heavy tanks deployed by France in Operation Daguet during the Gulf War.

A number of important caveats need to be applied to these forces. Firstly the figures should be read as maximums. The 420 heavy tanks, for example, earmarked for the heavy armoured and mechanised forces are in fact the total number of heavy tanks France will have in its arsenal when the defence reforms are fully implemented. The total projectable heavy tank levels are thus premised on France being able and willing to project every tank in its possession into a distant theatre.

More importantly, the potential Achilles' heel of these forces is the question of just how projectable they actually will be. In terms of command and control and inter-service and multinational interoperability there are likely, if the 1997–2002 *Loi-Militaire* is fully implemented, to be relatively few problems but in terms of moving the hardware and troops themselves to distant theatres there appear to be substantial difficulties. Before exploring these problems, it is important to state here that the suggestion of difficulties has been rejected by the French military. In June 1996 the then CEMA Jean-Philippe Douin argued that:

> France [has] enough air and sea-transport capacity to deal with both a major long-brewing crisis like the Gulf or an emergency in Africa...In the first instance we have time and can move our troops and equipment by ship. In the second, our airforce's fleet of transporters is sufficient to carry the troops we need to the spot[117].

Aside from the logical hole in this argument, namely that France doesn't have the capacity to respond to a quickly-brewing major crisis, there are other grounds for questioning Douin's assessment. Firstly the recent decision by France to cut its airforce transport capacity by 40 per cent to 52 aircraft as part of defence restructuring, and the decision not to fund FLA [see Chapter 3] point to a bottleneck in French airlift capabilities. France has been trying to respond to this by examining three main options: buying off-the-shelf US C130J or C17s, leasing Russian-Ukrainian Antonov's, or taking the European commercial track which will probably not be able to deliver a substantial enhancement of French capabilities before 2010–15[118], but has thus far taken no major decisions.

The argument that the 52 aircraft presaged by *Une Défense Nouvelle* will be enough to meet French needs is itself questionable. At present France has 67 Transall C160, 14 C130 and 8 CASA CN235 airlift aircraft and 14 refuelling aircraft. Despite this, and despite efforts to develop European pooling of airlift [see Chapter 4], France still needed in 1994, for example, to rent Russian transport aircraft to move little more than 2000 men into Rwanda, mainly from bases within Africa[119]. The idea that many thousands of troops and hundreds of pieces of heavy equipment could be rapidly projected from France by air is clearly far beyond present or likely French capabilities.

The alternative identified by Douin, the use of sea-transport, also needs to be questioned. At present France has just four heavy landing ships [*Transport de Chalands de Débarquement* – TCD], the two *Foudre*-class and two *Ouragan*-class vessels[120]. Though these are being updated, there are no plans to increase the overall size of the fleet. *Foudre* vessels can carry up to 450 troops and 30 tanks, and *Ouragan* vessels 350 troops and 25 tanks. France also has five of the smaller and shorter-range light landing ships [*Bâtiments de Transport Léger* – BATRAL] each capable of carrying up to 150 troops and a handful of tanks, though these are due out of service in 2010[121].

Thus generously France's landing vessels could carry up to 2500 troops and perhaps 150 to 200 tanks or armoured vehicles. The vessels are however deployed around the globe servicing the DOM-TOMS and France's military presence elsewhere, and it is unlikely that all could be made available for use in relation to a single time-sensitive operation. Even if they could, their combined carrying capacity is only sufficient to project less than one quarter of the troops and hardware of the heavy armoured group alone, leaving aside the other three projection forces.

At best France may be in essentially the same position it was in during the Gulf War when it took several months between September 1990 and January 1991 to deploy and logistically support the relatively modest forces of Operation Daguet using non-dedicated naval vessels and merchant shipping. By some senior military accounts France may, at least until 2002, be even worse off than in 1990/1 because of the upheavals caused by the defence reforms[122]. Either way it must be doubted that France could project perhaps three or four times the troops and military hardware in a similar timeframe. The essential point here is that the rapid expansion of France's projection forces is not being accompanied by any parallel expansion of the military airlift, or sea-lift, refuelling and logistics support capability.

It is difficult to see any way in which these shortfalls can be addressed in the medium term particularly in the context of defence budget constraints. France seems likely to continue to be effective in the exercise of force up to the 5000 troops level and may be able to purchase or lease adequate airlift to support these lower-level projections, though it is likely that some difficulties will persist through to at least 2015. France's capacity to project substantially greater force appears to be irresolvable nationally and thus points to either European or transatlantic solutions. This of course is in line with the assertion in official literature that a major engagement of French forces [up to 50 000 troops] would take place in the context of North Atlantic Alliance operations. It appears to be an acceptance that France would be dependent to a substantial degree on Europeans or, more likely in the medium term, the United States for major power projection in line with the multinational imperative shaping French defence policy as national options diminish.

The projection of nuclear power

The projection of French nuclear weapons outside Western Europe is a sensitive subject which has received relatively little attention in the open French literature and little official elaboration. The resolution of the more operational/less operational debate by 1994 and the ending of any prospect of France developing 'micro' or special-effects nuclear weapons following the CTBT agreement in 1996, appear to have ended speculation that France was developing 'useable' nuclear weapons, and pushed the issue of nuclear projection to the margins of debate.

In the context of this chapter, however, this issue is central to understanding the exercise of French military power outside Europe. It is important in terms of the evolution of nuclear posture and doctrine, in terms of the French response to the new threat context – particularly the proliferation of WMD and ballistic missiles – and in relation to the projection of conventional forces where issues of conventional-nuclear linkage arise.

One way into the issues is through the 1994 *Livre Blanc*. As was evident in the previous chapter, in defining the doctrine of *dissuasion* the defence white paper makes clear the latitude of French policy which has never been tied to a specified threat[123]. Two of the six 'scenarios' specified by the *Livre Blanc* implicate French nuclear weapons in dissuasive roles outside Europe. The first is in relation to 'national territory outside metropolitan France' [i.e., the DOM-TOMs] which are stated to be 'covered by *dissuasion*'[124]. The second and novel scenario is that of a 'regional conflict that may involve our vital interests'[125]. The latter, stated to be 'particularly relevant from the turn of [the] century', relates to threats on the European continent but 'equally' to threats in the longer-term from the Mediterranean, and the Near and Middle East 'involving a nuclear power' [*puissance nucléaire*]. In such a situation a 'deterrent manoeuvre, *adapted to the particular context*, could be necessary to accompany [the French] decision to engage [emphasis added]'[126]. The specification in the latter scenario that a nuclear power would have to be involved for French nuclear weapons to play a role is at odds with the allowance of the relevance of nuclear *dissuasion* to non-nuclear threats in the same publication[127].

In setting out new threats, including non-nuclear threats, which could justify the projection of nuclear power, and in suggesting the adaptation of the 'deterrent manoeuvre', the *Livre Blanc* is clearly reorienting French nuclear policy more overtly towards threats outside Europe. This is underlined by the assertion that: 'the credibility of [the French] deterrent posture rests on the availability of means that are sufficiently flexible and diversified to offer the President a variety of options as required'[128] As Bruno Tertrais points out, 'by introducing the words *flexible* and *diversified* into the doctrine, the *Livre Blanc* seemed to acknowledge that the future requirements of deterrence could include a more subtle approach than one reflected in the full-blown strategic strike apparently planned against the Soviet Union'[129].

As a Defence Ministry official Tertrais's careful wording is understandable, but the less cautious might find the words '... offer the President a *variety of options...*' evidence enough of doctrinal movement. There is evidence of doctrinal change too in the sensitivities exposed by the contradiction in the *Livre Blanc* between a definition of *dissuasion* which allows a French nuclear response to a non-nuclear threat and threat scenarios which evidently do not. These sensitivities centre around the assurances France has given to non-nuclear states in respect of its nuclear weapons. In the context of the NPT negotiations, France restated its Negative Security Assurance [NSA] to non-nuclear states first set out in 1982. In view of what follows part of this assurance is worth quoting. In it France gave the undertaking that it:

> would not use nuclear weapons against non-nuclear weapons states except in the case of an invasion or any other attack on France, its territory, its armed forces or other troops, or against its allies or a state toward which it has a security commitment, carried out or sustained by such a state in alliance or in association with a nuclear weapons state[130].

Similar NSAs were also given in the context of French agreement to legally binding Nuclear Weapon Free Zone [NWFZ] compliance in Africa [Pelindaba[131]], South America [Tlatelolco] and the South Pacific [Rarotonga].

Squaring the circle between the NSAs and the *dissuasion* statement in the *Livre Blanc* is a matter of understanding that the NSAs themselves are qualified by precedent and context. France, like others of the P-5 states, has insisted that certain issues condition adherence to NSAs. These include the UN Charter Article 51 right to self-defence, which overrides all other obligations, and the right to no longer be bound by legal norms if faced with an adversary which has itself first violated those norms. France has also indicated that NSAs are context-dependent, that is that they would be irrelevant if French vital interests were at stake and would be irrelevant also in relation to NPT treaty signatories if those signatories were not themselves observing their treaty obligations[132].

Disentangling this it is evident that, NSAs and NWFZs or no, France has considerable latitude to respond with nuclear weapons to non-nuclear threats given the enormous scope for interpretation around the issues of 'self-defence', adversarial 'non-observation of legal norms', 'vital interests', and 'states *in association* with a nuclear weapons state'. Nor is this scope necessarily only theoretical given the overlap of French interests, scenarios which could involve a French nuclear response, and French commitments in the NPT and NWFZ fora.

Accepting that French nuclear forces are now relevant to non-nuclear as well as nuclear threats to vital interests, including those outside France and Western Europe, raises the issue of how exactly the deterrent manoeuvre is being 'adapted to the particular context' outside Europe. One response to this can be made by considering trends in French nuclear hardware. The downsizing and simplification of the French nuclear arsenal, discussed in Chapter 3, has left France with a nuclear dyad, the submarine-based nuclear deterrent and an air-launched stand-off missile in air force and naval variants. *Une Défense Nouvelle* explains that this posture is to remain as the full complement of *Triomphant* class submarines are deployed and airforce and naval variants of Rafale equipped with the ASMP-amélioré replace the Mirage 2000N and ASMP[133]. The same source then adds: ' the air-launched component, based around the stand-off missiles, will bring flexibility and diversification to the means of penetration ['*modes de pénétration*']...indispensable...to dissuasion in the twenty-first century'[134].

Given this statement and the lack of evidence that France is preparing sub-strategic roles for its SSBNs, as Britain has[135], it is safe to ask exactly how the stand-off missile could provide the President with some of the 'variety of options' specified by the *Livre Blanc*. The existing ASMP missile, operational since 1986, has the small TN80/81 warhead, a range of up to 350 kilometres[136], and is reported to have an accuracy of around 350 metres[137]. Its eventual replacement, the ASMP-amélioré, due in service in 2008, although from the same missile family is reported to be a new development, the best compromise according to the *Assemblée Nationale* 'between operational requirements, technological constraints, and cost'[138]. The new missile will carry a variant of the 'robust' TNN warhead, heavier and less complex than the TN81, is expected to have a range in excess of 600 km, and is to have better penetration than ASMP [by virtue of stealth technologies] and to be more accurate[139].

While clearly not a precision 'micro' nuclear weapon with a claim to surgical nuclear strike capability, the range, accuracy and yield of the ASMP and particularly the ASMP-amélioré, projected from a French carrier or forward airbase, will give the French a range of nuclear options against any of the proliferating/unstable states threatening French interests in the DOM-TOMs or in relation to Mediterranean, Near and Middle East theatres suggested by the *Livre Blanc*. On this point it is pertinent that the target sub-set reported for the ASMP during the Cold War – which included 'hardened targets...targets vulnerable to nuclear strikes such as airfields and depots...air defence...and large naval targets'[140] – are not so far from the kind of hardened command post, weapons of mass destruction, and military infrastructure targets associated with the deterrence of emergent threats, a distinction eroded further by the improved specifications of the ASMP-*amélioré*.

While the press have for some years reported that France is adapting its nuclear forces 'unofficially' to the 'new role of dissuading rogue governments from [threatening] French territories or vital interests'[141] and even the IHEDN has speculated that 'the multiplication of threats and the variety of types of crises *may enlarge the spectrum of our [nuclear] targets ... [primarily] in relation to weaker but still military threatening countries ... either nuclear-armed or ready to use other weapons of mass destruction'* [emphasis added][142], there has been no move to formalise this in official doctrine. Exactly why these adjustments have not yet been followed by doctrinal clarity is a complex question.

One explanation for this can be found in the political sensitivities of the French response to proliferation in general and key potential proliferators in particular. The main thrust of the French response to proliferation risks has been diplomatic, centring on arms control agreements to constrain potential proliferators[143]. To strengthen these processes France [in addition to its claims about its own nuclear downsizing and pursuit of disarmament] has provided Negative Security Assurances, discussed above, and also given Positive Security Assurances [PSA] to the effect that:

> France, as a permanent member of the Security Council, pledges that in the event of attack with nuclear weapons, or the threat of such attack against a non-nuclear-weapons state party to the NPT, France will immediately inform the Security Council and act within the Council to ensure that the latter takes immediate steps to provide, in accordance with the Charter, necessary assistance to any state which is the victim of such an act of aggression[144].

The diplomatic approach to proliferation is informed by the view that France must avoid the perception that it is part of the 'North' arming itself against the 'South' and that France should ensure, where possible, that potential proliferators are not left vulnerable in insecure contexts to threats which could stimulate proliferation. On both counts official French statements of doctrinal adjustment and targeting expansion would undermine the thrust of French non-proliferation diplomacy in general[145] and complicate greatly French relations with some potentially proliferating states in the Maghreb and Middle East in particular.

A second explanation is that there is opposition within France to any adjustments which have the potential to erode the traditional emphasis on 'non-war' and 'non-use' in French strategy, a dynamic which was particularly evident in the 'more operational/less operational debate'[146]. This debate has not been ended by the CTBT, as some have supposed, but has rather been recast, in terms of target sub-sets rather than hardware, and temporarily subsumed[147].

A third explanation arises from the issue of France's security ambitions in Europe. As Pascal Boniface has pointed out:

there is a fundamental contradiction between the perspective of a Europeanisation of the French nuclear forces and the risk of sliding towards a nuclear policy that would no longer be solely deterrent. Our European partners will not follow us towards concepts of employment of nuclear weapons that we ourselves always rejected in the past[148].

This point has been echoed by Olivier Debouzy who has argued that there is little advantage for France in incorporating into its nuclear doctrine ideas which 'negatively colour' the perception of French nuclear forces by third parties and which risk permanent internal and external controversy[149].

Given that serious NBC and ballistic missile threats to France and French vital interests are emergent rather than existent, there are perhaps few reasons for France to spell out doctrinal changes now, and obviously quite a few reasons for not doing so. These concerns notwithstanding, doctrinal adjustment [such as ending the '*du faible au fort*' formulation, allowing that a nuclear response can be relevant to non-nuclear threats, and articulating a novel scenario for nuclear *dissuasion*] *is* underway. The extent to which the dissuasive 'variety of options' for the President implies a 'slide toward a nuclear policy which is not solely deterrent' will determine how quickly the suspended debate about doctrine and strategy re-emerges.

Reflecting this a longer-term problem for France [and indeed the other nuclear 'possessor' states] will be maintaining a climate of assurance and confidence amongst potential proliferators while simultaneously developing and deploying the nuclear hardware which is, however 'unofficially', being earmarked for power projection roles in relation to proliferation threats. For states potentially on the receiving end of what might be termed *dissuasion adaptée* [i.e., adapted to non-nuclear threats, weak states, new contexts and new targets] official French doctrinal imprecisions and assurances that separate capability from intention, may in the end matter rather less than aircraft carrier and stand-off nuclear missile deployments.

A final important question is that of conventional-nuclear linkage, that is the role of projected nuclear power in relation to projected conventional power. This issue arises because NBC proliferation raises the stakes for interventions by France. As one commentator put it: 'France [has] acute concerns that its ability to project power could be radically circumscribed by the spread of chemical and biological weapons and ballistic missile capabilities, particularly in Africa and the Middle East'[151]. Reflecting these concerns there appears to be an enhanced role for French nuclear weapons in deterring threats against French [and by extension allied] conventional forces deployed 'out of area'. A deterrent manoeuvre in this context would in essence be about projecting nuclear power to create strategic 'space' for conventional intervention by deterring NBC threats against the projected conventional forces.

The idea that French nuclear weapons 'back-up' projected French, European or even NATO forces in this way is, of course, implicit in *dissuasion par constat* because an aggressor must factor-in the possession of nuclear weapons by France at least in relation to Western vital interests. But it is given substance by the capacity to project nuclear weapons into remote theatres afforded by the ASMP [and follow-on] delivered from an aircraft carrier or from forward airbases.

Conventional-nuclear linkage of this sort, in the context of significant proliferation of NBC weapons, points to further pressure on French doctrine and perhaps even on the core concept of 'non-use'. It points also in the longer term, despite warnings to the contrary, to the requirement to reconcile these pressures with the Europeanisation of nuclear weapons.

Having said something about the projection of conventional and nuclear power it is now appropriate to reflect on the trends over the last decade and the implications for the future exercise of French military power outside Europe. It is evident that the defence reforms have orientated French defence towards 'out-of-area' missions since the end of the Cold War largely as a result of the 1990/1 Gulf War which set the precedent, and peacekeeping operations which have become the focus of much Western military activity in the post-Cold War era.

Trends in the French military presence in Africa, the DOM-TOMs and in *missions de paix* with the United Nations point to growing synergies between French forces in missions outside Europe. As a result of defence restructuring, professionalisation and a refined focus for conventional forces [*prévention* and *projection*] France seems likely to increase its military effectiveness in low-level operations across the globe and seems determined and able to underwrite its presence throughout the DOM-TOMs, a presence which yields much French influence for relatively little outlay.

The key novelty in the exercise of French conventional military power outside Europe has been the multilateralisation of French activity. France has lost much of its unilateral authority in Africa and in all but DOM-TOM mission areas French action is being shaped by a multilateral imperative, perhaps even a multilateral dependence, underpinned by the limits of French military power, not least in strategic air-lift, sea-lift and remote theatre support.

Finally, with respect to the projection of nuclear power it is evident that French doctrine is being adjusted to the new context and to threats seen as increasingly arising outside continental Europe. It is evident also that French nuclear hardware gives France the capability to support the traditional and novel roles implicit in doctrinal enlargement and adaptation.

Conclusions

Few things about France lend themselves to simple summary and matters of defence are no exception. The diversity of French interests, the scale of the French military diaspora, the nature of the French armed forces and the complexity of the French defence discourse require considerable caution in attempting to draw firm conclusions. Moreover, the evolution of French defence is ongoing and a study, such as this, which has reflected on the past, on a decade of transition and on the events, trends and processes which have shaped that transition must be cognisant of the fact that no end point has been reached. With these caveats in mind, what follows are some observations about the trajectory of French policy which, as the title of this work indicates, may be useful in thinking about French defence in the new millennium.

It is clear that France found adaptation to the post-Cold War context particularly demanding. The geopolitical freeze of the Cold War had suspended or resolved many of France's core security concerns [particularly the German question] and the stability of the bipolar order gave France the benefit of a strong Western alliance without subjecting it to the full costs of Western discipline, allowing France the latitude to cultivate its distinctiveness with few real risks. When the Cold War ended, France had to cope both with the geopolitical turbulence and with changes which were bound to be felt particularly acutely in Paris. The first of these was the renewed importance in Europe of an hegemonic United States, with which France has still fully to come to terms. The second was the unification of Germany which raised old ghosts about French security and continental stability and encouraged France to accelerate European union. These pressures were compounded by new demands on the defence budget as France squared up to the force requirements of the new context, shaped by the experience in the Gulf War of 1990/1 and by the wars in Yugoslavia, and by uncertainty about the continued relevance of Gaullist defence thinking in the changed geopolitical landscape.

In these circumstances it is not difficult to understand why France eschewed the defence budget cuts of most of its Western allies in order to keep its options open. Nor is it difficult to understand why the adaptation of defence was slow, cautious and incremental. The *Armées 2000/ORION* reforms of the late 1980s were cast in the Cold War mould and were quickly overtaken by events. The first genuinely post-Cold War reforms appeared in the 1992–4 *Loi-Militaire* but, despite the Gulf War and Maastricht, these were limited adjustments which preserved much of the Cold War emphasis on static national defence. Not until the 1994 *Livre Blanc sur la Défense* did France face up to the new agenda, 'denationalise' many aspects of defence and orientate French forces more fully towards European and international roles, and not until the 1996 *Une Défense Nouvelle* did France finally take the decisions of professionalisation, downsizing and reorganisation, implicit in the *Livre Blanc*, which will recast the French armed forces for the new millennium. That these reforms will not be completely implemented until 2015 means that France will have taken 25 years to adapt its defence.

Accompanying these reforms has been the issue of the continued relevance of Gaullist thinking about defence. It has been argued in this study that Gaullism acted to constrain French policy by insisting on orthodoxy and that part of the reason France had such difficulties in adapting to the new context was that it continued to steer by the Gaullist map and compass even as its relevance declined. It is clear that there is no longer a Gaullist consensus on defence, nor yet a post-Gaullist consensus. In certain areas of French military activity such as international peacekeeping Gaullism has little to say. In other areas such as European defence the Gaullist emphasis on the primacy of the nation-state, independence, national sanctuary and intergovernmentalism has long been eroded in the construction of Europe and European defence. Even nuclear independence is being compromised by the French desire to build a concerted deterrent in Europe as part of the construction of ESDI and even Gaullist analysts have for some time accepted that many, if not most, Gaullist guidelines for defence are no longer relevant in the post-Cold War context.

Despite this, Gaullism continues to exercise a powerful hold on certain elements of French defence. This is true of issues around the constitution which provides a context and means to perpetuate Gaullist ideas. In relation to this the argument has been made here that the centralisation of power in the presidential *domaine réservé* and even the cohabitation *domaine partagé* is anachronistic and undemocratic in the modern context because of the weaknesses of the checks and balances in the system. To respond to this France should develop further the steps the Socialist government has recently taken to subject defence policy to greater transparency and accountability, and eventually transform the presidential and governmental purview of defence into a *domaine ouvert*.

Nowhere, however, is the continued fidelity to Gaullist ideas in the post-Cold War era stronger than in relations with NATO, the *bête noir* of de Gaulle himself. With respect to NATO the relevance and application of

Gaullism is restated unambiguously in the 1994 *Livre Blanc sur la Défense*. It is evidenced in the suspension of President Chirac's bold conception of a return to the NATO fold, which ran aground in the shallow waters of Gaullism because NATO had not adapted to the degree required to meet the needs of a French strategic community [and political opposition] imbued with Gaullist thinking and still assessing the Franco-NATO relationship by the verities of 1966.

This study has argued that fidelity to Gaullism in relation to NATO is detrimental to French interests in the post-Cold War era because calculations of independence and influence have changed. France now has deep and unavoidable obligations to a renewed NATO, is less able to stand up to the United States, and has less scope for risk-free distinctiveness than in the Cold War. In the early 1990s French distance from NATO yielded only marginalisation yet France could not escape the consequences of NATO's evolution. Closing the gap between obligations and influence was arguably the main reason France moved back towards NATO after 1992 and this imperative has continued to strengthen as a result of the events in Bosnia and Kosovo. In the new context participation translates into influence and France should participate in NATO's fora in order to exercise influence commensurate with its obligations and the limitations of the new dispensation.

The case has been made here that the Gaullist anxiety about subordination to the United States, which is at the heart of arguments against reintegration with NATO, is flawed because it is premised on a misconception of the degree of subordination to the United States in the new context and because it overstates the degree of French autonomy which the Gaullist prescription seeks to defend. At the same time the observation has been made that these arguments are unlikely to prompt a French return to NATO in the near future, though the explanation for this lies less in the validity of the arguments shaping French policy than in the difficulties of overcoming the intellectual inertia of having looked at an issue in the same way for more than 30 years.

It seems more likely that France will find its way back to NATO sooner or later because the construction of ESDI demands it. Early French hopes for a uniquely European route to ESDI were abandoned as France came to terms with a renewed and dominant NATO and with European unwillingness to match the pace or scope of French ambitions for European security. In relation to its European partners France is a dynamic, *avant-garde* and uniquely capable state embedded largely, to subvert a NATO phrase, in a 'coalition of the unwilling'. With few like-minded partners and none who match French military clout, the obstacles to France leading the construction of a European defence and security competence are formidable indeed.

The EU has no defence competence and to develop one must address the issues of political will, decision-making, democratic accountability and budgetary constraints with which it has made little real progress over the past decade. Europeans continue to pursue national interest in matters of defence and security and share few core aspects of policy, being divided along Europeanist/Atlanticist, federalist/integrationalist, neutral/non-neutral, and northern European/southern European lines as well as being divided into nuclear and non-nuclear states and states with and without extra-European interests. Few European states share France's ambitions for a strategic Europe, a *l'Europe puissance*, and few Europeans seem willing to act in ways which threaten the dominance of the United States and NATO in Europe.

Narrowing the gap between the EU, WEU and NATO seems problematic because of the membership complications, the lack of an EU defence competence, the issue of EU expansion and Atlanticist resistance to the creation of a pan-European arrangement through the merger of the EU and WEU which would appear to exclude the North Americans. Finally, despite some progress in developing limited bilateral and *ad hoc* European defence co-operation, forging a uniquely European defence capability from the vast and disparate military resources available to Europe is also problematic because of the limited competence of those forces, duplication, and redundancy as well as the national emphasis in defence which means that NATO Europe's 17 nations are supporting 17 national armies rather than working towards a single European force. Attempts by France to overcome these obstacles through the co-operation of *'un cercle des solidarités renforcées'* face great difficulties because France ultimately has no stable 'inner core' around which to construct ESDI and because the multilateralisation required to realise ESDI will undermine the ability of France to pursue a national agenda.

By contrast the pursuit of ESDI within NATO would seem to be a far smoother and less fractious process. The evidence of the post-Cold War era is that the closer France has been to NATO the more willing the United States has been to endorse and support the development of ESDI and the more willing France's European partners in NATO have been to work with France to strengthen European defence co-operation. While France remains semi-detached there are limits to this engagement because the United States and some NATO Europeans will suspect French motives. Were France back fully in NATO, much of the allied reticence would evaporate and the construction of ESDI would become a proper reflection of the US-European balance and the limits of what the respective parties were willing to countenance and able in practice to undertake.

This appears to hold true also for France's return to NATO's Nuclear Planning Group which seems the natural forum to take forward the emergent trilateral nuclear relationship between France, the United States and Britain arising from the Franco-British JCNPD, US-French nuclear accords and technical co-operation in a 'no-test' context and the long-standing Anglo-American nuclear relationship. The NPG also seems the natural forum for the Europeanisation of France's [and perhaps Britain's] nuclear weapons in the pursuit of ESDI. Europeans – including France in the December 1996 Franco-German Nuremberg 'Common Concept' – have made clear the continued primacy of the US nuclear guarantee and President Chirac himself has argued that the Europeanisation of nuclear weapons should be about strengthening transatlantic deterrence. It seems clear from the statements of Britain, Germany, Spain and perhaps Italy as the most likely 'inner core' of an emergent *dissuasion concertée* that European co-operation on nuclear matters is most likely in a NATO context, and quite unlikely outside it. On all these grounds this study has argued that France should take its full place in NATO fora, including the NPG, in order to maximise its influence on the decisive processes shaping European security.

The longer term question of whether France will go on to lead the construction of a genuinely autonomous and perhaps one day free-standing ESDI through NATO will depend on what Europeans and the United States ultimately want and on the future international context which cannot be predicted with any certainty. It has however been argued here that France's present distance from NATO is only pushing back the day when Europeans do exercise meaningful defence autonomy.

Outside Europe the exercise of French military power in *missions de présence*, *missions de souveraineté* and *missions de paix* with the United Nations, have been subject to tremendous changes in the post-Cold War era. As a result of the catastrophe in Rwanda and French policy in Zaire, France now seems set on a policy trajectory which is more selective about French interests and obligations in Africa and which is increasingly multilateral in seeking pan-African, European or international responses to regional crises. The French military presence in Africa is much reduced and France seems most interested in exploiting its residual African assets in the pursuit of multilateral leadership rather than unilateral initiative.

Across the DOM-TOMs the downsizing and professionalisation of the armed forces in the 1996 defence reforms seem if anything likely to strengthen the French military presence across the world and enhance the effectiveness of French forces and the capacity to project limited force into regional theatres. This points to a French determination to underwrite its presence in the DOM-TOMs and to retain its global reach and relevance.

In a decade of participation in UN peacekeeping operations since the end of the Cold War, France has learned much from its experiences – in Cambodia, Somalia, Bosnia and Rwanda in particular – and developed a cohesive doctrine for *mission de paix* premised on the requirement for precision in UN mandates and robust rules of engagement. Where these doctrinal requirements are met, France seems increasingly effective in UN operations, and in co-operation with allies, and is likely to continue to uphold its rank as a UN Security Council permanent member and to continue to seek to manipulate the links between the UN, NATO and WEU to empower the UN by positing NATO/WEU as its military arm, and to constrain US hegemony and build European defence by embedding NATO and the WEU respectively in the context of UN authority.

Convergent trends in the role of French military forces in *missions de présence, missions de souveraineté* and *missions de paix* point to an enhanced capacity for France to exploit its limited military means, particularly in relation to the pre-emptive or low-level roles defined under the *prévention* rubric in the 1994 *Livre Blanc sur la Défense* and in *Une Défense Nouvelle*. While these actions are not risk-free they do offer France an opportunity to 'punch above its weight' by being able and willing to act where others are not.

Despite the restructuring of the French armed forces for the *projection* of up to 50 000 combat troops and large heavy armour formations for Gulf War-type crises, there will remain doubts about France's capability to project this level of force even after the defence reforms have been fully implemented, not least because France is not investing in commensurate air and sea-lift capabilities. Limited European means and the official French statement that large *projection* forces would be exercised in the context of North Atlantic Alliance operations point either to a reliance on the United States or to the projection of considerably fewer than 50 000 troops and the earmarked heavy armour in the context of any crisis which did not allow France many months, perhaps many more than it had in the Gulf War, to move its troops and hardware. France's limited capacity thereafter to sustain such a force out of area points to further dependence on the United States underscoring the limits of the unilateral exercise of French military power and the inevitable alignment of French and NATO operations.

Finally, France has begun the process of adjusting its nuclear weapons to the new context and new threats. While reaffirming the traditional elements of *dissuasion* which have not been eroded by Europeanisation, France has ended the uniquely prestrategic role of its short-range nuclear weapons and begun to enlarge and adapt its doctrine, in what has here been termed *dissuasion adaptée*, to include nuclear responses to non-nuclear threats, weak states, new contexts, new targets, and to support conventional power projection by creating 'strategic space' for French [and by implication allied] conventional forces by dissuading threats against those forces from the proliferation of weapons of mass destruction.

These changes have not yet been articulated in official doctrine because of the sensitivity of the issues within the French strategic community; anxiety that adapting doctrine may undermine French efforts to build European acceptance of nuclear weapons through *dissuasion concertée*; and tensions between the capacity to project nuclear forces and French diplomatic efforts to constrain proliferation; and tensions between the exercise of *dissuasion* and French commitments to the contrary in the context of the NPT and NFZ agreements it has signed and the negative and positive security assurances it has given. Despite this, change *is* underway and France seems unlikely to be able to postpone indefinitely official elaboration of the adaption of *dissuasion* nor the inevitable debates around Franco-NATO nuclear alignment, the enlargement and adaptation of doctrine, the resultant evolution of strategy, tensions between the Europeanisation and projection of nuclear weapons, and the conventional-nuclear linkage of projected military power.

While no definitive answers can be given at this point to the two final questions posed in the introduction – whether multilateralism will constrain or enhance French military power and whether France can resolve the tensions arising from the erosion of the distinctions between the '*trois cercles*' of French interests – some observations may be made. The evidence of the past decade suggests that European multilateralism constrains France because of different interests, perceptions, and capabilities and because of the inability or unwillingness of Europeans to act. By contrast Atlanticist multilateralism – from the Gulf War to Kosovo – in which France has co-operated with the United States and European allies like Britain which share France's willingness to act, its global outlook, extra-European interests and military competence, appears to empower France as one of the key players in world order roles. The difference would tend to suggest an Atlanticist emphasis in future French power projection, a point confirmed by French statements and by the 1996 defence reforms.

The construction of European defence would thus appear to pose two long-term challenges to France. One is the erosion of national autonomy as policy Europeanises and French national and European interests merge. The second is that deepening engagement with Europe seems likely to erode the French capacity to exercise military power on the international stage unless France can persuade at least some of its European partners, to a degree it has largely to date been unable to do, to enlarge their interests to include world order roles. The context in which the latter is most likely to happen is that of NATO providing one more reason why France should return fully to the North Atlantic Alliance.

It was a common theme amongst those from the French strategic community interviewed in Paris for this study that France has fewer and fewer cards to play in relation to NATO, the United States, and defence and security issues more generally. The case has been argued here that this is so only in terms of Gaullist thinking about defence. A France restored in NATO and freed from the constraints of Gaullism would be uniquely positioned to exploit its leadership in Europe, its range of military and military-related resources, its global reach, and its ability and willingness to act, in the construction of a renewed transatlantic partnership which recognised the real importance of the United States and which balanced European support for US world order roles with US support for greater European responsibility in Europe.

Appendix: Defence and Security Research in France

Finding contemporary French defence and security resources can be problematic if one is not familiar with the documentation centres, institutes, and libraries where material is held. This short appendix is intended as a guide to the most useful locations in Paris. The following list is certainly not exhaustive but there is little contemporary material the listed centres do not collectively hold:

Name: Centre de Documentation de l'Armement [CEDOCAR]
Correspondence: 32 boulevard Victor, 75015 Paris / CEDOCAR, 00460 Armées. Tel: +33(0)1 45526956 / Fax: +33(0)1 45328276.
Notes: Defence Ministry documentation centre with wide range of publications including many military-technical sources not found elsewhere. Tightly controlled access particularly for non-French nationals. Visit by appointment with librarian.
Metro: Balard.

* * *

Name: Direction des Journaux Officiels
Correspondence: 26 rue Desaix, 75727 Paris cedex 15. Tel: +33(0)1 40587878 / Fax: +33(0)1 45791784.
Notes: Centre for the consultation and purchase of official reports, debates, bulletins, cd-rom, etc. from the Assemblée Nationale, Sénat, and related official sources. Open access.
Metro: Dupleix.

* * *

Name: Délégation à l'Information et à la Communication de la Défense [DICOD]. Formerly: Services Information et Relations Publiques des Armées [SIRPA]
Correspondence: 14 avenue de Lowendahl, 75007 Paris. Tel: +33(0)1 45553650 / Fax: +33(0)1 47055585.
Notes: Public relations outlet for Defence Ministry and armed forces documentation. Useful source of official publications and data. Also has good press cutting service. Located within the Ecole Militaire. Access by identity papers / passport.
Metro: Ecole Militaire.

* * *

Name: La Documentation Française
Correspondence: 29 quai Voltaire, 75007 Paris. Tel: +33(0)1 40157000 & 40157272 / Fax: +33(0)1 40157230 / E-mail: postmaster@ladocfrancaise.gouv.fr / Website: http://www.ladocfrancaise.gouv.fr
Notes: Large library and bookshop complex holding primarily government and related official material. Open access.
Metro: Assemblée Nationale or Musée d'Orsay [RER].

* * *

Name: Fondation Nationale des Sciences Politiques [Sciences Po]
Correspondence: 27 & 30 rue Saint-Guillaume, 75337 Paris. Tel: +33(0)1 45495160 & 45495096.
Notes: Massive library and documentation centre holding a huge collection of books, journals, official documentation, press cuttings, microfiche, and so on. Access to resources by purchase of 'cartes de lecteur'.
Metro: Rue du Bac / Saint-Germain-des-Pres / Saint Sulpice / Sèvres-Babylone.

* * *

Name: Fondation pour les Recherches Stratégiques [FRS]
Formerly: FED and CREST.
Correspondence: 27 rue Damesme, 75013, Paris. Tel: +33(0)1 43137777 / Fax: +33(0)1 43137778. Website: http://www.frstrategie.org
Notes: Good library, particularly for journals. Visit preferably by appointment with librarian.
Metro: Maison Blanche.

* * *

Name: Institut Français des Relations Internationales [IFRI]
Correspondence: 27, rue de la Procession, 75740, Paris. Tel: +33(0)1 40616000 / Fax: +33(0)1 40616060. E-mail: ifri@ifri.org / Website: http://www.ifri.org
Notes: One of the best libraries, wide range of books and journals [notably IFRI publications], reference database unfortunately still on card system. Visit by appointment with the librarian.
Metro: Vaugirard.

* * *

Name: Institut des Hautes Etudes de Défense Nationale [IHEDN]
Correspondence: Centre de Documentation, 13/21 Place Joffre, BP 41, 00445 Armées. Tel: +33(0)1 44424347 / Fax: +33(0)1 44424218. Website: http://www.ihedn.fr
Notes: Excellent computer database, good centre for IHEDN and other institutes' working papers/conference papers, etc. Access by identity paper / passport.
Metro: Ecole Militaire.

* * *

Name: Institut de Relations Internationales et Stratégiques [IRIS]
Correspondence: 2 bis, rue Mercoeur, 75011 Paris. Tel: +33(0)1 53276060 / Fax: +33(0)1 53276070.
Notes: A small library, particularly useful for IRIS publications. By appointment.
Metro: Voltaire.

Notes and References

Introduction

1. *Livre Blanc sur la Défense 1994* (La Documentation Française, March 1994).
2. *Une Défense Nouvelle* (SIRPA/Ministère de la Défense, February 1996).
3. See for example: Pascal Boniface, *Vive la Bombe* (Editions 1, 1992) and *La Volonté d'Impuissance: la Fin des Ambitions Internationales et Stratégiques* (Editions de Seuil, 1996).
4. Pierre Lellouche, *Légitime Défense* (Patrick Banon, 1996).
5. Charles George Fricaud-Chagnaud, *Mourir pour la Roi de Prusse?: Choix Politique et Défense de la France* (Publisud, 1994).
6. Pierre M. Gallois, *Livre Noir sur la Défense* (Editions Payot et Rivages, 1994).
7. François Heisbourg, *Les Volontaires de l'An 2000: Pour une Nouvelle Politique de Défense* (Balland, 1995).
8. Frédéric Bozo, *La France et l'OTAN: De la Guerre Froide au Nouvel Ordre Européen* (Masson/IFRI, 1991).
9. Nicole Gnesotto, *L'Union et l'Alliance: les Dilemmas de la Défense Européenne*, Les Notes d'IFRI No 2 (July 1996).
10. Thierry Garcin, *La France dans le Nouveau Désordre International* (Bruylant, 1995).
11. Philip Gordon, *A Certain Idea of France: French Security Policy and the Gaullist Legacy* (Princeton University Press, 1993).
12. See for example, Jolyon Howorth and Patricia Chilton [eds], *Defence and Dissent in Contemporary France* (Croom Helm, 1984) and Jolyon Howorth and Anand Menon [eds], *The European Union and National Defence Policy* (Routledge, 1997).
13. Robbin Laird, *French Security Policy in Transition: Dynamics of Continuity and Change*, McNair Papers No 38 (INSS, 1995). See also the earlier Robbin Laird [ed] *French Security Policy: From Independence to Interdependence* (Westview Press, 1986).
14. Diego Ruiz Palmer, *French Strategic Options in the 1990s*, Adelphi Paper No 260 (IISS, 1991).
15. Amongst David Yost's many publications see: 'France in the new Europe', *Foreign Affairs* (Winter 1990/91) pp. 107–28 and 'Nuclear debates in France', *Survival* (Winter 1994/95) pp. 113–39.
16. John Chipman, *French Power in Africa* (Blackwell, 1989).
17. Robert Aldrich and John Connell [eds], *France in World Politics* (Routledge, 1989).
18. See: Bruno Colson, 'La culture stratégique française', *Stratégique* (Spring 1992) pp. 27–60 and François Léotard, 'Une nouvelle culture de la défense', *Défense Nationale* (July 1993) pp. 9–20.

1   French Defence in Context

1. John Chipman, *French Power in Africa* (Blackwell Press, 1989) p. 9.
2. Jolyon Howorth, 'The defence consensus and French political culture', in Michael Scriven and Peter Wagstaff [eds], *War and Society in Twentieth Century France*, (Berg Press, 1991) p. 167.

3. Quoted in John Keiger, 'France and international relations in the post-Cold War era: some lessons of the past', *Modern and Contemporary France*, (Autumn 1995) p. 265.
4. Haig Simonian, *The Privileged Partnership: Franco-German Relations in the European Community 1969–1984* (Clarendon Press, 1985) p. 181.
5. Quoted in Julius Friend, *The Lynchpin: French-German Relations 1950–1990*, Washington Papers No 154 (CSIS/Praeger Press, 1991) p. 85.
6. See for example: Raymond Poindevin and Jacques Bariéty, *Les Relations Franco-Allemandes 1815–1975* (Armand Colin, 1977); F. Roy Willis, *France, Germany and the New Europe 1945–1967* (Stanford University Press, 1968); Georges Valence, *France-Allemagne: Le Retour de Bismarck* (Flammarion, 1990); and Philip Gordon, *France, Germany and the Western Alliance* (Westview Press, 1995).
7. Nicholas Martin and Marc Créspin, *L'Armée Parle* (Fayard, 1983) pp. 143–70.
8. Jolyon Howorth, op. cit.
9. *Livre Blanc sur la Défense 1994* (La Documentation Française, 1994) p. 127.
10. Charles de Gaulle, *La France et son Armée* (Librairie Plon, 1938) p. 12.
11. On the history of French armed forces see: James S. Ambler, *The French Army in Politics 1945–1962* (Columbus Press, 1966); Jean Doise and Maurice Vaisse, *Diplomatie et Outil Militaire 1871–1991* (Editions du Seuil, 1992); Alistair Horne, *The French Army and Politics 1870–1970* (Macmillan Press, 1984); André Martel, *Histoire Militaire de la France* [Volume 4] (PUF, 1994); and Pierre Montagnon, *Histoire de l'Armée Française* (Pygmalion/Gerard Watelet, 1997).
12. Details taken from Dominique Fremy and Michele Fremy, *Le Quid* (Robert Laffont, 1983) pp. 843–54.
13. Perhaps the definitive, if largely uncritical, history of French colonialism is Jacques Thobie *et al.*, *Histoire de la France Colonial* [2 Vols] (Editions du Seuil, 1990). On the wars of colonial disengagement see: Anthony Clayton, *The Wars of French Decolonization* (Longman, 1994).
14. Quoted in Roy Macridis, 'French foreign policy', In Roy Macridis [ed], *Foreign Policy in World Politics* (Prentice-Hall, 1992) p. 38.
15. 'Entretien avec François Léotard', *Le Quotidien* (6 March 1994).
16. The trend is explored in: Henri Paris, *L'Arbalète, la Pierre au Fusil et l'Atome – La France va-t-elle Etre Encore en Retard d'une Guerre?*, (Albin Michel, 1997). For an excellent overview of the RMA see: Lawrence Freedman, *The Revolution in Strategic Affairs*, Adelphi Paper 318 (IISS/Oxford University Press, 1998).
17. André DePorte, 'The foreign policy of the Fifth Republic: between the nation state and the World', In: James Hollifield and George Ross [eds], *Searching for the New France* (Routledge, 1991) p. 250.
18. Charles de Gaulle, *Mémoires de Guerre Vol I* (Plon, 1954) p. 1.
19. André DePorte, op. cit., p. 252.
20. Dorothy Shipley White, *Black Africa and De Gaulle* (Pennsylvania State University Press, 1979) and Georges Lavroff [ed], *La Politique Africaine du Général de Gaulle* (Pédone, 1980).
21. Charles de Gaulle, *War Memoirs Vol III* (Simon and Schuster, 1960) p. 204.
22. Jean Doise and Maurice Vaisse, op. cit., p. 503.
23. Marc Theleri, *Initiation à la Force de Frappe Française* (Editions Stock, 1997) p. 294.
24. On France's first nuclear weapons test at Reggane see: Marcel Duval and Yves Baut, *L'Arme Nucléaire Française: Pouquoi et Comment?* (Kronos, 1992) pp. 139–54; and Bruno Barrillot, *Les Essais Nucléaires Français 1960–1996* (CDRPC, 1996) pp. 37–62.

25. On the history of French nuclear weapons, in addition to Marcel Duval and Yves Baut, ibid., see: Alexandre Sanguinetti, *La France et l'Arme Atomique* (René Julliard, 1964); Charles Ailleret, *L'Aventure Atomique Française* (Grasset, 1968); Alain Joxe, *Le Cycle de la Dissuasion 1945–1990* (La Découverte, 1990) and Marcel Duval and Dominique Mongin, *Histoire des Forces Nucléaires Françaises Depuis 1945* (Que sais-je?/ PUF, 1993). In English see: Robert Norris *et al.*, *Nuclear Weapons Databook, Vol V: British, French and Chinese Nuclear Weapons* (Westview Press, 1994) pp. 182–321.

26. Amongst Pierre Gallois's key works are: *Stratégie de l'Age Nucléaire* (Calmann-Lévy, 1960) and *Paradoxes de la Paix* (Presses de la Cité, 1967); amongst Lucien Poirier's are *Des Stratégies Nucléaires* (Hachette, 1977) and *Essais de Stratégie Théoretique* (Fondation pour les Etudes de Défense Nationale, 1982). For an excellent essay on the founding fathers of French nuclear strategy see: François Géré, 'Quatre généraux et l'apocalypse: Ailleret-Beaufre-Gallois-Poirier, *Stratégique* (Spring 1992) pp. 75–116.

27. Charles Ailleret, ' Défense dirigée ou défense tous azimuts?', *Revue de Défense Nationale* (December 1967) pp. 1923–32.

28. François Maurin, 'L'originalité français et le commandement', *Défense Nationale* (July 1989) pp. 45–57.

29. See: Frédéric Bozo, *La France et l'OTAN: De la Guerre Froide au Nouvel Ordre Européen* (Masson/IFRI, 1991) pp. 68–70.

30. Cyril Black *et al.*, *Rebirth: A History of Europe Since World War II* (Westview Press, 1992) p. 339.

31. Charles de Gaulle (1954) op. cit., p. 70.

32. Matthew Parris, *Scorn* (Penguin, 1995) p. 144 [attrib. Richard Wilson].

33. Michael Harrison, *The Reluctant Ally: France and Atlantic Security* (Johns Hopkins University, 1981).

34. Ibid., p. 52.

35. Ibid.

36. See Maurice Vaisse, *Alger: Le Putsch* (Editions Complexe, 1983) and Alistair Horne, *A Savage War of Peace* (Penguin/Elisabeth Sifton, 1987) pp. 436–60.

37. For a superb study of France and arms control see: Jean-Luc Marret, *La France et le Désarmement* (L'Harmattan, 1997). For French approaches to arms control see pp. 39–62 and for de Gaulle's arms control 'refusal' and the 'empty chair' see pp. 259–330.

38. For example, Michael Harrison, op. cit. and Philip Gordon, *A Certain Idea of France: French Security Policy and the Gaullist Legacy* (Princeton University Press, 1993). These books repay reading in sequence.

39. The best overview of this transformation is probably: Patrice Buffotot, *Le Parti Socialist et la Défense* (Institute de Politique Internationale et Européenne, 1982).

40. Cited in: Olivier Duhamel, 'The fifth republic under François Mitterrand: evolutions and perspectives', In: Stanley Hoffman [ed], *The Mitterrand Experiment* (Polity Press, 1987) p. 143.

41. *The Military Balance 1981–82* (IISS, 1981) p. 31 and *The Military Balance 1986–87* (IISS, 1986) p. 63.

42. See Mitterrand's restatement of Gaullist nuclear verities in: *Intervention de M. François Mitterrand, President de la République sur la Theme de la Dissuasion*, Service de Presse, Palais de l'Elysée (5 May 1991).

43. *Discours prononce par M. François Mitterrand, President de la République Française, devant le Bundestag à l'occasion du 20ème Anniversaire du Traité de Coopération*

Franco-Allemand, Service de Presse, Palais de l'Elysée (10 January 1983). The speech is reprinted in François Mitterrand, *Réflexions sur la Politique Extérieure de la France* (Fayard, 1986) pp. 183–208.

44. For contemporary accounts see: Xavier Luccioni, *L'Affaire Greenpeace* (Payot, 1986) and Richard Shears and Isobelle Gidley, *The Rainbow Warrior Affair* (Counterpoint, 1986). For Lacoste's view see: Pierre Lacoste and Alain-Gilles Minella, *Amiral Lacoste: Un Amiral au Secrèt* (Flammarian, 1997). On the French intelligence services in general, with further insight about the Rainbow Warrior affair, see: Jean-Jacques Cécile, *Le Renseignement Français à l'Aube du XXIᵉ Siecle* (Lavauzelle, 1998); Pascal Krop, *Les Secrèts de l'Espionage Français* (JCL, 1996); Pierre Lacoste [ed], *Les Renseignement à la Française* (Economica, 1998); Douglas Porch, *Histoire des Services Secrèts Françaises* [2 vols] (Albin Michel, 1997); and Dominique Prieur, *Agent Secrèt* (Fayard, 1995).

45. Dominique Moisi, 'French foreign policy: the challenge of adaptation', *Foreign Affairs* (Autumn 1988) p. 164.

46. Philip Gordon, *A Certain Idea of France: French Security Policy and the Gaullist Legacy* (Princeton University Press, 1993) p. 3.

47. Bruno Tertrais, *The French Nuclear Deterrent After the Cold War*, Report No P-8012 (RAND, 1998) p. 4.

48. Anand Menon, ' The "consensus" on defence policy and the end of the Cold War', In: Tony Chafer and Brian Jenkins [eds], *France: from the Cold War to the New World Order* (Macmillan, 1996).

49. Philip Gordon, op. cit., p. 164.

50. Official translation. See: *La Constitution Française* [Français-Anglais] (Assemblée Nationale, 1997) pp. 14–15.

51. Ibid.

52. Ibid, pp. 16–17.

53. This point is made by Jolyon Howorth in 'Foreign and defence policy; from independence to interdependence', In: Peter Hall *et al.*, [eds], *Developments in French Politics* (Macmillan Press, 1990) p. 210. For a detailed discussion of presidential control over nuclear weapons see: Samy Cohen, *La Monarchie Nucléaire* (Fayard, 1986).

54. On de Gaulle's nuclear emphasis see: Alexandre Sanguinetti, *La France et l'Arme Atomique* (René Juillard, 1964) particularly chapter 2; on withdrawal from NATO see Frédéric Bozo, *La France et l'Otan* (IFRI/Maison, 1991), particularly chapters 3 and 4.

55. For a good overview of Giscard's foreign policy see: Samy Cohen and Marie-Claude Smouts [eds], *La Politique Extérieure de Valéry Giscard d'Estaing* (Presse de la Fondation Nationale des Sciences Politiques, 1985).

56. For a discussion of the acceptance of a more 'flexible' strategy under Giscard see Parti Républicain, *Le Projet Républicain* (Flammarion, 1978) pp. 92–3.

57. Perhaps Mitterrand's clearest statement on this issue was the *Intervention de M. François Mitterrand, President de la République, sur la Theme de la Dissuasion*, op. cit.

58. This point is mentioned in Vincent Wright, *The Government and Politics of France* (Open University Press, 1989) p. 58. I confirmed the point with Wright.

59. The military and political reaction to Mitterrand's 1992 testing moratorium decision is discussed in my 'France and nuclear weapons testing', *Bulletin of Arms Control*, No 19 (August 1995) pp. 14–25.

60. This somewhat simplified summation draws on many sources. It can be more fully explored in Philip Gordon, op. cit., pp. 81–105.

61. James McMillan, *Twentieth Century France* (Edward Arnold, 1992) pp. 217–23.

62. For an excellent overview of foreign policy-making in the first period of cohabitation see: Samy Cohen, 'La politique étrangère entre l'Elysée et Matignon', *Politique Etrangère* (Autumn 1989) pp. 487–503. See also: Jolyon Howorth, 'François Mitterrand and the "domaine réservé" from cohabitation to the Gulf War', *French Politics and Society* (Winter 1992) pp. 43–58; Stephanie Mesnier, 'Le rôle du Quai d'Orsay de mai 1986 à mai 1988', *Revue Administrative* (December 1990) pp. 489–98; and Philippe le Prestre, 'The lessons of cohabitation', In: Philippe le Prestre [ed], *French Security Policy in a Disarming World* (Lynne Reinner, 1989) pp. 15–48.

63. On the second period of cohabitation and the distribution of power see: Jean-Claude Zarka, 'Le "domaine réservé" à l'épreuve de la seconde cohabitation', *Revue Politique et Parlementaire* (1994) pp. 40–4.

64. Gérard Courtais, 'Les français face à l'enigme Chirac', *Le Monde* (10 May 1996).

65. On Chirac, Jospin and defence in the third period of cohabitation see: Franck Laffaille, 'Défense nationale et cohabitation', *Revue Droit et Défense* (Autumn 1997) pp. 32–7.

66. For a discussion of Jospin's defence initiatives after May 1997 see Chapter 3.

67. Dominique David [ed], *La Politique de Défense de la France: Textes et Documents* (FEDN, 1989) pp. 35–42.

68. For details of these arrangements see: Patrice Dubos, 'Qui fait la politique de défense de la France?', *Les Cahiers de Mars* (Spring 1995) pp. 31–42; Jean-Luc Mathieu, *La Défense Nationale* (Que Sais-je?/PUF, 1996); *France and the Armed Forces of France* (SIRPA, 1998); and SGDN [ed], *L'Organisation Générale de la Défense de la France* (La Documentation Française, 1990).

69. For example: David Garraud, 'Essais nucléaire: Juppé et Léotard de demarquent de Mitterrand', *Libération* (8 May 1994) p. 4.

70. For a useful overview of the 1972 and 1994 *Livres Blancs*, see: Jean-Luc Mathieu, *La Défense Nationale*, op. cit., pp. 38–55.

71. *Projet de Loi Relatif à la Programmation Militaire pour les Années 1997–2002* (Ministère de la Défense, May 1996).

72. *Projet de Budget de la Défense pour 1999* (Ministère de la Défense, September 1998). For a useful discussion of the relationship between the *Loi-Militaire* and Budget see: Yves Artru, 'La programmation militaire et l'annualité budgétaire: le mariage impossible?', *Défense Nationale* (March 1997) pp. 57–68.

73. *Une Défense Nouvelle* (Minstère de la Défense/SIRPA, February 1996).

74. Good general studies of the French political process in English include Vincent Wright, *The Government and Politics of France*, op. cit. and Anne Stevens, *The Government and Politics of France* (Macmillan, 1992).

75. A nuanced account of Parliament's role in defence policy can be found in 'Entretien avec Xavier de Villepin', *Relations Internationales et Stratégiques* (Spring 1994) pp. 7–19.

76. On war declaration and other aspects of parliamentary influence over defence policy see: Simon Cohen, 'Le control parlementaire de la politique de défense', *Revue du Parliament* (Spring 1977) pp. 378–446.

2 Adjusting to Change: French Defence Policy, 1989–94

1. See for example Jean Gallois [pseudonym], *Notre Défense en Mal d'un Politique* (Paris: Economica, 1988) and François Heisbourg, 'Défense français: l'impossible status quo', *Politique Internationale* (Summer 1987) pp. 137–53.

2. François Heisbourg, 'Réflexions sur la politique de défense de la France', *Politique Etrangère* (Spring 1990) pp. 157–69; Jacques Lanxade, 'Quelle armée pour demain?', *Politique Internationale* (Summer 1991) pp. 119–26; Jean d'Albion, *La France Sans Défense* (Calmann-Levy, 1991); and Lucien Poirier, 'La crise des fondaments', *Stratègique* (Spring 1992) pp. 117–54.

3. Diego Ruiz Palmer, *French Strategic Options in the 1990s*, Adelphi Paper 260 (IISS, Summer 1991) pp. 34–8.

4. Philip Gordon, *A Certain Idea of France: French Security Policy and the Gaullist Legacy* (Princeton University Press, 1993) 158–9.

5. *Projet de Loi Relatif à la Programmation Militaire pour les Années 1990–1993* (Ministère de la Défense, May 1989). See also Rocard's speech on the *Loi* at the IHEDN in September 1989 reproduced in: Michel Rocard, 'Les orientations de la politique de défense de la France', *Défense Nationale* (November 1989) pp. 13–30. For a useful critique see: Xavier de Villepin, 'Les hommes sont-ils plus sage?: reflexions sur la revision de la loi de programmation militaire', *Défense Nationale* (August/September 1989) pp. 13–22.

6. Jean-François Lazerges, 'Armées 2000', *Défense Nationale* (May 1991) pp. 25–40.

7. Arséne Orain, 'The defence posture of France: an overview', *Military Technology* 5 (Spring 1990) p. 34.

8. Henri Paris, 'Les armées de l'an 2000', *Défense Nationale* (November 1989) pp. 31–42. See also Diego Ruiz Palmer, op. cit., pp. 59–65.

9. 'France's power base', *Jane's Defence Weekly* (23 June 1990) p. 1252.

10. Jean-Pierre Chevènement, 'Confidence and perseverance', *Military Technology* (Spring 1990) p. 32.

11. Hubert Védrine, *Les Mondes de François Mitterrand* (Paris, Fayard, 1996) p. 452.

12. Quoted in: Julius Friend, *The Lynchpin: Franco-German Relations 1950–1990*, Washington Papers No 154 (Washington, CSIS /Praeger 1991) p. 81.

13. Jacques Amalric, 'Les inquiétudes de M. Mitterrand', *Le Monde* (24 November 1989).

14. Christian Millotat and Jean-Claude Philippot, 'Le jumelage Franco-Allemand pour la sécurité de l'Europe', *Défense Nationale* (October 1990) pp. 67–80 and Hubert Védrine and Jean Musitelli, 'Les changements des années 1989–1990 et l'Europe de la prochaine décennie', *Politique Etrangère* (Spring 1991) pp. 165–78.

15. See for example: Frédéric Bozo, *La Politique Etrangère de la France Depuis 1945* (Editions La Découverte, 1997) p. 97.

16. Daniel Colard, 'L'Allemagne unie et le couple franco-allemand', *Défense Nationale* (May 1991) pp. 99–114.

17. *Text of the statement by Mitterrand and Kohl on European Community Union*, Reuters [Transcript] (19 April 1990).

18. Reproduced as 'La lettre commune de MM Kohl et Mitterrand', *Le Monde* (December 9–10, 1990).

19. See for example Jean-Pierre Chevènement's statement on NATO indispensability in Minstère de la Défense, *Propos Sur la Défense*, 15 (May/June 1990) 112. See also Hubert Védrine's comments in 'Paris tente de reassurer Washington sur la pérenité de l'Alliance Atlantique, *Le Monde* (29 May 1991) p. 4.

20. Ministère de le Défense, *Propos sur la Défense*, No 20 (March/April 1991) pp. 85–6.

21. Philip Gordon, op. cit., p. 166.

22. On French thinking about NATO at the time see: Frédéric Bozo, 'La France et l'OTAN: vers une nouvelle alliance', *Défense Nationale* (January 1991), pp. 19–34.

23. Frédéric Bozo [1997], op. cit., p. 99.
24. Steven Philip Kramer, *Does France Still Count?: The French Role in the New Europe*, Washington Papers No 164 (Washington, CSIS/Praeger, 1994) pp. 22–3.
25. *NATO Facts and Figures*, (NATO Information Service, 1989) p. 377.
26. On French public opinion on the Gulf War see: 'Les français face à la crise du Golfe', *Figaro* (14 January 1991).
27. Hubert Védrine, *Les Mondes de François Mitterrand*, op. cit., p. 528.
28. François Heisbourg, 'France and the Gulf crisis', In: Nicole Gnesotto and John Roper, *Western Europe and the Gulf*, ISS/WEU, 1992, pp. 21–2.
29. Philip Gordon [1993], op. cit., pp. 180–1.
30. *The Military Balance 1991–92* (London: IISS/Brassey's, 1991) pp. 55–8.
31. 'Daguet' is an interesting appellation for the French light armoured division. Its literal translation is Brocket, a young stag in its first or second year at the stage when its antlers are just beginning to develop. As a metaphor of inexperience, impotence, and subordination to the bigger stags in the group, it is difficult to think of anything which more succinctly captures the essence of the French role in the Gulf War.
32. These figures are put together from the *The Military Balance 1991–92*, op. cit. pp. 238–42; Jeffrey McCausland, *The Gulf Conflict: A Military Analysis*, Adephi Paper 282 (London, IISS/Brassey's, November 1994); Dominique and Michele Fremy, *Le Quid* (Paris, Editions Robert Laffont, 1993) pp. 1005–8; and Jean-Jacques Langendorf, *Le Bouclier at la Tempête: Aspects Militaires de la Guerre du Golfe* (Georg, 1995).
33. Frédéric Praeter, 'La France et la crise du Golfe', *Politique Etrangère*, (Summer 1991) pp. 450–1.
34. 'La participation française aux opérations', *Les Cahiers de Mars* (Autumn 1991) pp. 36–49.
35. Erwan Bergot, *L'Opération Daguet: les Français dans la Guerre du Golfe* (Presses de la Cité, 1991) pp. 189–239.
36. 'L'Opération Daguet: organisation du commandement et cooperation interalliée', *Les Cahiers de Mars* (Autumn 1991) pp. 30–5.
37. François Heisbourg [1992], op. cit., p. 25.
38. Jean-Marie Colombani, 'Le "rang" de la France', *Le Monde* (5 March 1991).
39. On the French combat experience in the war see: *La France en Guerre* (PUF, 1991); Roland Passevant, *Le Golfe: Tempête pour la Paix* (Messidor, 1991); and David Reyrat, *L'Engagement de la France dans la Guerre du Golfe* (Jean Picollec, 1992).
40. Author's interview at DICOD/SIRPA, 9 March 1999.
41. See for example 'Un orchestre où chacun joue son rôle?', *Le Monde* (22 September 1990) and François Cornu, M. Chevènement a donné le coup d'envoi de l'Opération "Daguet"', *Le Monde* (27 September 1990).
42. Hubert Védrine, op. cit., p. 530.
43. Maurice Schmitt, *De Diên Biên Phu à Koweit Cité* (Editions Grasset, 1992) p. 211.
44. Jean-Pierre Chevènement was the founding president of the France-Iraq friendship society.
45. One of the most important elements of this was the political furore around the targets of French air strikes in the early phases of the air war. Anxious to minimise problems with Iraq, Chevènement persuaded the French war cabinet to confine early operations to targets in Kuwait. In the political storm which followed, the government was widely attacked for refusing to strike targets in Iraq thereby sparing the aggressor. Mitterrand was prompted to refute this in a press

conference on 20 January and the French began attacks shortly after against targets in Iraq. See: Josette Alia and Christian Clerc, *La Guerre de Mitterrand: la Dernière Illusion* (Olivier Orban, 1991) pp. 303–6 and Clare Trean, 'La malaise de M. Chevènement', *Le Monde* (21 January 1991).

46. Chevènement's discomfort is evident in his letter to President Mitterrand spelling out his opposition to the government's position. This letter is reproduced in: Brigitte Stern, *Guerre du Golfe: Le Dossier d'une Crise Internationale 1990–1992* (Documentation Française, 1993) pp. 376–7. For Chevènement's *cri du coeur* see: Jean-Pierre Chevènement, *Une Certain Idée de la République m'Amène à . . .* (Albin Michel, 1992).

47. Simon Doux, *L'Opposition Française à la Guerre du Golfe 1990–1991* (Institute d'Etudes Politique, 1992).

48. François Heisbourg [1992], op. cit., p. 28.

49. For an overview of France's main defence procurement programmes in the early 1990s see: Daniel Coulmy, 'Les grands programmes d'armement', *Stratégique* (Spring 1992) pp. 167–98. On constraints see: Jacques Isnard, 'France: la difficile équation budgétaire', *DAH* (September 1990) pp. 26–30 and Jean de Galard, 'Le combat est engagé entre les finances et la défense', *Air et Cosmos* (26 August– 8 September 1991) pp. 46–7.

50. Figures compiled from: Patrick Saint-Exupéry, 'Mitterrand, L'Africain amer', *Le Monde* (9 November 1994), and 'Les précédentes interventions militaires françaises', *Le Monde* (24 May 1996).

51. Philip Gordon, French Security Policy After the Cold War: Continuity, Change and Implications for the United States, Report No P–4229-A (RAND, 1992), p. 36.

52. On the British problems in the Gulf War see: Lousie Fawcett and Robert O'Neill, 'Britain, the Gulf crisis and European security', In: Nicole Gnesotto and John Roper, op. cit., pp. 141–58.

53. Patrice Langereux, 'Spot a bien servi les alliés dans la tempête du désert', *Air et Cosmos* (22 April 1991) pp. 36–7 and Jean-Paul Dufour, 'Le satellite français SPOT a guidé les raids américains dans le Golfe', *Le Monde* (26 May 1991).

54. On allied intelligence in the Gulf War see: Desmond Ball, *The Intelligence War in the Gulf*, Canberra Papers No 78 (Australian National University, 1991).

55. Pierre Joxe, *Transcript of Speech to the Assemblée Nationale* (6 Jan 1991).

56. For more on this see my *Command, Control, Communications and Intelligence in the Gulf War*, Canberra Working Papers No 238 (Australian National University, 1991).

57. See for example: Pascal Boniface and Jean Golliet, *Les Nouvelles Pathologies des Etats dans les Relations Internationales* (Paris, Dunod, 1993); Pierre Lellouche, *Le Nouveau Monde: De l'Ordre de Yalta au Désordre de Nations* (Paris, Editions Grasset 1992); and Pierre M. Gallois, 'Les changements stratègiques dans le monde', *Défense Nationale* (June 1992) pp. 23–30.

58. For insightful overviews of these issues see: Alain Dumoulin, 'Les forces multinationales', *Défense Nationale* (August 1991) pp. 67–85; François Heisbourg, 'L'après Guerre du Golfe', *Politique Etrangère* (Summer 1991) pp. 411–64; Jacques Lanxade, 'Quelle armée pour demain?' *Politique Internationale* (Summer 1991) pp. 117–26; and, Jacques Touvenin, 'La condition militaire aujourd'hui', *Défense Nationale* (February 1992) pp. 13–28.

59. This debate was not new. De Gaulle himself had argued in favour of professionalisation of the armed forces in his *Vers l'Armée de Metier* (Paris, Librarie Berger-

Levrault, 1934) though in office he had failed to implement change. As the pressures on the defence budget increased through the 1980s and after the problems of the Gulf War, the issue came once again to the fore as for example in: François Cailleteau, 'La conscription: les elements du problème', *Défense Nationale* (January 1990) pp. 13–27; Bernard Boene and Michel Martin, *Conscription et Armée de Metier* (Paris, FEDN/Documentation Français, 1991); and, Henri Paris, 'Armée de metier où de conscription?', *Défense Nationale* (May 1993) pp. 89–96.

60. A useful overview of French intelligence problems and responses can be found in two linked articles: Maurice Faivre, 'Les renseignement dans la Guerre du Golfe', *Stratégique* (Winter 1991) pp. 373–401, and Maurice Faivre, ' Les renseignement après la Guerre du Golfe, *Stratégique* (Spring 1992) pp. 257–74.

61. *The Military Balance 1991–1992*, (London, IISS/Brassey's, 1992) p. 242.

62. For an analysis of the German role in the Gulf War see: Karl Kaiser and Klaus Becher, 'Germany and the Iraq Conflict', In: Nicole Gnesotto and John Roper, *Western Europe and the Gulf* (Paris: ISS/WEU 1992) pp. 39–69.

63. Philip Gordon, France, Germany and the Western Alliance (Westview Press, 1995) pp. 38–9.

64. *The Military Balance 1991–1992*, op. cit., 238–42.

65. Cited in Lawrence Freedman and Efraim Karsh, *The Gulf Conflict 1990–91* (Faber and Faber, 1993) p. 358.

66. His speech is reproduced as Jacques Delors, 'European Integration and Security', *Survival* (March/April 1991) pp. 99–109.

67. On German security 'pulls' see: Lothar Gutjahr, *German Foreign and Defence Policy after Unification* (Pinter, 1994) and Wolfgang F. Schlör, *German Security Policy: an Examination of the Trends in German Security Policy in a New European and Global Context*, Adelphi Paper 277 (IISS/Brassey's, 1993).

68. *The Military Balance 1991–1992*, op. cit., p. 242.

69. One of the most comprehensive anaylses of this view of the post-Cold War world and the place of France within it can be found in Pierre Lellouche, *Le Nouveau Monde: de l'Ordre de Yalta au Dèsordre des Nations* (Paris: Grasset 1992).

70. On NATO adaptation immediately after the end of the Cold War see: David Yost, *NATO Transformed: The Alliance's New Roles in International Security* (USIP, 1998) pp. 72–90.

71. This valuable point is made in Margaret Blunden, 'France after the Cold War: inching towards the Alliance', *Defense Analysis*, (Autumn 1993) p. 262.

72. Frédéric Bozo [January 1991], op. cit., p. 30.

73. Anand Menon, Anthony Forster and William Wallace, 'A common European defence', *Survival* (Autumn 1992) p. 107.

74. Jacques Amalric and Jean-Pierre Langellier, 'M. Mitterrand et Kohl proposent de renforcer les responsabilités européennes en matière de défense', *Le Monde* (17 October 1991).

75. Anand Menon, Anthony Forster and William Wallace, 'A common European defence', op. cit., p. 110.

76. Hartmut Bühl, 'Le corps Européen', *Les Cahiers de Mars* (Summer 1996) pp. 35–45.

77. Article B, Title 1, *Treaty on European Union* (European Community, 1992).

78. There are numerous general studies of the wars in Yugoslavia. See, for example: Mark Almond, *Europe's Backyard War* (Heinemann, 1994); Christopher Bennett, *Yugoslavia's Bloody Collapse: Causes, Courses and Consequences* (Hurst 1995)); Laura Silber and Allan Little, *The Death of Yugoslavia* (Penguin 1995).

79. John Zametica, *The Yugoslav Conflict*, Adephi Paper 270 (IISS Summer 1992) 59–63.
80. Ibid., pp. 34–40.
81. See: Marie-Janine Calic, 'German Perspectives', In: Alex Danchev and Thomas Halverston [eds], *International Perspectives on the Yugoslav Conflict* (Macmillan/St Martin's Press 1996) pp. 52–75; Heinz Junen Axt, 'Hat genscher Jugoslawien entzweit? mythen und fakten zur außenpolitik des vereiten Deutschland', *Europa Archiv* (Spring 1993) pp. 351–60; Peter Viggo Jakobsen, 'Myth-making and Germany's unilateral recognition of Croatia and Slovenia', *European Security* (Autumn 1995) pp. 400–16; Hanns W. Maull, 'Germany in the Yugoslav crisis', *Survival* (Winter 1995/6), pp. 99–130. I am indebted to Wibke Hansen for assistance with these sources.
82. Hubert Védrine, op. cit., pp. 615–6.
83. *La Politique Etrangère de la France: Textes et Documents* (Ministre des Affaires Etrangères, May 1991) p. 19.
84. For insightful analysis of the impact of the Yugoslavian conflicts on the Franco-German relationship see: Henri Starck, 'Dissonances Franco-Allemandes sur fond de guerre Serbo-Croat', *Politique Etrangère* (Summer 1992), pp. 339–47; and Henri Starck, 'France-Allemagne: entente et mésentente', *Politique Etrangère* (Winter 1993/4), pp. 989–1001.
85. Pia Christina Wood, 'France and the post Cold War order: the case of Yugoslavia', *European Security* (Spring 1994) pp. 537–43.
86. Jacques Poos at the Luxembourg summit quoted in: Joel Haveman, 'EC urges end to Yugoslav violence, threatens aid cut', *Los Angeles Times* (29 June 1991).
87. John Zametica, op. cit., pp. 65–6.
88. *The Military Balance 1993–94* (IISS/Brassey's, 1993) p. 256; and, 'Entre faucons et colombes', *Le Point* (22–28 August 1992) p. 23.
89. Olivier Lepik, 'French perspectives', In: Alex Danchev and Thomas Halverston (eds), op. cit., p. 80.
90. Jim Hogland, 'Sarajevo: a bold visit none could ignore', *International Herald Tribune* (1 July 1992) p. 6.
91. See for example, William Drozdiak, 'French President's secret mission: generous but solitary', *International Herald Tribune* (30 June 1992) pp. 1–2.
92. An interesting discussion of the factors bearing on French policy at the time can be found in Steven Kramer, *Does France Still Count?*, Washington Paper No 164 (CSIS/Praeger, 1994) pp. 50–4. See also: Joseph Fitchett, 'Mitterrand's panache gives him stature', *International Herald Tribune* (29 June 1992).
93. Annka Savill, 'France marches out of step on world sanctions', *The Independent* (1 June 1992).
94. For a personal account see: Rezak Hukanovic, *The Tenth Circle of Hell* (Abacus, 1998).
95. Velika Kladusa, 'Percée humanitarian d'un bataillon français', *Libération* (1 November 1992).
96. *The Military Balance 1993–94*, op. cit., p. 257.
97. William Droziak, 'Pressure for Bosnian intervention rises in France', *International Herald Tribune* (24–25 December1992).
98. Dominque Moisi, 'Behind the tougher French stance', *International Herald Tribune* (18 January 1993).
99. Patrice-Henry Desaubliaux, '5793 Casques Bleues français en ex-Yugoslavie', *Figaro* (10 February 1994); and, Jacques Isnard, 'Le coup de semonce de Léotard sur la Bosnie', *Le Monde* (18 May 1994).

100. For a stinging French assessment of collective Western incompetence see: Michel Jobert, 'Serbs are right to laugh at the West's posturing', *International Herald Tribune* (9 June 1993); and on the French role in particular see: Michel Bombier, 'Les errements de la politique française', *Le Monde* (27 March 1993).
101. See Silber and Little, op. cit., pp. 309–11.
102. David Buchan, 'Friction with the UN fails to deter French', *Financial Times* (21 January 1994).
103. For a robust defence of French policy and a sharp exposition of the dilemma 'trapping' France in Bosnia see: Hubert Védrine, 'Non, la France n'a pas à rougir', *Le Nouvel Observateur* (17–23 February 1994) pp. 42–5; see also: 'Bosnie: Mitterrand défend la politique de la France', *Figaro* (26 May 1994).

3 The Reform of Defence

1. François Cailleteau, 'Quelles menaces?', *Relations Internationales et Stratégiques* (Winter 1993) p. 90.
2. Pierre Lellouche, *L'Europe et sa Sécurité*, Rapport d'Information 1294, Assemble Nationale (May 1994) pp. 62–71 and Marcel Duval, 'Risques et options pour la défense', *Défense Nationale* (May 1992) pp. 59–63.
3. See: Jean-François Bureau, 'La réforme militaire en France: une mutation identitaire', *Politique Etrangère* (Spring 1997) pp. 73–6 and Paul-Marie de la Gorce, 'Contexte international et défense', *Défense Nationale* (December 1993) pp. 9–22.
4. Jacques Lanxade, 'La doctrine militaire français face aux bouleversements stratégiques en Europe', *Relations Internationales et Stratégiques* (Winter 1991) p. 36.
5. *Livre Blanc Sur La Défense 1994*, (La Documentation Française, March 1994) p. 22.
6. On the utility of the French armed forces in the post-Cold War context see: Jacques Lanxade, ' Quelle armée pour demain?', *Politique International* (Summer 1991) pp. 119–26 and Jacques Lanxade, ' La défense française dans le nouveau cadre géostratégique', *Défense Nationale* (August/September 1995) pp. 7–22.
7. For an insightful overview of this evolution see: Paul R.S. Gebhard, *The United States and European Security*, Adelphi Paper 286, IISS/Brassey's (February 1994).
8. This point is made by Stanley Sloan in Robert Grant [ed], *Changing French-American Security Relations*, US-CREST (December 1993) pp. 50–2.
9. Figures for US conventional downsizing in Europe are taken from the *The Military Balance 1989–90* (IISS/Brassey's, 1989–90) pp. 26–7 and *The Military Balance 1994–95*, (IISS/Oxford University Press, 1994–95) pp. 31. In 1995 unofficial estimates suggested that as few as 480 US nuclear weapons were left in Europe. See: Robert Norris and William Arkin, 'US nuclear weapons locations', *Bulletin of the Atomic Scientists* (November/December 1995) pp. 74–5.
10. Joseph Fitchett, 'Paris concedes to NATO on Franco-German corps', *International Herald Tribune* (1 December 1992) p. 1. The caveats are: the use of the Eurocorps is subject to French and German agreement; the corps is to be used only for defined and specified missions; and the corps is to be used as a unit [i.e., not broken up or redistributed] and under the title of the Eurocorps.
11. Pascal Boniface, 'La France', In Pascal Boniface [ed], *L'Année Stratégiques 1995: Les Equilibres Militaires* (Dunod/IRIS 1995) p. 12.
12. Robert Grant, 'France's new relationship with NATO', *Survival*, 38(1) (Spring 1996) p. 63. This excellent article repays reading in its entirety, pp. 58–80.
13. Ibid.
14. Ibid., p. 61.

15. Typical of this is François Fillon, 'Le domaine partagé de la défense', *Le Monde* (5 March 1993).
16. This point is made by Paul Gebhard, *The United States and European Security*, Adelphi Paper No 286 (IISS/Brassey's, February 1994) p. 13.
17. François Heisbourg, 'Sécurité: l'Europe livrée à elle-même', *Politique Etrangère* (Spring 1994) pp. 247–60 and Frédéric Bozo, 'Organisations de sécurité et insécurité en Europe', *Politique Etrangère* (Summer 1993) pp. 447–58.
18. Paul Cornish, 'European Security: the end of architecture and the new Europe', *International Affairs* (October 1996) pp. 756–7.
19. For the text of the summit declaration see: *NATO Handbook* (NATO Information and Press Office, 1995) pp. 269–77.
20. Cited in David Buchan, 'France goes on the defence offensive', *Financial Times* (24 January 1994).
21. Claude Monier, 'Orientation de l'alliance atlantique vers le partenariat pour la paix', *Défense Nationale* (March 1995) pp. 166–9.
22. An excellent overview of this aspect of policy can be found in Camille Grand, 'La politique français et le non-prolifération nucléaire', *Défense Nationale* (August/September 1994) pp. 99–112.
23. Jean-Luc Marret, *La France et le Désarmement* (L'Harmattan, 1997). See also: André Collet, 'Le règlementation des armements: de nouvelles voies', *Défense Nationale* (May 1994) pp. 99–107; Jean Compagnon, 'La convention sur les armes chimiques remise sur rails', *Défense Nationale* (July 1997) pp. 29–36; and Daniel Pichard, 'Le désarmement', *Les Cahiers du Chear* (December 1994) pp. 89–103.
24. The moratorium was a complex decision with domestic as well as international political intent. These issues are explored in my 'France and Nuclear Weapons Testing', *Bulletin of Arms Control*, No 19 (August 1995) pp. 14–25.
25. The principles set out by Jules Moch in the Fourth Republic to guide French participation in arms control were adapted under de Gaulle to define a set of ideas which kept France out of most international arms control fora until the mid-to late-1970s and informed policy up until the end of the Cold War. The principles were that France should participate only in meaningful arms control aimed at disarmament, that arms control processes should be open, inclusive and non-discriminatory and that arms control should be verifiable. Processes which accorded special status to the US and USSR, which constrained French autonomy or which were not verifiable were consequently rejected. See Jules Moch, *La Folie des Hommes* (Laffont, 1954); *En Retard d'une Paix* (Laffont, 1958) and *Non à la Force de Frappe* (Laffont, 1963). For French arms control history see: Jean-Luc Marret, op. cit., and David Yost, *France's Deterrent Posture and Security in Europe [Vol II]*, Adelphi Papers No 195 (IISS, Winter 1984/5) pp. 35–60.
26. A useful overview of France's historical view of the NPT and of its role as a distributor of civil nuclear technology can be found in Harald Muller, *Falling into line?: France and the NPT*, PPNN Occasional Papers No 6, University of Southampton (May 1990). On US-French policy convergence on nuclear arms control see: Robert Grant, *Counterproliferation and International Security: The Report of the French-American Working Group*, US-CREST, June 1995.
27. Perhaps the most thoughtful and influential of these was Pascal Boniface, *Vive la Bombe*, (Editions 1, 1992).
28. Marcel Duval and Yves le Baut, *L'Arme Nucléaire Française: Pourquoi et Comment?* (Kronos, 1992) pp. 247–8.

29. On the 1990–3 *Loi-Militaire* see: Marcel Duval and Dominque Mongin, *Histories des Forces Nucléaires Française depuis 1945* (Que Sais-Je?/ PUF, 1993) pp. 118–20.

30. On the nuclear upgrades see: 'L'environment de la force nucléaire est renforce', *Air et Cosmos* (16–21 December 1991) p. 40. The command and control systems themselves are described in Marc Theleri, *Initiation à la Force de Frappe Française 1945–2010* (Editions Stock, 1997) pp. 257–90.

31. For a definitive overview of the Iraqi WMD programmes found by the UN see: *Report of the Secretary-General on the Status of the Implementation of the Special Commission's Plan for the Ongoing Monitoring and Verification of Iraq's Compliance with the Relevant Parts of Section C of Security Council Resolution 687 [1991]*, UN Security Council Report S/1995/864 (11 October 1995).

32. André Dumoulin, 'L'inexorable adaptation de la doctrine nucléaire française', In: *L'Europe et la Sécurité*, (Les Dossiers du Grip, 1994–5), pp. 297–9.

33. David Yost, 'Nuclear debates in France', *Survival* (Winter 1994/5) pp. 114–5.

34. The 'more operational/less operational' division in the strategic community was in fact a simplification of a range of strategy options. For an analysis of the complexity of these issues see: Bruno Tertrais, *The French Nuclear Deterrent After the Cold War*, Report No P-8012 (RAND, 1998) pp. 22–5.

35. 'Intervention de M. Joxe', *Figaro* (19 May 1992).

36. Pascal Boniface [1992], op. cit., pp. 155–7.

37. See: Pascal Boniface, *Contre le Revisionisme Nucléaire* (Ellipses/IRIS, 1994). The relevant chapters are 4–7, pp. 41–75.

38. Quoted in Alain Joxe's insightful assessment of Mitterrand's statement in *Le Débat Stratégique* (May 1994) p. 1.

39. *Intervention de M. François Mitterrand, President de la République, sur la Theme de la Dissuasion*, Service de Presse, Palais de l'Elysée (5 May 1994).

40. See: François de Rose, 'L'avenir de la dissuasion', *Relations Internationales et Stratégiques* (Summer 1992) pp. 92–3.

41. 'France, Britain to Join in Study of their A-Forces', *International Herald Tribune* (10 April 1987).

42. Ibid.

43. Giovanni de Briganti, 'France, Britain agree to closer military links', *Defense News* (8 February 1988).

44. 'Paris offers UK nuclear targetting talks', *Financial Times* (19 December 1987) p. 6.

45. These arguments are made in François Heisbourg, 'The British and French Nuclear Forces', *Survival* (July/August 1989) pp. 316–7. For a prescient view of these issues see: Edward Kolodzeij, 'British-French nuclearisation and European denuclearisation: challenge for US policy', *Atlantic Community Quarterly* (Fall 1988) pp. 305–11.

46. Pascal Boniface, 'French nuclear strategy and European deterrence: les rendez-vous manqués', *Contemporary Security Policy*, (August 1996) pp. 227–9.

47. For a discussion of German flirtations with nuclear weapons technology see: Mark Hibbs, 'Tomorrow a Eurobomb?', *The Bulletin of Atomic Scientists* (January/February 1996) pp. 16–23.

48. David Yost, 'Europe and nuclear deterrence', *Survival* (Autumn 1993) p. 112.

49. Jerome Paolini, 'La France, l'Europe et la bombe', *Commentaire*, No 54 (Summer 1991) p. 253.

50. Philip Gordon [1992], op. cit., pp. 26–7.

51. 'Britain and France to study nuclear links', *International Herald Tribune* (5/6 May 1990) and Ian Kemp, 'France/UK seek closer ties', *Jane's Defence Weekly* (21 April 1990) p. 738.

52. 'Paris et Londres sur le sentier de la bombe commune', *Libèration* (4 October 1992). Julian Nundy, 'France seeks closer link with Britain', *Independent* (2 October 1992).
53. 'Franco-British Nuclear Cooperation', *Atlantic News*, No 2466 (23 October 1992) p. 2.
54. John Ridding and Philip Stephens, 'France and UK signal improved relations at summit', *Financial Times* (27 July 1993) and, Stuart Croft, 'European integration, nuclear deterrence and Franco-British nuclear cooperation', *International Affairs* (October 1996) pp. 777–9.
55. Jacques Isnard, 'Programmation militaire française: un pari sur l'avenir', *Défense et Armament International* (September/October 1992) p. 18.
56. Jean de Galard, 'Budget 1992 de la défense: un net rallentissement du nucléaire', *Air et Cosmos*, (25 November–1 December 1991) p. 42. See also: 'Le Budget de la Défense 1992 [Dossier]', *Armées d'Aujourd'hui* (February 1992) pp. 39–51 and 'Le nucléaire dans la projet de Loi de Programmation', *Air et Cosmos*, (9–15 November 1992) p. 48.
57. Jean de Galard, 'De 1992 à 1997, 620 milliard de francs de credits budgetaires pour la défense', *Air et Cosmos*, (7–15 December 1992) p. 38.
58. That the new *Loi-Militaire* was premised on changing assumptions and an emergent agenda is made clear in Jean-Michel Boucheron's extraordinarily detailed report on the *Loi-Militaire*. See his *Paix et Défense* (Dunod, 1992) pp. 199–267 on the changed thinking and pp. 268–467 on the modular *Loi* itself.
59. A good overview of these plans is provided by Arthur Paecht, *Une Nouvelle Donne Pour l'Espace Militaire*, Rapport 1892, Assemblée Nationale (January 1995).
60. 'Le renseignement et la communication dans la Loi de Programmation 1992–1994', *Air et Cosmos* (23–29 November 1992) p. 44. On the Sarigue DC-8 intelligence aircraft see Boucheron, op. cit., p. 106 and pp. 325–6.
61. 'Le programme spatial militaire dans la Loi de Programmation 92–94', *Air et Cosmos* (30 November–6 December 1992) p. 46.
62. Boucheron, op. cit., pp. 343–414.
63. Jacques Isnard, op. cit., pp. 20–1.
64. The members of the *Livre Blanc* commission were: Marceau Long [Commission President]; from the Prime Minister's office: Maurice Schmitt, Philippe Marland, Patrick Lecointre, Achille Lerche, Bernard de Monferrand; from the Defence Ministry: Renaud Donnedieu de Vabres, Jacques Lanxade, Henri Conze, François Roussely, Jean Rannou and Jean-Claude Mallet; from the Foreign Ministry: Bruno Racine, Jean-Marie Guehenno and François Barry-Delongchamp; from Budget Isabelle Bouillet; from the Interior Ministry Jean Riolacci; from Recherche Jérôme Paolini; from Industry Didier Lombard; from the Cooperation Ministry Antoine Pouillieute; from the CEA Roger Baleras; from the SGDN Eric de la Moisonneuse; from DAS Philippe Mallard; from the Conseil d'Etat Marc Guillaume; and six 'personalités: Thierry de Montbrial, Raymond Levy, Jean Prada, Gabriel Robin, Mary-Jean Voinot and Michel Alliot. See: *Livre Blanc sur la Défense 1994*, op. cit., pp. 203–5.
65. Jean de Galard, 'Le Livre Blanc de la Défense: a peine paru, déjà conteste', *Air et Cosmos* (28 February–6 March 1994) p. 37.
66. *Livre Blanc sur la Défense 1994*, op. cit., p. 11.
67. Ibid., p. 13.
68. I am indebted to Pascal Boniface for this insight.
69. For a few years France looked at the idea of developing a multinational BMD system with the US, Italy and Germany, before pulling out in 1996 on grounds of

cost and the risk of undermining *dissuasion*. See: Jean-Luc Prome, 'Quelle choix pour une défense antimissile?', *Science et Vie* (June 1993) pp. 122–7; Jacques Dupont, 'Coopération américo-européenne en vue dans les systèmes de défense antibalistique', *Air et Cosmos* (14 October 1994) pp. 34–5; and, James Atkinson, *La Coopération Transaltantique dans la Domaine de la Défense Antimissile Européenne*, Report No 1588 (WEU, 4 November 1997).

70. *Livre Blanc sur la Défense 1994*, op. cit., pp. 87–98.
71. Ibid., p. 127.
72. 'Long-range vision: procurement chief grapples with structural disarmament', *Armed Forces Journal International* (June 1984) p. 42.
73. *Livre Blanc sur la Défense 1994*, p. 199.
74. 'Long-range vision: procurement chief grapples with structural disarmament', op. cit.
75. For an insightful analysis of this see: Henri Paris, 'Domaine réservé: la conspiration du silence en matière de défense', *Revue Politique et Parlementaire*, 976 (March/April 1995) pp. 45–9.
76. François Heisbourg, *Les Volontaires de l'An 2000* (Balland, 1995), particularly pp. 125–46 and 177–204.
77. Dominique Moisi, 'De Mitterrand à Chirac', *Politique Etrangère* (Autumn 1995) pp. 853–4.
78. Xavier Gautier, 'La France, première cible', *Figaro* (24 July 1995).
79. 'Chirac prone la reprise des tirs', *Figaro* (8 October 1993).
80. Charles Rebais, 'Retour à la doctrine Gaullist', *Figaro* (14 June 1995).
81. Michael Sutton, 'Chirac's foreign policy: continuity with adjustment', *The World Today* (July 1995) p. 137.
82. Sylvie Pierre-Brossolette, 'Chirac sur deux fronts', *L'Express* (15 June 1995) pp. 8–9.
83. Vincent Hugeaux and Sylvaine Pasquier, 'Bosnie: Chirac dans la guerre', *L'Express* (27 July 1995) pp. 18–21.
84. Jean Dupont, 'La France a tire des AS30 Laser en Bosnie', *Air et Comsos* (9 February 1996) p. 38.
85. Many in the US were critical that France usurped the US role in Bosnia by insisting that the Dayton Accords be formally signed in Paris and that they be known as the Elysée Peace Treaty. See: Sarah Helm, 'Paris tries to steal US peace laurels', *The Independent* (6 December 1995).
86. Eric Biegala, 'Déjà, en Mai, le calvaire des casques bleus', *Figaro* (28 December 1995) p. 1.
87. 'Paris denies cover-up on pilot's treatment', *International Herald Tribune* (28 December 1995) p. 6.
88. James Geary, 'Politics and Massacres', *Time* (24 June 1996) p. 24.
89. Partrice-Henry Desaubliaux, 'Bosnie: la France en premier ligne', *Figaro* (18 December 1995). For figures see *The Military Balance 1996–1997* (IISS/Oxford University Press 1996) pp. 303–4.
90. Jean-François Mancel, 'Bosnie: le poids de la France', *Figaro* (14 December 1995).
91. The latter point is made in Dominique Moisi, 'Chirac of France', *Foreign Affairs*, (November/December 1995) p. 10.
92. See for example: Jacques Baumel, 'Les essais sont indispensables', *Figaro* (13 October 1993); Pierre Lellouche, 'La suspension des essais nucléaires: une decision choquante et dangereuse', *Figaro* (16 April 1992); and Pierre Lellouche, 'Défense: le danger de l'obsolescence', *Figaro* (13 October 1993).
93. 'Querelle sur le nucléaire', *L'Express* (7–13 October 1993) pp. 52–3.

214 *Notes and References*

94. René Galy-Dejean *et al.*, *La Simulation des Essais Nucléaires*, Rapport d'Information 847, Commission de la Défense, Assemblée Nationale (15 December 1993); and Patrice-Henry Desaubliaux, 'La neccessaire reprise des essais nucléaire', *Figaro* (17 December 1993).
95. *Intervention de M. François Mitterrand, President de la République, sur la Theme de la Dissuasion*, Service de Presse, Palais de l'Elysée (5 May 1994) pp. 9–14.
96. See: Dominique Garraud, 'Essais nucléaires: Juppé et Léotard de demarquent de Mitterrand', *Libération* (8 May 1994) and Patrice-Henry Desaubliaux, 'Essais nucléaires: Balladur en desaccord avec Mitterrand', *Figaro* (11 May 1994).
97. Charles Rebois, ' Retour à la doctrine gaulliste', *Figaro* (14 June 1995).
98. On the outcry against France see: Ramesh Thakur, *The Last Bang before a Total Ban: French Nuclear Testing in the Pacific*, PRC Working Paper 159 (September 1995). and Russell Johnston, *Resumption of French Nuclear Testing in the Pacific*, WEU Report 1488, WEU (7 November 1995). On the German reaction see: Holger H. Mey and Andrew Denison, 'View From Germany: France's nuclear tests and Germany's nuclear interests', *Comparative Strategy*, No 15 (Autumn 1996) pp. 169–71.
99. William Pfaff, 'The French feel safer with a nuclear deterrent of their own', *International Herald Tribune* (11 September 1995).
100. David Buchan, 'Chirac looks for stronger UK links', *Financial Times* (14 May 1996).
101. On this point see: Pierre Lellouche, 'Essais nucléaires: leçons d'un psycho-drame', *Politique Etrangère* (Winter 1995/6) pp. 95–111; Michel Tatu, 'Après Mururoa', *Politique Internationale* (Autumn 1995) pp. 143–60; and a remarkable report from the right-dominated Assemblée Nationale, Aymeri de Montesquiou, *Les Reactions Internationales à la Reprise des Essais Nucléaires: Beaucoup de Bruit pour Rien?'*, Assemblée Nationale Rapport 2946 (28 June 1996).
102. On the final six tests see: *Fin des Essais Nucléaires* (SIRPA Dossier, January 1996) and Bruno Barrillot, *Les Essais Nucléaires Français 1960–1996* (CDRPC, 1996) pp. 177–223.
103. See my 'French nuclear weapons testing and the CTBT process' *Bulletin of Arms Control* (December 1996) pp. 18–23.
104. On the zero option decision see ibid., pp. 18–19.
105. Pascal Boniface, 'Essais et doctrine nucléaires: la victoire à la Pyrrhus des revi-sionnistes', *Relations Internationales et Stratégiques* (Winter 1995) p. 49. The whole article, pp. 44–9, repays reading.
106. Stephen Hoadly, 'France mends fences with Rarotonga accord', *Jane's Defence Weekly* (10 April 1996) p. 18.
107. 'L'heure de choix: entretien avec M. Charles Millon, Ministre de la Défense', *Armées d'Aujourd'hui* (October 1995) pp. 10–15.
108. 'Composition, organisation et fonction du Comité Stratégique', *Air et Cosmos* (24 November 1995) p. 36.
109. 'France Faces Budget Cutbacks', *International Defence Review* (October 1995) p. 5.
110. Peter Lewis, 'French security policy: the year of the disappearing budget', *Jane's Defence 1996: The World in Conflict* (1996) pp. 42–3.
111. 'Editorial: la nouvelle compagne de M. Chirac', *Le Monde* (24 February 1996).
112. Eric de la Maisonneuve, 'L'armée professionelle et la nation', *La Croix l'Evéne-ment* (9 March 1996); 'Chevènement: on a baisse la France', *Nouvel Observateur* (24 February 1996) p. 12; Thomas Sanction, 'Farewell to some arms', *Time* (4 March 1996) p. 18. See also: Eric de la Maisonneuve, 'Le soldat et le politique',

Défense Nationale (June 1996) pp. 61–72 and Alain Faure-Durfourmantelle, 'Défense de la nation et défense militaire', ibid., pp. 73–86.

113. J.A.C. Lewis, 'French government looks ahead to long-term change', *Jane's Defence Weekly* (17 April 1993) p. 14.

114. '*Services Communes*' are the armed forces joint or shared support services including health services, fuel, some communication and information assets, and materials/infrastructure.

115. Volunteers being those who neither joined the armed forces on professional terms nor were conscripted, but rather those who volunteered to undertake a period of national service.

116. See: Dominique Conort, 'Les ressources humaines: les défis de la professionalisation', *Défense Nationale* (July 1996) p. 92.

117. The reductions set out in *Une Défense Nouvelle* as being planned over 18 years [i.e., 1997–2015] were in fact intended to be implemented in the first five years to 2002 and thereafter sustained.

118. David Buchan, 'A 21st century army', *Financial Times* (26 February 1996) p. 12.

119. 'Les chefs militaire et la Loi de Programmation', *Air et Cosmos* (31 May 1996) p. 24.

120. J.A.C. Lewis, 'Fitter, leaner, forces for a multi-polar world', *Jane's Defence Weekly* (11 June 1997) p. 71.

121. *Une Défense Nouvelle*, op. cit., p. 6. In his television interview Chirac created some confusion by stating that the new policy called for up to 50–60 000 personnel to be projected, which appeared to be at odds with the proposal for a main projection force of up to 30 000.

122. These figures were taken from: Pierre Langereux *et al.*, 'Un nouveau modèle de défense à l'horizon 2015', *Air et Comsos* (1 March 1996) p. 24; 'A year of blue water fleet modernisation', *Jane's Defence Weekly* (11 June 1997) p. 80; 'Air force support hit by personnel cuts', ibid., p. 83; and, J.A.C. Lewis, 'All change for the future: how the big shake-out will shape up', *Jane's Defence Weekly* (13 March 1996) pp. 19–20.

123. 'Trois ans pour demanteler le Plateau d'Albion', *Air et Cosmos* (3 May 1996) p. 33; and J.A.C. Lewis, 'Fitter, leaner, forces for a multi-polar world', op. cit., p. 69.

124. 'La dissuasion, d'une ère à l'autre', *Armées d'Aujourd'hui* (March 1996) p. 14.

125. 'Lancement du missile stratégique M51', *Air et Cosmos* (3 May 1996) p. 34.

126. The information presented here is based on *Une Défense Nouvelle*, op. cit., pp. 20–1; 'Une Défense Nouvelle', *Armées d'Aujourd'hui* [Special Issue] (March 1996) p. 14; 'Une nouvelle modèle de défense à l'horizon 2015' op. cit., pp. 10–13; 'France to cut nuclear arms in sweeping defence review', *Financial Times* (23 February 1996); Giovanni de Briganti, 'France continues to pare down nuclear forces', *Defense News* (14–20 October 1996) p. 40; and J.A.C. Lewis, 'All change for the future: how the big shake-out will shape up', op. cit., pp. 19–20.

127. *Une Défense Nouvelle*, op. cit., p. 18.

128. *Livre Blanc sur la Défense 1994*, op. cit., pp. 76–86.

129. *SIRPA Actualité* (9 March 1996) p. 34.

130. Rapport Annexe of the *Débat Sur Une Défense Nouvelle*, Journal Officiel de la Republique Française (3 July 1996) p. 9996.

131. Jacques Isnard, 'Le budget militaire sera réduit de 100 milliards de francs en cinq ans', *Le Monde* (24 February 1996) p. 1.

132. *Une Défense Nouvelle*, op. cit., p. 24.

133. Ibid., pp. 24–5, and Rapport Annexe of the *Debat Sur Une Défense Nouvelle*, op. cit., p. 9990.

134. *Une Défense Nouvelle*, op. cit., pp. 42–50.
135. Jean-Pierre Casamayou, 'Une pole mariant Aérospatiale et Dassault', *Air et Cosmos* (1 March 1996) p. 17.
136. Jean Dupont, 'Créer un grande pole d'electronique', *Air et Cosmos* (1 March 1996) p. 18.
137. Restructuring DCN was part of a wider restructuring of the DGA earmarked for 30 per cent cuts in programme budgets from 1997–2002.
138. Jean Pierre Casamayou, 'Le pole nucléaire: CEA/DAM revu à la baisse', *Air et Cosmos* (1 March 1996) p. 19.
139. Edwina Campbell, *France's Defence Reforms: The Challenge of Empiricism*, London Defence Studies Papers No 36 (CDS/Brassey's, 1997) p. 21
140. On the problems behind the defence industry reforms see: ibid., pp. 21–30; 'Nouveau depart pour l'industrie de défense', *Air et Cosmos* (1 March 1996) pp. 14–19; and Paul Greenish, *Difficult Times in France: A Study of French Defence Industrial Policy, Practice and Prospects*, Seaford House Papers (Royal College of Defence Studies, 1997) pp. 47–66.
141. David Buchan, op. cit., p. 12.
142. François Heisbourg, 'Nouvelle politique de défense: de la coupe aux terres', *l'Echos* (18 March 1996).
143. Gérard Courtais, 'Les Français face à l'enigme Chirac', *Le Monde* (10 May 1996).
144. John Vinoair, 'Nice but inadequate, French say of Chirac', *International Herald Tribune* (3 February 1997).
145. See: 'Défense: Alain Richard', *Figaro* (5 June 1997); 'Affaires étrangère: Hubert Védrine', *Figaro* (5 June 1997); Joseph Fitchett, 'Jospin fills cabinet with a new generation', *International Herald Tribune* (5 June 1997); Jacques Isnard, 'Diplomatie et défense: conflits en perspective avec l'Elysée', *Le Monde* (6 June 1997).
146. The clearest statements of this general acceptance of the Chirac's reforms are to be found in Jospin's annual lectures to the IHEDN in September 1997 and September 1998. See respectively: *Discours du Premier Ministre M. Lionel Jospin à l'Institut des Hautes Etudes de Défense Nationale*, DPIC Bulletin, (4 September 1997) pp. 2–11; and ibid., (3 September 1998) pp. 3–15.
147. 'Lionel Jospin affirme face à Jacques Chirac son autorité dans la conduite de la diplomatie', *Le Monde* (30 June 1997).
148. Jean-Dominique Merchet, 'Jospin en vol de reconnaissance', *Libération* (29 July 1997).
149. Jean-Dominique Merchet, 'Le Premier Ministre balise son territoire militaire', *Libération* (16 September 1997).
150. Jacques Baumel, 'La défense éternelle sacrifiée', *Figaro* (22 August 1997).
151. Alexandra Schwartzbrod, 'Défense: un budget qui sent la poudre', *Libération* (20 August 1997).
152. 'Les inquiétudes du Général Douin', *Figaro* (14 October 1997).
153. Jacques Isnard, 'Les Chefs d'Etat-Major s'inquiétent du sort de la programmation militaire', *Le Monde* (17 September 1997).
154. Jacques Isnard, 'Des généraux inquiéts de leur perte d'influence', *Le Monde* (13 June 1998).
155. Patrice-Henry Desaubliaux, 'Un contrat de quatre ans pour la défense', *Figaro* (4 April 1998).
156. Jacques Isnard, 'Nouveaux chefs militaire à Matignôn et à la défense', *Le Monde*, (14–15 June 1998).
157. Patrice-Henry Desaubliaux, 'Grandes manoeuvres à la défense', *Figaro* (28 July 1998).

158. Jacques Isnard, 'Nomination d'un nouveau Secrétaire Général de la Défense Nationale', *Le Monde* (10 July 1998).

159. Patrice-Henry Desaubliaux, 'La leçon d'Alain Richard aux élèves officiers', *Figaro* (2 September 1998) and Jacques Isnard, 'Alain Richard confronté à une Grand Muette qui rouspète', *Le Monde* (29 October 1998).

160. In 1994 Samy Cohen published an important – if contentious – analysis arguing that the French military had, particularly since the end of the Cold War, been steadily losing power and influence in French politics and society. See: Samy Cohen, *La Défaite des Généraux*, (Fayard, 1994). See also: Pascal Boniface, *L'Armée: Enquête sur 300,000 Soldats Méconnus* (Editions 1, 1990).

161. 'L'Opposition s'inquiète du risque de repli de l'industrie de défense', *Le Monde*, (15 July 1997).

162. For more on the Europeanisation of defence industries and the implications for France see: Pierre Dabezies and Jean Klein, *La Reforme de la Politique Française de Défense* (Economica, 1998) pp. 55–74; Pierre Dussauge and Christophe Cornu, *L'Industrie Française de l'Armement: Coopérations, Restructurations et Intégration Européenne* (Economica, 1998); Martene Lignières-Cassou, *Rèussir la Diversification des Industries de Défense*, Report No 911 (Assemblée Nationale, May 1998); Paul Martin-Lalande, *Restructurer l'Industrie de Défense: Un Defi Economique et Social*, Report No 2823 (Assemblée Nationale, May 1996); Paul Quiles and Guy-Michel Chauveau, *L'Industrie Française de Défense: Quel Avenir?*, Report No 203 (Assemblée Nationale, September 1997); Jean-Louis Scaringella, *Les Industries de Défense en Europe* (Economica, 1998).

163. *La Gestion Budgétaire et la Programmation Militaire au Ministère de la Défense*, Cour des Comptes, Rapport Public Particulier (June 1997). See also: Bernard Bombeau, 'La défense sans la loupe de la Cour des Comptes', *Air et Cosmos* (27 June 1997) pp. 46–7; and, Jacques Isnard, 'La Cour des Comptes dénonce la dérive des coûts dans des grands programmes d'armement', *Le Monde* (25 September 1997).

164. Interview with Alain Richard, 'Je lancerai dès cet automne une revue générale des programmes', *Le Monde* (25 September 1997).

165. Jacques Isnard, 'Alain Richard confie à un civil la communication des armées', *Le Monde* (10 July 1998); and, Patrice-Henry Desaubliaux, 'La défense réorganise as communication', *Figaro* (23 July 1998). For Bureau's views on the defence reforms see: Jean-François Bureau, 'La réforme militaire en France une mutation identitaire', *Politique Etrangère* (Spring 1997), pp. 69–81.

166. J.A.C. Lewis, 'France to outline procurement plans', *Jane's Defence Weekly* (2 September 1998) p. 13.

167. See: Franck Laffaile, 'Défense nationale et cohabitation', *Revue Droit et Défense* (Autumn 1997) pp. 32–8.

168. On this 'malaise' and the response see: Dominque Moisi, 'The trouble with France', *Foreign Affairs* (May/June 1998) pp. 94–104; and Jonathan Fenby, *On the Brink: the Trouble with France* (Little, Brown and Company, 1998), particularly pp. 18–53 and 420–34.

169. Jacques Boyon, 'Défense nationale et cohabitation', *Figaro* (26 December 1997).

4 French Defence Policy in the New Europe

1. *Livre Blanc sur la Défense*, (La Documentation Française, March 1994), p. 56.

2. Ibid., p. 57.

3. Pascal Boniface, 'France' In: Pascal Boniface, *L'Année Stratégique 1995* (Dunod/ IRIS, 1995) pp. 11–12.
4. Olivier Debouzy, 'France-OTAN: la fin de l'autre guerre froide', *Commentaire* (Summer 1996) pp. 349–52.
5. Frédéric Bozo, 'France' In Michael Brenner [ed], *NATO and Collective Security* (Macmillan, 1998), p. 58.
6. Ibid.
7. Ibid., p. 62.
8. For a discussion of the IMS and Article 5/non Article 5 mission issue see: Gabriel Robin, 'A quoi sert l'Otan?', *Politique Etrangère* (Spring 1995) pp. 171–80.
9. On this formal versus actual role of the Military Committee and SACEUR see: Roy W. Stafford, 'Defence planning in NATO: a consensual decision-making process', In: Robert Pfaltzgraff and Uri Ra'anan [eds], *National Security Policy: The Decision Making Process* (Handon, Con, Archon Books, 1984), particularly pp. 155–7.
10. Philip Gordon, *US and ESDI in the New Europe*, Les Notes de l'IFRI, No 4 (IFRI, 1998) pp. 33–4.
11. Paul Gebhard, *The United States and European Security*, Adelphi Paper 286 (IISS/ Brassey's, (February 1994) p. 22.
12. Frédéric Bozo[1998], op. cit., p. 53.
13. Robert Grant, 'France's new relationship with NATO', *Survival*, 38(1) (Spring 1996) p. 62.
14. Speech of Hervé de Charette, North Atlantic Council, 5 December 1995.
15. Charles Millon, 'La France et la rénovation de l'Alliance Atlantique', *Revue de l'OTAN* (May 1996) p. 13.
16. Robert Grant [1996], op. cit., p. 62.
17. For Chirac's upbeat assessment of the Europeanising trends in NATO and the implications for France see: Jacques Chirac, 'La politique de défense de la France', *Défense Nationale* (August/September 1996) pp. 13–14. The 'NATOisa-tion' of French defence policy has received far less attention than the European-isation of NATO in the French literature. In opposition in early 1997 Lionel Jospin himself warned of the 'NATOisation' of Europe, see: Stanley R. Sloan, 'French defense policy: Gaullism meets the post Cold War World', *Arms Control Today* (April 1997) p. 6; while more specifically see: Anne-Marie le Gloannec, 'Europe by other means?', *International Affairs* (January 1997) pp. 87.
18. Luc Rosenzweig, 'Les responsables de l'Otan ne croient plus dans un retour prochain de la France', *Le Monde* (7 June 1997).
19. Brain Knowlton, 'Chirac calls for revamping NATO', *International Herald Tribune* (2 February 1996).
20. Tom Rhodes, 'Chirac pushes for more balanced Alliance', *The Times* (1 February 1996).
21. See for example: Paul Quilès, 'Défense Européenne et Otan: la dérive', *Le Monde* (11 June 1996) and Philippe Delmas, 'Quatre questions sur un gambit', *Le Monde* (11 June 1996).
22. Michael Meimeth, 'Germany', In Michael Brenner [ed], *NATO and Collective Security* (Macmillan, 1998), p. 97.
23. Philip Gordon [1998], op. cit., pp. 15–20.
24. For a French view of the compromises see: Hervé de Charette, 'France for a streamlined NATO: setting the record straight', *International Herald Tribune* (10 December 1996); see also: Alain Frachon, 'Commandement sud de l'Otan: la 'ba-

taille' de Naples', *Le Monde* (31 January 1997); 'Riling NATO', *The Economist* (21 June 1997); 'War over Naples', *The Economist* (30 November–6 December 1996).

25. Philip Gordon, op. cit., pp. 36–7.

26. Numerous versions of who wrote what and whether Chirac's critical observation that AFSOUTH was 'of capital importance' to France was in English or French can be found in: Thomas Friedman, 'Look here, France, you're not going to get the Sixth Fleet', *International Herald Tribune* (2 December 1996); Joseph Fitchett, 'US tries to fend off Paris on NATO post', *International Herald Tribune* (4 December 1996); Daniel Vernet, 'Jacques Chirac a engagé son autorité sur l'affaire du commandement sud', *Le Monde* (12 December 1996); Alain Frachon, 'Le France recule sur le commandement sud de l'Otan', *Le Monde* (24 January 1997); and Jacques Amalric, 'Otan: comment Washington a coulé Paris', *Libération* (27 February 1997).

27. 'Le ministre de la défense allemand critique la France sur l'Otan', *Le Monde* (2 October 1997).

28. French arguments for 'Southern enlargement' were supported by Belgium, Canada, Italy, Luxembourg, Portugal, and Turkey, while the British reportedly favoured Slovenian accession and Germany was 'sensitive' to the Romanian and Slovenian case. See: Peter Wise, 'Nato divisions emerge over new member states', *Financial Times* (30 May 1997) and 'Who will join the club?', *The Economist* (7 June 1997) pp. 6–7.

29. See: Pascal Boniface, 'Le débat française sur l'élargissement de l'Otan', *Relations Internationales at Stratégiques* (Autumn 1997) pp. 33–47. Another excellent article though more general in nature is: Jean-François Guilhaudis, 'Considérations sur l'élargissement de l'Otan', *Défense Nationale* (November 1995) pp. 57–66.

30. Boniface, op. cit., pp. 40–1.

31. Romanian and Hungarian accession would, for example, have placed the Hungarian minority in Transylvania question in a wholly NATO context. The prospect alone of first wave entry was enough for both parties to sign a joint accord on the issue. See: Craig Whitney, 'Nato expansion: symbol and substance', *International Herald Tribune* (18 May 1997).

32. Laurent Zechini, 'Washington récuse formellement la candidature de la Roumanie et de la Slovénie', *Le Monde* (14 June 1997).

33. On the arguments for US choices see: Madeleine Albright, 'Enlarging NATO: why bigger in better', *The Economist* (15 February 1997) pp. 19–21; on French perceptions see: Daniel Vernet, 'Otan: l'alliance des paradoxes', *Le Monde* (14 July 1997) and Fabien Roland-Lévy, 'Grogne anti-américaine à l'Assemblée Nationale', *Le Monde* (12 July 1997).

34. See: Charles Trueheart, 'On the eve of NATO summit, France calls diplomatic ceasefire', *International Herald Tribune* (4 July 1997) and Daniel Vernet, 'La France risque de se trouver isolée lors du sommet atlantique de Madrid', *Le Monde* (4 July 1997).

35. Hubert Védrine, quoted in: Daniel Vernet, 'France-l'Otan: une bonne idée en panne', *Le Monde* (29/30 June 1997).

36. An early attempt to assess the cost of expansion which raises, though does not definitively answer, the main issues is: Ronald Asmus *et al.*, 'What will NATO enlargement cost?', *Survival* (Autumn 1996) pp. 5–26.

37. For a critical French analysis of the evolution of Franco-NATO rapprochement between December 1995 and the election of the Jospin administration in May 1997 see: Pascal Boniface, 'La France, l'autonomie stratégique européenne et

l'Otan', *Relations Internationales et Stratégiques* (Winter 1997) pp. 20–6. See also the thoughtful analysis by Adrian Treacher, 'New tactics, same objectives: France's relationship with NATO', *Contemporary Security Policy* (August 1998) pp. 91–110.

38. Alain Joxe, 'L'Otan peut attendre', *Le Monde* (13 May 1997).
39. Klaus Naumann, 'NATO's new military structure', *NATO Review* (Spring 1998) pp. 10–14.
40. On the Congressional pressure to withdraw see: 'Will Congress force the US out of Bosnia?', *The Economist* (25 October 1997) p. 25.
41. Ivo Daalder, 'Bosnia after SFOR: options for continued engagement', *Survival* 39(4) (Winter 1997/8) p. 16. The whole article repays reading, pp. 5–18.
42. Brian Knowlton, 'France backs a US call to keep NATO in Bosnia', *International Herald Tribune* (25 September 1997).
43. Philip Gordon, *US and ESDI in the New NATO*, op. cit., pp. 40–3.
44. 'La France à toujours d'un arrangement avec l'Otan', *Le Monde* (2 December 1997).
45. Jacques Isnard, 'Solitaire mais solidaire', *Le Monde* (4 December 1997).
46. Ibid.
47. *The Military Balance 1998/99* (IISS/Oxford University Press, 1999) p. 31.
48. Christian Lecomte, 'L'état-major de l'Otan se méfierait des militaires français', *Le Monde* (16 December 1997).
49. Jacques Isnard, 'Le commandant Gourmelon, poisson pilote du renseignement militaire français en Bosnie', *Le Monde* (6 May 1998).
50. 'Une taupe française démasquée à l'Otan', *Libération* (3 November 1998).
51. Charles Trueheart, 'France says spy incident will not hurt NATO ties', *International Herald Tribune* (4 November 1998).
52. Another article in *Libération* denied this, arguing that 'without doubt' Brunel knew NATO's targets. See: Jean-Dominique Merchet, 'Des renseignements essentiels', *Libération* (3 November 1998).
53. Quoted in Charles Trueheart, op. cit.
54. For a prescient overview see: Jean-Arnault Derens, 'Kosovo: la guerre inévitable?', *Etudes* (June 1998) pp. 739–49.
55. Gilles Andréani, 'Old French problem – or new transatlantic debate?', *RUSI Journal* (February/March 1999) p. 24.
56. Gérald Dubos, 'A deux pas du Kosovo', *Armées d'Aujourd'hui* (February 1999) pp. 14–15.
57. *Strategic Survey 1998/99* (IISS/Oxford University Press, 1999) pp. 119–23.
58. 'Les forces alliées et adverses', *Le Monde* (25 March 1999).
59. Jean-Michel Aphatie, 'Jean-Pierre Chevènement dit "sa preférence pour un solution politique" du conflit', *Le Monde* (26 March 1999).
60. 'M. Séguin quitte la présidence du RPR et les européennes', *Le Monde* (17 April 1999). It should be noted that Séguin did not cite opposition to the war as his reason for quitting but the timing of his departure carried that implication.
61. Ariane Chemin, 'Robert Hue: "cette guerre, c'est une connerie"', *Le Monde* (27 March 1999).
62. Jean-Michel Aphatie, 'Une parti des responsables français contestent le cadre et les modalités de l'action de l'Otan', *Le Monde* (27 March 1999) and Pierre Bourchieu *et al.*, 'Arrêt des bombardements, autodétermination', *Le Monde* (31 March 1999).
63. Ariane Chemin, 'A Paris, dissonances chez les manifestants hostiles à la guerre', *Le Monde* (29 March 1999).

64. Jean Cot, 'Quand les moyens tuent le fin', *Le Monde* (27 March 1999).
65. *The Alliance's Strategic Concept*, Press Communiqué NAC-S(99) 65, 24 April 1999. For a French perspective see: Alain Richard, 'The future of the Atlantic Alliance: a French view', In: *NATO 50: Mapping the Future*, The Washington Summit, 25–27 April 1999, pp. 22–4.
66. John Ruggie, 'Consolidating the European pillar: the key to NATO's future' , *The Washington Quarterly* (Winter 1997) p. 115.
67. Jean-Claude Casanova, 'Dissuasion Concertée', *L'Express* (28 September 1995).
68. François Heisbourg, 'La réintégration de la France dans l'Otan', *Echos* (4 January 1996).
69. Quoted in Michael Sutton, 'Chirac's foreign policy: continuity – with adjustment', *The World Today* (July 1995) p. 137.
70. The idea of this new transatlantic bargain is set out in: David Gompert and Stephen Larrabee, *America and Europe: a Partnership for a New Era* (Cambridge University Press, 1997). The possibility of this type of new bargain materialising in the near to medium term is assessed in: Frédéric Bozo, op. cit., pp. 45–60.
71. On Europe as *l'Europe-puissance* see: Yves Boyer, 'France and the European project: internal and external issues', In: Kjell A. Eliassen, *Foreign and Security Policy in the European Union* (Sage Press, 1998) p. 94.
72. This long-term ambition for a strategic Europe is not, of course, shared by all in France, nor is it usually articulated in such stark terms. Nevertheless it is consistent with French official statements on the European project, CFSP, ESDI, the future of NATO, European military autonomy and a European nuclear deterrent. See for example: Jacques Chirac, 'Construire une architecture Européenne de sécurité', *Regard Européen* (Summer 1997) pp. 50–5.
73. *The Alliance's Strategic Concept*, NAC-S(99)65, 24 April 1995, in particular articles 13, 14, 17, 18, 30, 42, 52(c) and 61.
74. Martin Holland [ed.], *Common Foreign and Security Policy: the Record and Reforms* (Pinter 1997) and Elfriede Regelsberger *et al.*, *The Foreign Policy of the European Union: From EPC to CFSP and Beyond* (Lynne Reinner, 1997).
75. Cited in Steven Kramer, *Does France Still Count?*, Washington Papers No 164 (Praeger/CSIS, 1994) p. 54.
76. This point is discussed in Geoffrey Howe, 'Bearing more of the burden: in search of a European foreign and security policy', *The World Today* (January 1996) pp. 24–5.
77. Some of the most thoughtful work on EU accountability, defence and security is that by Patricia Chilton. See for example: Patricia Chilton, *The Defence Dimension of the IGC: an Alternative Agenda* (ISIS, March 1996).
78. Jacques Santer, 'The European Union's security and defence policy', *NATO Review* (November 1995) p. 7.
79. See: Catriona Gourlay and Eric Remarcle, 'The 1996 IGC: the actors and their interaction', In: Kjell A. Eliassen, op. cit., pp. 59–93.
80. For a superb study of the implications of the EU for the national defence policies of member states see: Jolyon Howorth and Anand Menon [eds], *The European Union and National Defence Policy* (Routledge, 1997).
81. On Germany's thinking about CFSP and ESDI see: Alfred Frisch, 'L'Allemagne et la défense européenne', *Documents: Revue des Questions Allemandes* (Spring 1996) pp. 30–6 and FED, *Les Politiques de Défense Franco-Allemandes: Etude Comparée* (FED, 1997). For useful background see Henri Ménudier, *Le Couple Franco-Allemand en Europe* (Institut d'Allemand d'Asnières, 1993).

82. For discussions of these obstacles see: Nicole Gnesotto, *L'Union et l'Alliance: les Dilemmas de la Défense Européenne*, Les Notes de l'IFRI (IFRI, July 1996) pp. 27–34; Robin Niblett, 'The European disunion: competing visions of integration', *Washington Quarterly* (Winter 1997) pp. 94–107; and Francis Gutmann, 'Après Madrid, Amsterdam, Luxembourg... La France, L'Europe et l'Otan', *Défense Nationale* (February 1998).
83. 'Entretien avec Hubert Védrine', *Le Monde* (29 August 1997).
84. For a realistic but upbeat assessment of the WEU at the time of the NATO summit see: Jean-Marie Guehenno, 'L'avenir de l'UEO', *Les Cahiers de Chear* (September 1994) pp. 11–24. Guehenno was French ambassador to the WEU at the time.
85. *European Security: a Common Concept of the 27 WEU Countries*, WEU Council of Ministers, Madrid (14 November 1995).
86. Russell Johnson, *WEU: Information Report* (WEU, March 1995) pp. 47–55.
87. Charles Millon, 'Towards a European defence pillar', *Defence Review* (Annual 1996) pp. 18–19.
88. Some of the most insightful work about the British position on the EU, WEU and NATO is that of Alyson Bailes. See for example her 'European defence and security: the role of NATO, WEU and EU', *Security Dialogue* (Spring 1996) pp. 55–64 and 'Europe's defence challenge: reinventing the Atlantic Alliance', *Foreign Affairs* (January/February 1997) pp. 15–20. For useful background see David Chuter, 'The United Kingdom', In: Howorth and Menon, op. cit., pp. 105–20.
89. These points are discussed in Philip Gordon, 'Does the WEU have a role?', *Washington Quarterly* (Winter 1997) pp. 135–7.
90. Peter Norman, 'Paris and Bonn in foreign policy pact', *Financial Times* (28 February 1996).
91. Millon, op. cit., p. 18.
92. Catriona Gourlay and Eric Remarcle, op. cit., p. 90.
93. Tony Blair, speech to the House of Commons, 18 June 1997 quoted in Boyer, op. cit., p. 105.
94. *Discours du Ministre de la Défense, M. Alain Richard, à l'Institut des Hautes Etudes de la Défense Nationale*, DPIC Bulletin (10 February 1998) pp. 19–20.
95. Germany was, for example, with the UK, one of the states which blocked the use of the WEU to provide aid and support stabilisation in 1997 in Operation Alba. On the complexities of the Franco-German relationship on an 'inner core' see: Robin Niblett, 'The European disunion: competing visions of integration', *Washington Quarterly* (Winter 1997) pp. 105–7.
96. Figures taken from *The Military Balance 1998/99* (IISS/Oxford University Press, 1999, and Charles Millon, op. cit., p. 19.
97. Michael O'Hanlon, 'Transforming NATO: the role of European forces', *Survival* (Autumn 1997) pp. 8–10.
98. Eric Revel, 'La future armée européenne compterait 300,000 hommes', *La Tribune* (14 March 1996).
99. Diamantis Vacalopoulos, 'Pour une force armée européenne de métier', *Le Monde* (5 July 1996).
100. Hartmut Bühl, 'Le corps européen', *Les Cahiers de Mars* (Summer 1996) p. 38.
101. Jean-François Durand, 'Les forces multinationales européennes', *Les Cahiers de Mars*, (Summer 1996) pp. 21–3.
102. Ibid., pp. 23–4.
103. Klaus-Uwe Wolff, 'La force navale franco-allemande', *Les Cahiers de Mars*, (Summer 1996) pp. 46–9.

104. John Lichfield and Christopher Bellamy, 'Britain sets sail with an old naval foe', *The Independent* (2 May 1996).

105. Ibid.

106. 'Premier exercise du groupe aérien franco-britannique', *Le Monde* (26 September 1996).

107. Millon, op. cit., p. 19.

108. Jean-Pierre Casamayou, 'Vers l'Europe du transport aérien militaire', *Air et Cosmos*, (10 January 1997) p. 35.

109. Mary Dejevsky, 'Chirac says Europe needs its own spy satellite, free of US', *The Independent* (4 December 1996).

110. Shaun Gregory, *French Military Satellite Systems: Implications for European Security*, ISIS Report (Spring 1996) p. 1.

111. 'Helios, les yeux d'espions de la France', *Libération* (7 July 1995).

112. Jacques Isnard, 'Le satellite Helios-I assure l'autonomie stratégique de la France en Irak', *Le Monde* (19 September 1996).

113. 'Une défense en commun', *Libération* (10 December 1996) and Jacques Isnard, 'Discorde entre Paris et Bonn sur le satellite-espion Helios 2', *Le Monde* (15 October 1996).

114. On the Alert system see: Serge Brosselin, 'Vers une système d'alerte satellitaire européenne', *Défense* (September 1995) pp. 30–2. On France and MEADS/BMD see: WEU, *La Coopération Transatlantique dans la Domaine de la Défense Antimissile Européenne*, WEU Report 1588 (November 1997) and Claude Roche, 'Une pays nucléaire comme la France doit-il se doter d'une défense active contre les missiles balistiques?', *Les Cahiers de la Fondation* (September 1998) pp. 13–18.

115. Pierre Langereux, 'Le centre satellitaire de l'UEO commencera a fonctionner en mars', *Air et Cosmos* (7 February 1993) p. 42.

116. Christian Lardier, 'Première visite au centre de Torrejon', *Air et Cosmos* (31 January 1997) p. 40.

117. Ministère de la Défense, *Projet de Budget de la Défense pour 1999*, (September 1998) p. 29.

118. Marc Prevot, 'L'OCCAR: une approche pragmatique pour améliorer la coopération européenne', *Relations Internationales et Stratégiques* (Autumn 1997) pp. 48–52.

119. See in this respect Nicole Forgeard, 'Industrie de défense français: regroupement et projection européenne', *Les Cahiers du Chear* (March 1996) pp. 9–25 and Christine Holzbauer-Madison *et al.*, *L'Année Européenne 1997* (Association Belles Feuilles, 1997) pp. 48–61.

120. Barbara Starr, 'USA warns of three-tier NATO technology rift', *Jane's Defence Weekly* (1 October 1997), p. 15.

121. On the French debate about the RMA see: David Yost, 'La France, les Etats-Unis et la révolution militaro-technique', *L'Armement* (May/June 1994) pp. 135–41; Alain Joxe, 'Etats des lieux de la RAM [Révolution dans les Affairs Militaires]' *L'Armement* (March 1996) pp. 137–42; François Géré, *Demain la Guerre: une Visite Guidée* (Calmann-Levy, 1997); and Bruno Tertrais, 'Faut-il croire à la révolution dans les affaires militaires?', *Politique Etrangère* (Autumn 1998) pp. 611–30.

122. Marcel Duval, 'La dissuasion française et l'Europe: une approche historique', *Relations Internationales et Stratégiques* (Spring 1996) pp. 86–7.

123. Michael Mazarr, 'Virtual nuclear arsenals', *Survival* (Autumn 1995) p. 7.

124. Frédéric Bozo, 'Le nucléaire entre marginalisation et banalisation', *Politique Etrangère* (Spring 1995) pp. 195–204.

125.   François Heisbourg, *Les Volontaires de l'An 2000* (Balland, 1995) pp. 147–75. In this study Heisbourg was referring to the internal defence debates in France, but the point is also relevant to the relations between France and its European allies.

126.   *Livre Blanc sur la Défense 1994*, op. cit., p. 51.

127.   'Accord franco-américain sur les arsenaux nucléaires', *Le Monde* (18 June 1996) and Nicola Butler, 'Sharing secrets', *Bulletin of the Atomic Scientists* (January/February 1997) pp. 11–12.

128.   Bruno Tertrais, *The French Nuclear Deterrent After the Cold War*, Report No P8012 (Rand, 1998) pp. 32–3.

129.   *Livre Blanc sur la Défense1994*, op. cit., p. 83.

130.   Bruno Tertrais [1998], op. cit., p. 8.

131.   *Une Défense Nouvelle*, op. cit., p. 20.

132.   Pascal Boniface, 'French nuclear strategy and European deterrence: les rendez-vous manqués', *Contemporary Security Policy* (August 1996) p. 230.

133.   Bruno Tertrais [1998], op. cit., p. 8.

134.   British-French Joint Statement on Nuclear Cooperation, UK-French Summit, 29–30 October 1995.

135.   'La dissuasion européenne', *Le Monde* (15 April 1995).

136.   Pascal Boniface [1996], op. cit., p. 235.

137.   Ibid.

138.   Alain Juppé, 'Quel horizon pour la politique étrangère de la France?', *Politique Etrangère* (Spring 1996) p. 147.

139.   Chirac's speech to the IHEDN on 8 June 1996 is reproduced in: Jacques Chirac, 'La politique de défense de la France', *Défense Nationale* (August/September 1996) pp. 7–18.

140.   *Discours du Premier Ministre, M. Lionel Jospin, à l'Institut des Hautes Etudes de Défense Nationale*, DPIC Bulletin (4 September 1997).

141.   David Yost, *The United States and Nuclear Deterrence in Europe*, Adelphi Paper 326 (IISS/Oxford, March 1999) p. 36.

142.   Stuart Croft, 'European integration, nuclear deterrence and Franco-British nuclear cooperation', *International Affairs* (October 1996) pp. 777–80.

143.   See Matthias Küntzel, *Bonn und die Bombe: Deutsche Atomwaffenpolitik von Adenauer bis Brandt* (Campus Verlag, 1992) p. 29. This was later published in English as *Bonn and the Bomb* (Pluto Press, 1995). See also Beatrice Heuser's detailed discussions in: *NATO, Britain, France and the FRG: Nuclear Strategies in Europe 1945–2000* (Macmillan, 1997) and *Nuclear Mentalities?: Strategies and Beliefs in Britain, France and the FRG* (Macmillan, 1998).

144.   Quoted in 'Bonn nimmt Juppé's atom-vorschägesehr zuruckhaltend auf', *Frankfurther Allegemeine Zeitung* (13 September 1995).

145.   This point is made in Pascal Boniface, 'La dissuasion nucléaire dans la relation franco-allemande', *Relations Internationales et Stratégiques* (Summer 1993) p. 23.

146.   'Concept commun franco-allemand en matière de sécurité et de défense', *Le Monde* (30 January 1997). The key parts of the Nuremberg Accord were leaked by *Le Monde* on 25 January 1997. See: 'Défense: l'accord confidentiel Kohl-Chirac', *Le Monde* (25 January 1997).

147.   Arnaud Leparmentier, 'Joshka Fischer plaide pour une "politique intérieure Européenne"', *Le Monde* (28 October 1998).

148.   For arguments against bringing the Franco-British and Franco-German dialogues together see: Frédéric Bozo, 'Dissuasion concertée: le sens de la formule', *Rela-*

tions Internationales et Stratégiques (Spring 1996), p. 99. The whole article repays reading pp. 93–100.

149. On Spain and nuclear weapons issues see: Carlos Miranda, 'La position de l'Espagne sur les questions nucléaires', *Relations Internationales et Stratégiques* (Spring 1996) pp. 121–5.

150. 'M. Chirac annonce la suppression du service militaire dans six ans'', *Le Monde* (24 February 1996).

151. Philippe Séguin, 'A rebuttal: why France's nuclear plan is serious', *International Herald Tribune* (6 September 1995).

152. Bruno Tertrais, *Nuclear Policies in Europe*, Adelphi Paper 327 (IISS/Oxford University Press, March 1999) pp. 59–60.

153. 'Concept commun franco-allemand en matière de sécurité et de défense', op. cit. That France too accepted this wording created a political furore in France because of the primacy given to NATO and the US nuclear guarantee. See: '"Défense concertée" Franco-Allemande: réactions réservées', *Le Monde* (26/7 January 1997); Elisabeth Auvillain, 'Le débat français sur la défense laisse l'Allemagne perplexe', *La Croix* (1 February 1997); and 'Otanising', *The Economist* (8 February 1997).

154. 'M.Mellick recense les différentes formules d'une doctrine nucléaire Européenne', *Le Monde* (4 February 1992).

155. André Dumoulin, 'Vers une dissuasion nucléaire Européenne?', *Le Monde Atlantique* (Spring 1996) p. 33.

156. Bruno Tertrais [1999], op. cit., pp. 66–71.

157. Ibid., p. 69.

158. Jacques Chirac, 'La politique de défense de la France', op. cit., pp. 7–18.

159. David Omand, 'Nuclear deterrence in a changing world: the view from the UK perspective', *RUSI Journal* (June 1996) p. 19.

160. Holger Hey and Andrew Dennison, 'France's nuclear tests and Germany's nuclear interests', Comparative Strategy (Autumn 1996), p. 170. See also: Harald Müller, 'L'Europe, l'Allemagne et le débat sur le désarmement nucléaire', *Relations Internationales et Stratégiques* (Summer 1998) pp. 112–34.

161. Carlos Miranda, op. cit., pp. 123–4. See also: *NATO Facts and Figures*, (Nato Information Service, 1989) pp. 146–7.

162. David Yost [March 1999], op. cit., pp. 38–9.

163. Shaun Gregory, *Nuclear Command and Control in NATO*, (Macmillan Press, 1996) pp. 103–4

164. David Yost [1999], op. cit., pp. 64–7.

5 French Defence Policy in a Global Context

1. *Livre Blanc sur la Défense1994*, (La Documentation Française, March 1994) p. 102.
2. *The Military Balance 1994–95* (IISS/Oxford University Press, October 1994–95) p. 45.
3. *Livre Blanc sur la Défense1994*, op. cit. I have placed the 2510 '*missions de presence*' troops based in the former Yugolsavia on this map under the '*missions de paix*' rubric for narrative convenience.
4. See: Georges Lavroff [ed], *La Politique Africaine du Général de Gaulle* (Pédone, 1990). In English a useful if uncritical analysis of de Gaulle and this transition can be found in Dorothy Shipley White, *Black Africa and de Gaulle* (Pennsylvania State University Press, 1979).

5. 'Resolution on Neo-colonialism', quoted in Colin Legum, *Pan Africanism: a Short Political Guide* (Pall Mall Press, 1962) p. 254.
6. See Dominique Moisi and Pierre Lellouche, 'French policy in Africa: the lonely battle against destabilisation', *International Security* (Spring 1979) pp. 108–33.
7. On contemporary French-African relations see: Roland Louvel, *Quelle Afrique pour Quelle Cooperation?* (L'Harmattan, 1994); Serge Michailof, *La France et l'Afrique* (Karthala, 1993); Pierre Péan, *Affaires Africaines* (Fayard, 1983); and François-Xavier Verschave, *La Francafrique* (Editions Stock, 1998). The definitive study of security relations in English is John Chipman's *French Power in Africa* (Blackwell, 1989).
8. See for example: Guy Martin, 'France and Africa' in: Robert Aldrich and John Connell [eds], *France in World Politics* (Routledge, 1989) pp. 103–5.
9. On Foccart see: Pierre Péan, *L'Homme et l'Ombre: Eléments d'Enquête Autour de Jacques Foccart, l'Homme le Plus Mystérieux et le Plus Puissant de la 5e République* (Fayard, 1990) and Jacques Foccart and Phillipe Gaillard, *Foccart Parle: Entretien Avec Jacques Foccart* (Fayard, Vol I [1995] and Vol II [1997]).
10. An excellent overview of this role is provided by André Dumoulin, *La France Militaire et l'Afrique* (GRIP/Editions Complexe, 1997).
11. Ibid., pp. 123–5. See also: 'Les précédentes interventions militaires françaises', *Le Monde* (24 May 1996) p. 2.
12. Amongst these were the self-styled Emperor Jean-Bedel Bokassa of the Central African Republic and the President of Zaire, Mobutu Sese Seko. See respectively: Charles Onana, *Bokassa: Ascension et Chute d'un Empereur* (Editions Duboirris, 1998); N'Gbanda Atumbu, *Les Derniers Jours du Maréchal Mobutu* (Gideppe, 1998); and Colette Braekman, *Le Dinosaure: Le Zaire de Mobutu* (Fayard, 1992).
13. There are some annual variations for these figures: the *Livre Blanc 1994* has 8600 personnel in 1994; *Figaro* has 8200 for 1996 ['Afrique; la France ne baisse pas la garde' (20 March 1996) p. 2]; and Dumoulin has 8400 for 1997 [op. cit., pp. 113–14].
14. Guy Martin, op. cit., pp. 114–15.
15. Robin Luckham, 'French militarism in Africa', *Review of African Political Economy* (May 1982) p. 56. The whole article repays reading pp. 55–84.
16. For a range of views, the balance of which implicate France in the genocide, see: Medhi Ba, *Un Génocide Français* (L'Esprit Frappeur, 1997); Colette Braeckman, *Rwanda: Histoire d'un Génocide* (Fayard, 1994); Jacques Castonguay, *Les Casques Bleus au Rwanda* (L'Harmattan, 1998); Alain Destexhe, *Rwanda: Essai sur le Génocide* (Editions Complexe, 1994); Jean-Paul Gouteux, *Un Génocide Secrèt d'Etat: La France et le Rwanda 1990–97* (Editions Sociales, 1998); Gérard Prunier's two-volumed study of the Rwandan genocide and its aftermath published in English as: *The Rwanda Crisis, 1959–94: History of a Genocide* (C. Hurst & Co Press, 1995) [particularly good on the French role in the 1994 catastrophe] and *Rwanda in Zaire: From Genocide to Continental War* (C. Hurst & Co Press, 1999); and Paul Quiles [ed], *Enquete sur la Tragédie Rwandaise* [4 Vols], Report No 1271 (Assemblée Nationale, 1998).
17. Theirry Garcin, 'L'intervention française au Rwanda', *Le Trimestre du Monde*, (Summer 1996) pp. 67–9. See also: Louis Balmand, *Les Interventions Militaires Françaises en Afrique* (Pédone, 1998) pp. 99–108.
18. J.A.C. Lewis, 'New mission for France as it re-enters Rwandan conflict', *Jane's Defence Weekly* (9 July 1994) p. 19.
19. 'La France et la tragédie rwandaise au histoire devenu plus lisible', *Marchés Tropicaux* (23 January 1998) pp. 156–8.

20. Thierry Garcin, op. cit., pp. 63–72.
21. See for example: Philippe Richard, 'Afrique: quelle politique de sécurité?', *Damoclès* (Spring 1997) pp. 8–9. For the French role in arming and training the Rwandan government see: Colette Braeckman, *Qui a Armé le Rwanda?: Chronique d'une Tragédie Annoncée* (Les Dossiers du GRIP, 1994).
22. *Strategic Survey 1994–95* (IISS/Oxford University Press, May 1995) p. 209.
23. See for example: François Jolivald, 'Regards sur la politique africaine de François Mitterrand', *Marchés Tropicaux* (4 July 1997) pp. 1474–5; Claude Wauthier, 'La politique africaine de Jacques Chirac', *Relations Internationales et Stratégiques*, (Spring 1997) pp. 121–8; and, François Gaulme, 'La France et l'Afrique – de François Mitterrand à Jacques Chirac', *Marchés Tropicaux* (26 May 1995) pp. 1112–14. On Foccart's reappearance at the Elysée see: 'Why were they there?', *The Economist* (11–17 January 1997), p. 12.
24. Paul-Marie de la Gorce, 'La deuxième mort de Jacques Foccart', *Jeune Afrique*, (13–16 August 1997) pp. 40–2.
25. David Buchan, 'France broadens African policy focus', *Financial Times* (8 April 1997).
26. 'La France sonne la retraite a Bangui', *Libération* (24 July 1997).
27. José Garçon, 'L'onde de choc rwandaise', *Libération* (3 July 1997) p. 5; Pierre Prier, 'La France veut réviser sa politique africaine', *Figaro* (5 July 1997); and Joseph Fitchett, 'New policy for Africa: don't rely on troops from France', *International Herald Tribune* (11 August 1997).
28. Dumoulin, op. cit., pp. 89–110.
29. J.A.C. Lewis, 'France looks for ways to keep its African influence intact', *Jane's Defence Weekly* (23 October 1996) p. 23: Thomas Lippman, 'US revises plans for Africa force', *International Herald Tribune* (10 February 1997) and, 'America loses its Afrophobia', *The Economist* (26 April 1997) pp. 47–8.
30. A Franco-German brigade did however exercise in Gabon in autumn 1997. See: Jean-Dominique Merche, 'Ne plus arbiter entre forces rivales', *Libération* (4 August 1997).
31. Dumoulin, op. cit., pp. 89–110.
32. See for example: Dominique de Combles de Nayves, 'La nouvelle politique militaire française en Afrique', *Défense Nationale* (August/September 1998) pp. 12–16; Albert Bourgi, 'La fin de l'épopée coloniale?', *Jeune Afrique* (13–26 August 1997) p. 3; Mireille Duteil, 'La France va-t-elle perdre l'Afrique?', *Le Point* (July 1997) pp. 38–41; 'Non, la France n'a pas "perdu" l'Afrique', *Marchés Tropicaux* (23 May 1997) p. 1071; and, Hugo Sada, 'Réexamen de la politique militaire française en Afrique', *Défense Nationale* (June 1997) pp. 183–5.
33. Marcel Maymil, 'Avenir de la présence française à Djibouti', *Bulletin d'études de la marine* (July 1996) pp. 43–4. For an excellent political history of Djibouti and a critique of the French presence see: Ali Coubba, *Djibouti: Une Nation en Otage* (L'Harmattan, 1993).
34. 'Les FFDj: forces françaises stationnées à Djibouti', *CID Tribune* (June 1996) pp. 214–15. For example in Operation Godoria in May 1991.
35. 'L'armée française à Djibouti', *Marchés Tropicaux* (15 December 1995) pp. 2772–3.
36. 'Les forces françaises de Djibouti [FFDj]: un emplacement exceptionnel', *Armées d'Aujourd'hui* (April 1996) pp. 24–5.
37. Troops from Djibouti did not participate directly in the Gulf War itself, remaining on stand-by due to their specialised roles, but Djibouti played a major part in the French operations throughout the war and subsequent to it.

38. Quoted in John Connell and Robert Aldrich, 'Remnants of Empire: France's overseas departments and territories', In: John Connell and Robert Aldrich [eds], *France in World Politics* (Routledge, 1989) p. 148.
39. Jean-Marie Lemoine, 'Les forces terrestres dans les DOM-TOMs', *Les Cahiers de Mars* (Summer 1990) p. 23.
40. The best study of these issues in English is: John Connell and Robert Aldrich, *France's Overseas Frontiers: Départements et Territoires d'Outre-Mer* (Cambridge University Press, 1992). See also: Jean-Luc Mathieu, *Histoire des DOM-TOM* (Que Sais-Je?/PUF, 1993).
41. John Connell and Robert Aldrich [1989], op. cit., p. 66.
42. For indigenous voices of opposition to the French possession of the DOM-TOMs see: Elie Castor and Georges Othily, *La Guyane: Les Grandes Problems* (Editions Caribéennes, 1984); Jean-Marie Colombani, *Double Calédonie: d'Une Utopie à l'Autre* (Denoel, 1999); Guy Numa, *Avenir de Antilles-Guyane* (L'Harmattan, 1986); Yves Salesse, *Mayotte: L'illusion de la France* (L'Harmattan, 1995); Jean-Marie Tjibaou, *La Présence Kanak* (Editions Odile Jacobs, 1990); Pieter de Vries and Hon Seur, *Mururoa et Nous: Experiences des Polynésiens au Cours des 30 Années d'Essais Nucléaires dans la Pacifique Sud* (CDRPC, 1997).
43. *La Réunion*, IHEDN Report No 124 (January 1996).
44. *Livre Blanc sur la Défense 1994*, op. cit., p. 103.
45. *Les DOM-TOM: Une Vulnerabilité Pour la France?*, IHEDN Report [unnumbered] (May 1998) p. 12.
46. Jean Fasquel, *Mayotte, La France et les Comores* (L'Harmattan, 1991); Bernard Lugan and François-Xavier Rocchi, 'Comores', *L'Afrique Reelle*, (Summer 1997) pp. 34–52; and Pascal Perri, *Comores: les Nouveaux Mercenaries* (L'Harmattan, 1994).
47. See for example: Jean-Marc Balancie, 'La presence française dans l'Océan Indien', *Stratégique* (Spring 1993) pp. 203–33 and Henri Labrousse, 'La stratégie française dans l'Océan Indien', *Stratégique* (Winter 1992) pp. 227–56.
48. Olivier Gohin, 'La défense de la France Outre-Mer', *Droit et Défense* (January 1994) p. 9. Gohin, based at the Université de Réunion, is one of the most insightful of French writers on the region's security issues.
49. Quoted in ibid.
50. *Les DOM-TOM: Une Vulnerabilité Pour la France?*, op. cit., pp. 14–17.
51. On this history and regional French opposition to France see: John Connell and Robert Aldrich [1989], op. cit., pp. 269–71; Ramesh Thakur, *The Last Bang before a Total Ban: French Nuclear Testing in the Pacific*, Working Paper 159, Peace Research Centre (Australian National University, September 1995) pp. 32–3; and Pieter de Vries and Hon Seur, op. cit.
52. Jean-Yves Faberon, 'L'accord de Nouméa du 21 Avril 1998', *Regards sur l'Actualité* (May 1998) pp. 19–31.
53. For an excellent geopolitical overview of France's role in the South Pacific, see: François Doumenge, *Géopolitique du Pacifique Sud* (CRET, 1990); on the ZNC region see pp. 113–40.
54. See: *Guide des Forces Nucléaires Françaises*, Damoclès (January 1992) pp. 48–9; Robert S. Norris et al., *Nuclear Weapons Databook, Volume V: British, French and Chinese Nuclear Weapons* (Westview Press, 1994) pp. 205–11.
55. On the move to the South Pacific see: Marcel Duval and Yves le Baut, *L'Arme Nucléaire Française: Pourquoi et Comment?* (Kronos, 1992) pp. 205–15. On the total number of tests see Bruno Barrillot, *Les Essais Nucléaires Français 1960–1996*, (Etudes de CDRPC, 1996) pp. 367–74.

56. See for example the French parliament's defence of the French action and analysis of the international reaction in: Aymeri de Montesquiou, *Les Réactions Internationles à la Reprise des Essais Nucléaires: Beaucoup de Bruit Pour Rien?*, Report No 2946, Assemblée Nationale (June 1996).

57. For a thoughtful overview of French policy in the South Pacific see: Nic Maclellan and Jean Chesneaux, *After Moruroa: France in the South Pacific* (Ocean Press, 1998).

58. *Restructurations de la Défense 2000–2002: Dossier de l'Information* (Ministère de la Défense/SIRPA, March 1999). The Dossier is not paginated but the relevant details can be found in a brochure within the Dossier entitled *'Les mesures d'adaptation'* .

59. 'L'après CEP: quel avenir pour la Polynésie Française?', *Bulletin d'Etudes de la Marine* (July 1996) pp. 48–53.

60. Dominique Fremy and Michele Fremy, *Quid 1993* (Editions Robert Laffont, 1993) p. 852.

61. Jean Labrousse, 'Guyane Française et le Centre Spatial Guyanaise', *Défense* (December 1997) pp. 17–19.

62. Ibid., p. 17.

63. *Livre Blanc sur la Défense 1994*, op. cit., p. 103.

64. Jean Labrousse, op. cit., p. 18.

65. Ibid.

66. Richard Dubos, 'Le rôle des forces prépositionnées', *Les Cahiers de Mars* (Summer 1994) p. 45. The whole article is very useful pp. 41–7.

67. *Livre Blanc sur la Défense 1994*, op. cit., p. 103.

68. Dominque Fremy and Michele Fremy, op. cit., pp. 846, 848.

69. Jean-Marie Lemoine, op. cit., p. 23.

70. 'Spécial Caraïbe: Les tensions dans la "Mediterranée américaine"', *Revue Politique et Parliamentaire* (July 1986, pp. 59–60). This special issue contains four valuable, if now dated, articles about the French presence in the Caribbean and regional relations, pp. 10–66.

71. For an updated look at regional issues and the place of the French Antilles see: Yves Salkin, 'Regards sur les petites Antilles', *Défense Nationale* (January 1997) pp. 135–44.

72. Henri Motte, 'Saint-Pierre et Miquelon: l'espoir d'être au le neant', *La Revue Maritime* (June 1995) pp. 33–47.

73. Olivier Bertaux, 'Outre-mer: le service militaire adapté', *Défense Nationale* (August/September 1994) pp. 71–8.

74. *Restructurations de la Défense 2000–2002: Dossier de l'Information* (Ministère de la Défense/SIRPA, March 1999). The Dossier is not paginated but the relevant details can be found in two of the brochures within the Dossier entitled *'Les mesures d'adaptation'* and *'La professionnalisation'*.

75. In 1994 France deployed around 300 000 professional military personnel as part of an armed force of approximately 560 000. At the time 19 390 [and 2150 naval forces] were deployed in the DOM-TOMs of which most of the naval forces and around one third of the standing forces were professional, in all around 8500. Thus in 1994 approximately 3 per cent of all professional forces were deployed in the DOM-TOMs. In 2002 the French armed forces will number 434 000 of whom 326 000 [75 per cent] will be professional. Professional standing forces in the DOM-TOMs may be as high as 17 500 or 5.5 per cent of the total.

76. The best overview of these trends, though now a little out of date, is the IRIS collection of essays: *Les Dom-Toms dans la Politique de la Défense: le Nouveau Monde* (IRIS/Dunod, 1992).

77. 'Forces françaises deployées hors du territoire métropolitain', *Projet de Budget de la Défense pour 1999* (Ministère de la Défense, September 1998) p. 31. The figures do not include deployments in Macedonia and Kosovo since 1998. For 1994–5 figures see *Livre Blanc sur la Défense 1994*, op. cit., p. 102.

78. See for example the florid language of François Trucy in *Rapport au Premier Ministre: Participation de la France aux Opérations de Maintien de la Paix* (February 1994), pp. 69–70.

79. Marie-Claude Smouts, 'Political Aspects of Peacekeeping', In: Brigitte Stern (ed), *United Nations Peacekeeping Operations: A Guide to French Policies*, (United Nations University Press, 1998) p. 8. This was published in French as *La Vision Française des Opérations de Maintien de la Paix* (United Nations University Press, 1997).

80. Three excellent overviews are: Marie-Claude Smouts, *La France et l'ONU* (Presses de la Fondation Nationale des Sciences Politique, 1979); the high-level collection of essays in André Levin, *La France et l'ONU 1945–1995* (Arléa-Corlet, 1995); and Charles Zorgbibe, *La France, l'ONU et le Maintien de la Paix* (PUF, 1996).

81. Philippe Guillot, 'France, peacekeeping and humanitarian intervention', *International Peacekeeping* (Spring 1994) p. 30.

82. Thierry Tardy, 'La France et l'ONU: 50 ans de relations contrastées', *Regards sur l'Actualité* (November 1995) pp. 7–9.

83. On de Gaulle and Hammarskjöld see: Jean Lacouture, *De Gaulle: The Ruler: 1945–1970* (Collins Harvell, 1991) pp. 245–6.

84. Ramesh Thakur, *International Peacekeeping in Lebanon* (Westview Press, 1987).

85. Brigitte Stern, op. cit., p. 129.

86. Georges Abi-Saab, 'Le deuxième génération des opérations de maintien de la paix', *Le Trimestre du Monde* (Winter 1994) pp. 87–97.

87. Thierry Tardy, op. cit., p. 14.

88. Pierre le Peillet, 'Les treize missions militaires de l'ONU', *Défense Nationale* (April 1993) pp. 165–9. See also Brigitte Stern, op. cit., Appendix 1 pp. 126–7.

89. Michel Loridon, 'L'Armée française an Cambodge', *Les Cahiers de Mars* (Autumn 1992) pp. 78–84.

90. See: Maurice Faivre, 'Participation française aux missions de paix de l'ONU', *Défense Nationale* (November 1992), pp. 189–91; 'Cambodge: mission accomplie', *SIRPA Actualité* (November 1993) pp. 23–35; and, Jean Benet, 'Aspects majeurs de l'intervention française au Cambodge', *Objectif 21* (Winter 1997) pp. 19–43

91. Marie-Claude Smouts, In Brigitte Stern, op. cit., p. 20.

92. Philippe Guillot, op. cit., pp. 30–43.

93. Thierry Tardy, op. cit., p. 19.

94. Ibid.

95. See for example: Mario Bettati, *Le Droit d'Ingérence: Mutation de l'Ordre Internationale* (Odile Jacobs, 1996); Pascal Boniface, 'L'intervention militaire entre intérêts, morale, volonté et réticences', *Relations Internationales et Stratégiques* (Winter 1996) pp. 26–36; Marie-Claude Delpal, *Politique Extérieure et Diplomatie Morale – le Droit d'Ingérence Humanitaire en Question*, FEDN Dossier No 50, (FEDN, April 1993); Alain Pellet, *Droit d'Ingérence ou Devoir d'Assistance Humanitaire* (La Documentation Française, 1995); and, Charles Zorgbibe, 'Condamner l'action humanitaire?', *Revue Politique et Parlementaire* (July 1996) pp. 62–6.

96. Michel Klen, 'L'enfer somalien', *Défense Nationale* (February 1993) pp. 135–43; and, Roland Marchal, 'Somalie: autopsie d'une intervention', *Politique Internationale* (Spring 1993) pp. 121–208.

97. For operational details and a wider discussion of the French role see: 'L'intervention française en Somalie', *Objectif 21* (Spring 1995) pp. 12–36; and Gérard Prunier, 'L'inconcevable aveuglement de l'ONU', *Le Monde Diplomatique* (November 1993) p. 7.

98. One of the most remarkable essays anywhere on the role of the UN in Bosnia can be found in Jean Franchet, *Casque Bleu Pour Rien: Ce Que J'ai Vraiment Vu en Bosnie* (J.C. Lattes, 1995).

99. See Xavier Gautier, *Morillon et les Casques Bleus: Mission Impossible?* (Editions 1, 1993) pp. 215–26. Morillon's own views on the early UN presence in Bosnia can be found in Philippe Morillon, *Croire et Oser: Chroniques de Sarajevo* (Grasset, 1993). It is also well worth reading the reflections of another French commander in Bosnia: Jean Cot, *Dernière Guerre Balkanique?: Ex-Yugolslavie: Temoignage, Analyses, Prespectives* (L'Harmattan/FED, 1996).

100. Philippe Morillon, 'The military aspects of field operations', In: Brigitte Stern, op. cit., p. 98.

101. See: Jacques Lanxade, 'L'opération Turquoise', *Défense Nationale* (February 1995) pp. 7–15. In seeking to spread the blame around a few years after the event the French *Assemblée Nationale* report into the 'tragedie rawandaise' was highly critical of the impotence of UNAMIR. See: 'Enquête sur un génocide', *Armées Aujourd'hui* (March 1999), p. 24. For details of the report itself see: *Enquête sur la Tragedie Rwandaise 1990–94*, op. cit.

102. Boutros Boutros-Ghali, *An Agenda for Peace*, United Nations A/46/277–S/24111, (UN, 17 June 1992).

103. François Chauvancy, 'La doctrine militaire française pour les opérations de maintien de la paix', Défense Nationale (June 1995) pp. 144–5.

104. Trucy had in fact made related remarks in a Sénat finance report in November 1992, but the Trucy report represented the working through of these ideas.

105. Ministère de la Défense, *Livre Blanc sur la Défense 1994* (La Documentation Française, March 1994) p. 61.

106. For excellent discussions of the UN-NATO-WEU interface see Patricia Chilton, 'Maintien de la paix: les nouveaux concepts stratégiques et les organisations internationales', *Relations Internationales et Stratégiques* (Spring 1995) pp. 15–28; and, Jean Klein, 'Interface between NATO/WEU and UN/OSCE', In: Michael Brenner [ed.], *NATO and Collective Security* (Macmillan, 1998) pp. 249–77.

107. Jean-Bernard Raimont, *La Politique d'Intervention dans les Conflits: Eléments de Doctrine pour la France*, Report No 1950 (*Assemblée Nationale*, 23 February 1995).

108. Some of these issues are explored in: Yves-Marie Laulan, 'L'armée française entre l'humanitaire et l'opérationnel', *Défense Nationale* (August 1996) pp. 75–81.

109. See: Jacques Lanxade, 'Orientations par la conception, la préparation, la planification, le commandement et l'emploi des forces françaises dans l'opérations militaires fondées sur un résolution du Conseil de Sécurité des Nations Unis', *Objectif 21* (Spring 1995) pp. 4–9. See also: Charles Zorgbibe, 'La France et le maintien de la paix: propositions', *Les Cahiers du Cedsi* (October 1998) p. 59.

110. For overviews of the French emphasis on the projection of conventional force see: Pierre Pascallon, 'L'evolution de nos forces classiques d'intervention', *Revue Politique et Parlementaire* (September/October 1993) pp. 47–54; *L'Ation par la Projection: un Nouvea Concept au Coeur de Notre Politique de Défense*, IHEDN Report No 48–1 (IHEDN, 1995); Raymond Bassac *et al.*, 'Les interventions extérieures: pourquoi? comment?', *Défense* (September 1996) pp. 31–66; and

Pierre Pascallon [ed], *Les Interventions Extérieures de l'Armée Française* (Bruylant Bruxelles, 1997).

111. 'La prévention, une nouvelle priorité', *Armées d'Aujourd'hui* (March 1996) p. 15.
112. 'La programmation militaire 1997–2002', *La Journel Officiel de la République Française* (3 July 1996) p. 9991.
113. 'L'adaptation du déploiement outre mer', *CID Tribune* (June 1997) pp. 106–9.
114. See the dossier entitled 'Les forces préposionnées de l'armée de terre', *Armées d'Aujourd'hui* (December 1993/January 1994) pp. 36–65.
115. *Defense and the Armed Forces of France* [English language version], (SIRPA, June 1998) p. 21 and 'La programmation militaire 1997–2002', op. cit., pp. 9988–9.
116. *Défense: Forces Terrestres*, Loi de Finances pour 1997, Assemblée Nationale Report No 2993 (10 October 1996) p. 19.
117. 'Interview with Jean-Philippe Douin', *Jane's Defence Weekly* (19 June 1996) p. 112.
118. *Défense: Credits d'Equipement*, Loi de Finances pour 1999, Assemblée Nationale Report No 1078 (8 October 1998).
119. Philippe Morillon, 'Military aspects of field operations', In Brigitte Stern, op. cit., p. 99.
120. Jean-Michel Boucheron, *Paix et Défense* (Dunod, 1992) p̄. 367.
121. *The Military Balance 1998–99* (IISS/Oxford University Press, 1999) p. 51.
122. Jacques Isnard quotes senior military personnel to this effect in: 'L'armée de terre réduit sa force de "projection" extérieure', *Le Monde* (2 October 1996).
123. *Livre Blanc sur la Défense 1994*, op. cit., p. 82.
124. Ibid., p. 93.
125. Ibid., p. 91.
126. Ibid.
127. It should be recalled that Mitterrand expressly ruled out the use of French nuclear weapons in relation to non-nuclear threats during the Gulf War and that he was consequently critiqued for undermining deterrence. Given however that there is no constraint on the President with respect to what constitutes 'vital interests' this decision could have been reversed.
128. *Livre Blanc sur la Défense 1994*, op. cit., p. 83.
129. Bruno Tertrais, *The French Nuclear Deterrent After the Cold War*, Report No P–8012 (RAND,1998) p. 26.
130. Ibid., p. 40, note 3.
131. France signed the Pelindaba protocols on 11 April 1996. For a useful discussion see: Camille Grand, *A French Nuclear Exception?*, Stimpson Paper No 38 (January 1998) pp. 26–30.
132. Tertrais, op. cit., pp. 40–1.
133. For a less than flattering overview of *Rafale* as a legacy of France's 'independence' in arms procurement and vis-à-vis US and European rivals see: 'Une folie nommée Rafale', *L'Expansion* (30 October–9 November 1995) pp. 58–64.
134. *Une Défense Nouvelle*, SIRPA/Ministère de la Défense (February 1996) p. 20.
135. On the UK's substrategic use of Trident, see the remarkably open article by British civil servant David Omand, ' Nuclear deterrence in a changing world: the view from a UK perspective', *RUSI Journal* (June 1996) pp. 15–22.
136. Jean-Michel Boucheron, *Paix et Défense* (IRIS/Dunod, 1992) p. 556. See also: Daniel Reydellet, 'Histoire des missiles nucléaires françaises', *L'Armement* (December 1996/January 1997) pp. 132–5.
137. Robert Norris *et al.*, *Nuclear Weapons Databook, Volume V: British, French and Chinese Nuclear Weapons* (Westview Press, 1994) p. 288.

138. Jacques Baumel, *Dissuasion Nucléaire*, Tome IV, Commission de la Défense Nationale et des Forces Armées, Projet de Loi de Finances 1997, No 3033 (10 October 1996) p. 26.
139. Marcel Duval, 'Perspectives d'avenir de la dissuasion française', *Défense Nationale* (December 1996) p. 21.
140. Robert Norris *et al.*, op. cit., pp. 288.
141. See for example: Giovanni de Briganti, 'France continues to pare down nuclear forces', *Defense News* (14–20 October 1996) p. 40.
142. *Quel Avenir Pour la Dissuasion Nucléaire Française?*, IHEDN, 123 Session Regional [Grenoble] (September–November 1995) pp. 21–2.
143. The best French study of non-proliferation is Thérèse Delpech, *L'Héritage Nucléaire* (Editions Complexe, 1997).
144. Quoted in Camille Grand, op. cit., p. 29.
145. On this point see Olivier Debouzy, 'L'avenir de la dissuasion nucléaire française', *Commentaire* (Winter 1994/5) pp. 871–2.
146. See for example David Yost, 'Nuclear debates in France', *Survival* (Winter 1994/5) pp. 113–16 and David Yost, 'France's nuclear dilemmas', *Foreign Affairs*, (January/February 1996) pp. 114–17.
147. The debate is ongoing, see for example: Marcel Duval, op. cit., pp. 22–3.
148. Pascal Boniface, 'Dénucléarisation rampante ou retour à la suffisance nucléaire?', In: Pierre Pascallon [ed], *Quelle Défense Pour La France?* (IRIS/DUNOD, 1993) p. 189.
149. Olivier Debouzy, 'L'avenir de la dissuasion nucléaire française', *Commentaire* (Winter 1994/5) p. 872.
150. Robert Joseph, 'Proliferation, counter-proliferation and NATO', *Survival*, (Spring 1996) p. 117.

Bibliography

Official documents

The Alliance's Strategic Concept, NAC–S(99)65, (Washington, 24 April 1995).

Jacques Baumel, *Dissuasion Nucléaire*, Tome IV, Commission de la Défense Nationale et des Forces Armées, Projet de Loi de Finances 1997, No 3033 (Paris, 10 October 1996).

Boutros Boutros-Ghali, *An Agenda for Peace*, United Nations A/46/277–5/24111, (New York: United Nations, 17 June 1992).

British-French Joint Statement on Nuclear Cooperation, UK-French Summit, 29–30 October 1995.

Defense and the Armed Forces of France [English language version], (Paris: SIRPA, June 1998).

Défense: Credits d'Equipement, Loi de Finances pour 1999, Assemblée Nationale Report No 1078 (Paris: 8 October 1998).

Défense: Forces Terrestres, Loi de Finances pour 1997, Assemblée Nationale Report No 2993 (Paris: 10 October 1996).

Discours du Ministre de la Défense, M. Alain Richard, à l'Institut des Hautes Etudes de la Défense Nationale, DPIC Bulletin (Paris: 10 February 1998).

Discours du Premier Ministre, M. Lionel Jospin, à l'Institut des Hautes Etudes de Défense Nationale, DPIC Bulletin, (Paris: 4 September 1997) pp. 2–11.

Discours du Premier Ministre, M. Lionel Jospin, à l'Institut des Hautes Etudes de Défense Nationale, DPIC Bulletin (Paris: 3 September 1998) pp. 3–15.

Une Défense Nouvelle (Paris: Minstère de la Défense, February 1996).

Discours Prononce par M. François Mitterrand, President de la République Française, devant le Bundestag à l'occasion du 20ème Anniversaire du Traite de Coopération Franco-Allemand, Service de Presse, Palais de l'Elysée (Paris: 10 January 1983).

European Security: a Common Concept of the 27 WEU Countries, WEU Council of Ministers, Madrid (Paris: 14 November 1995).

Intervention de M. François Mitterrand, President de la République, sur la Theme de la Dissuasion, Service de Presse, Palais de l'Elysée (Paris: 5 May 1994).

Livre Blanc sur la Défense 1972 (Paris: Ministère de la Défense Nationale, 1972).

Livre Blanc sur la Défense 1994 (Paris: La Documentation Française, 1994).

Projet de Budget de la Défense pour 1999 (Paris: Ministère de la Défense, September 1998).

Projet de Loi Relatif à la Programmation Militaire pour l'Années 1990–1993 (Paris: Ministère de la Défense, May 1989).

Projet de Loi Relatif à la Programmation Militaire pour les Années 1997–2002 (Paris: Ministère de la Défense, May 1996).

Jean-Bernard Raimont, *La Politique d'Intervention dans les Conflits: Eléments de Doctrine pour la France*, Report No 1950, Assemblée Nationale, (Paris: 23 February 1995).

Report of the Secretary-General on the Status of the Implementation of the Special Commission's Plan for the Ongoing Monitoring and Verification of Iraq's Compliance with the Relevant Parts of Section C of Security Council Resolution 687 [1991], UN Security Council Report S/1995/864 (New York: 11 October 1995).

Restructurations de la Défense 2000–2002: Dossier de l'Information (Paris: Ministère de la Défense/SIRPA, March 1999).

SGDN, *L'Organisation Générale de la Défense* (Paris: La Documentation Française, 1990).
François Trucy, *Rapport au Premier Ministre: Participation de la France aux Opérations de Maintien de la Paix* (Paris: February 1994).

Books/monographs

L'Action par la Projection: un Nouvea Concept au Coeur de Notre Politique de Défense, IHEDN Report No 48–1 (Paris: IHEDN, 1995).

Jean d'Albion, *Une France Sans Défense* (Paris: Calmann-Levy, 1991).

Robert Aldrich and John Connell [eds], *France in World Politics* (London: Routledge, 1989).

Josette Alia and Christian Clerc, *La Guerre de Mitterrand: la Dernière Illusion* (Paris: Olivier Orban, 1991).

Mark Almond, *Europe's Backyard War* (London: Heinemann, 1994).

Robert Anderson and Arthur Hartman [eds], *France and the United States: Charting a New Course* (Washington: The Atlantic Council of the United States, February 1995).

N'Gbanda Atumbu, *Les Derniers Jours du Maréchal Mobutu* (Paris: Gideppe, 1998).

Medhi Ba, *Un Génocide Français* (Paris: L'Esprit Frappeur, 1997).

Desmond Ball and Jeffrey Richelson, *Strategic Nuclear Targeting* (Syracuse: Cornell University Press, 1986).

Desmond Ball, *The Intelligence War in the Gulf*, Canberra Papers No 78 (Canberra: Australian National University, 1991).

Louis Balmand, *Les Interventions Militaires Françaises en Afrique* (Paris: Pédone, 1998).

Bruno Barrillot, *Les Essais Nucléaires Français 1960–1996*, (Lyon: Etudes de CDRPC, 1996).

Jacques Baumel and Jean-Paul Pigasse, *La France et sa Défense* (Paris: Editions de Forgues, 1994).

Christopher Bennett, *Yugoslavia's Bloody Collapse: Causes, Courses and Consequences* (London: Hurst 1995).

Erwan Bergot, L'Operation Daguet: *Les Français dans la Guerre du Golfe* (Paris: Presses de la Cité, 1991).

Mario Bettati, *Le Droit d'Ingérence: Mutation de l'Ordre Internationale* (Paris: Odile Jacobs, 1996).

Cyril Black *et al.*, *Rebirth: A History of Europe Since World War II* (Boulder: Westview Press, 1992).

Bernard Boene and Michel Martin, *Conscription et Armée de Metier* (Paris, FEDN/Documentation Français, 1991).

Pascal Boniface, *L'Armée: Enquête sur 300,000 Soldats Méconnus* (Paris: Editions 1, 1990).

Pascal Boniface, *Vive la Bombe*, (Paris: Editions 1, 1992).

Pascal Boniface and Jean Golliet, *Les Nouvelles Pathologies des Etats dans les Relations Internationales* (Paris, Dunod, 1993).

Pascal Boniface, *Contre le Revisionisme Nucléaire* (Paris: Ellipses/IRIS, 1994).

Pascal Boniface [ed], *L'Année Stratégique 1995: Les Equilibres Militaires* (Paris: Dunod/IRIS 1995).

Yves Bonnet, *Mission ou Démission: le Prix de la Défense* (Paris: Jean Picollec, 1996).

Jean-Michel Boucheron, *Paix et Défense* (Paris: Dunod, 1992).

Frédéric Bozo, *La France et l'OTAN: De la Guerre Froide au Nouvel Ordre Européen* (Paris: IFRI/Masson, 1991).

Frédéric Bozo, *La Politique Etrangère de la France Depuis 1945* (Paris: Editions La Découverte, 1997).

Colette Braeckman, *Le Dinosaure: Le Zaire de Mobutu* (Paris: Fayard, 1992).

236   *Bibliography*

Colette Braeckman, *Qui a Armé le Rwanda?: Chronique d'une Tragédie Annoncée* (Brussels: Les Dossiers du GRIP, 1994).
Colette Braeckman, *Rwanda: Histoire d'un Génocide* (Paris: Fayard, 1994).
Michael Brenner [ed], *NATO and Collective Security* (London: Macmillan, 1998).
Leeanne Broadhead [ed], *Issues in Peace Research 1995–1996* (Bradford: University of Bradford Press, 1996).
Patrice Buffotot, *Le Parti Socialist et la Défense* (Paris: Institute de Politique Internationale et Européenne, 1982).
Edwina Campbell, *France's Defence Reforms: The Challenge of Empiricism*, London Defence Studies Papers No 36 (London: CDS/Brassey's, 1997).
Jacques Castonguay, *Les Casques Bleus au Rwanda* (Paris: L'Harmattan, 1998).
Jean-Jacques Cécile, *Le Renseignement Français à l'Aube du XXIe Siecle* (Paris: Lavauzelle, 1998).
Tony Chafer and Brian Jenkins [eds], *France: From the Cold War to the New World Order* (London: Macmillan, 1996).
Jean-Pierre Chevènement, *Une Certain Idée de la République m'Amène à ...* (Paris: Albin Michel, 1992).
Patricia Chilton, *The Defence Dimension of the IGC: an Alternative Agenda* (London: ISIS, March 1996).
John Chipman, *French Power in Africa* (London: Blackwell Press, 1989).
Anthony Clayton, *The Wars of French Decolonization* (London: Longman, 1994).
Samy Cohen and Marie-Claude Smouts [eds], *La Politique Extérieure de Valéry Giscard d'Estaing* (Paris: Presse de la Fondation Nationale des Sciences Politiques, 1985).
Samy Cohen, *La Monarchie Nucléaire* (Paris: Fayard, 1986).
Samy Cohen, *La Défaite des Généraux* (Paris: Fayard, 1994).
Jean-Marie Colombani, *Double Calédonie: d'Une Utopie à l'Autre* (Paris: Denoel, 1999).
Comite 5, *L'Action par la Projection: un Nouveau Concept au Coeur de Notre Politique de Défense*, Report No 48–1, (Paris: IHEDN, 1995).
John Connell and Robert Aldrich, *France's Overseas Frontiers: Départements et Territoires d'Outre-Mer* (Cambridge: Cambridge University Press, 1992).
La Constitution Française [Français-Anglais] (Paris: Assemblée Nationale, 1997).
Jean Cot, *Dernière Guerre Balkanique?: Ex-Yugolslavie: Temoignage, Analyses, Prespectives* (Paris: L'Harmattan/FED, 1996).
Ali Coubba, *Djibouti: Une Nation en Otage* (Paris: L'Harmattan, 1993).
Pierre Dabezies and Jean Klein, *La Reforme de la Politique Française de Défense* (Paris: Economica, 1998).
Alex Danchev and Thomas Halverston [eds], *International Perspectives on the Yugoslav Conflict* (London: Macmillan/St Martin's Press 1996).
Dominique David, *La Politique de Défense de la France: Textes et Documents* (Paris: FEDN, 1989).
Dominique David, *Conflits, Puissances et Stratégies en Europe: le Dégel d'un Continent* (Brussels: Bruylant, 1992).
Marie-Claude Delpal, *Politique Extérieure et Diplomatie Morale – le Droit d'Ingérence Humanitaire en Question*, FEDN Dossier No 50, (Paris: FEDN, April 1993).
Thérèse Delpech, *L'Héritage Nucléaire* (Brussels: Editions Complexe, 1997).
Alain Destexhe, *Rwanda: Essai sur le Génocide* (Brussels: Editions Complexe, 1994).
Les Dom-Toms dans la Politique de la Défense: le Nouveau Monde (Paris: IRIS/Dunod, 1992).
Les DOM-TOM: Une Vulnerabilité Pour la France?, IHEDN Report [unnumbered] (Paris: May 1998).

Jean Doise and Maurice Vaisse, *Diplomatie et Outil Militaire 1871–1991* (Paris: Editions du Seuil, 1992).

François Doumenge, *Géopolitique du Pacifique Sud* (Paris: CRET, 1990).

Simon Doux, *L'Opposition Française à la Guerre du Golfe 1990–91* (Paris: Institut d'Etude Politique, 1992).

André Dumoulin, *La France Militaire et l'Afrique* (Brussels: GRIP/Editions Complexe, 1997).

Pierre Dussauge and Christophe Cornu, *L'Industrie Française de l'Armement: Coopérations, Restructurations et Intégration Européenne* (Paris: Economica, 1998).

Marcel Duval and Yves le Baut, *L'Arme Nucléaire Française: Pouquoi et Comment?* (Paris: Kronos, 1992).

Marcel Duval and Dominque Mongin, *Histoires des Forces Nucléaires Française depuis 1945* (Paris: Que sais-je?/PUF, 1993).

Kjell A. Eliassen [ed], *Foreign and Security Policy in the European Union* (London: Sage Press, 1998).

Jean Fasquel, *Mayotte, La France et les Comores* (Paris: L'Harmattan, 1991).

Jonathan Fenby, *On the Brink: the Trouble with France* (London: Little, Brown and Company, 1998).

Gregory Flynn, *French NATO Policy: The Next Five Years*, Report No N–2995–AF (Washington: RAND, June 1990).

Jacques Foccart and Phillipe Gaillard, *Foccart Parle: Entretien Avec Jacques Foccart* (Paris: Fayard, Vol I [1995] and Vol II [1997]).

La France dans l'Océan Indien, Report No 22 (Paris: IHEDN, 1991).

Jean Franchet, *Casque Bleu Pour Rien: Ce Que J'ai Vraiment Vu en Bosnie* (Paris: J.C. Lattes, 1995).

Lawrence Freedman and Efraim Karsh, *The Gulf Conflict 1990–91* (London: Faber and Faber, 1993).

Lawrence Freedman, *The Revolution in Strategic Affairs*, Adelphi Paper 318 (Oxford: IISS/Oxford University Press, 1998).

Julius Friend, *The Lynchpin: Franco-German Relations 1950–1990*, Washington Papers No 154 (Washington: CSIS/Praeger Press, 1991).

Jean Gallois [pseudonym] *Notre Défense en Mal d'un Politique* (Paris: Economica, 1988).

Pierre M. Gallois, *Stratégie de l'Age Nucléaire* (Paris: Calmann-Lévy, 1960).

Pierre M. Gallois, *Paradoxes de la Paix* (Paris: Presses de la Cité, 1967).

René Galy-Dejean *et al*, *La Simulation des Essais Nucléaires*, Rapport d'Information 847, Commission de la Défense, Assemblée Nationale (Paris: 15 December 1993).

Charles de Gaulle, *Vers l'Armée de Metier* (Paris, Librarie Berger-Levrault, 1934).

Charles de Gaulle, *La France et son Armée* (Paris: Librairie Plon, 1938).

Xavier Gautier, *Morillon et les Casques Bleus: Mission Impossible?* (Paris: Editions 1, 1993).

Paul R.S. Gebhard, *The United States and European Security*, Adelphi Paper 286, (London: IISS/Brassey's, 1994).

François Gere, *Demain la Guerre: une Visite Guidée* (Paris: Calmann-Levy, 1997).

La Gestion Budgétaire et la Programmation Militaire au Ministère de la Défense, Cour des Comptes, Rapport Public Particulier (Paris: June 1997).

Franz-Olivier Giesbert, *François Mitterrand: Une Vie* (Paris: Seuil, 1996).

Robert Gildea, *France Since 1945* (Oxford: Oxford University Press, 1996).

Valerie Giscard d'Estaing, *Le Pouvoir et la Vie* (Paris: Compagnie, 1988).

Nicole Gnesotto and John Roper, *Western Europe and the Gulf*, (Paris: ISS/WEU, 1992).

Nicole Gnesotto, *L'Union et l'Alliance: les Dilemmas de la Défense Européenne*, Les Notes de l'IFRI (Paris: IFRI, July 1996).

David Gompert and Stephen Larrabee, *America and Europe: a Partnership for a New Era* (Cambridge: Cambridge University Press, 1997).

Philip Gordon, *A Certain Idea of France: French Security Policy and the Gaullist Legacy* (Princeton: Princeton University Press, 1993).

Philip Gordon, *France, Germany and the Western Alliance* (Boulder: Westview Press, 1995).

Philip Gordon, *US and ESDI in the New Europe*, Les Notes de l'IFRI, No 4 (Paris: IFRI, 1998).

Jean-Paul Gouteux, *Un Génocide Secrèt d'Etat: La France et le Rwanda 1990–97* (Paris: Editions Sociales, 1998).

Camille Grand, *A French Nuclear Exception?*, Stimpson Paper No 38 (Washington: January 1998).

Robert Grant [ed], *Changing French-American Security Relations*, US-CREST (Arlington: December 1993).

Robert Grant, *Counterproliferation and International Security: The Report of the French-American Working Group*, US-CREST (Arlington: June 1995).

Paul Greenish, *Difficult Times in France: A Study of French Defence Industrial Policy, Practice and Prospects*, Seaford House Papers (London: Royal College of Defence Studies, 1997).

Shaun Gregory, *Command, Control, Communications and Intelligence in the Gulf War*, Canberra Working Papers No 238 (Canberra: Australian National University, 1991).

Shaun Gregory, *Nuclear Command and Control in NATO*, (London: Macmillan/St Martin's Press, 1996).

Shaun Gregory, *French Military Satellite Systems: Implications for European Security* (London: ISIS, Spring 1996).

Alfred Grosser, *Affaires Exterieures* (Paris: Champs/Flammarion, 1984).

Guide des Forces Nucléaires Françaises (Lyon: Damoclès Special Issue, January 1992).

Lothar Gutjahr, *German Foreign and Defence Policy after Unification* (Pinter, 1994).

Peter Hall *et al* [eds], *Developments in French Politics* (London: Macmillan Press, 1990).

Michael Harrison, *The Reluctant Ally: France and Atlantic Security* (London: Johns Hopkins University, 1981).

François Heisbourg, *Les Volontaires de l'An 2000: Pour une Nouvelle Politique de Défense* (Paris: Balland, 1995).

Bertel Heurlin [ed], *Germany in Europe in the Nineties* (London: Macmillan, 1996).

Beatrice Heuser, *NATO, Britain, France and the FRG: Nuclear Strategies in Europe 1945–2000* (London: Macmillan, 1997).

Beatrice Heuser, *Nuclear Mentalities?: Strategies and Beliefs in Britain, France and the FRG* (London: Macmillan, 1998).

Stanley Hoffman [ed], *The Mitterrand Experiment* (Cambridge: Polity Press, 1987).

Martin Holland [ed], *Common Foreign and Security Policy: The Record and Reforms* (London: Pinter, 1997).

James Hollifield and George Ross [eds], *Searching for the New France* (London: Routledge, 1991).

Christine Holzbauer-Madison *et al*, *L'Année Européenne 1997* (Paris: Association Belles Feuilles, 1997).

Alistair Horne, *The French Army and Politics 1870–1970* (London: Macmillan Press, 1984).

Alistair Horne, *A Savage War of Peace* (London: Penguin/Elisabeth Sifton, 1987).

Jolyon Howorth and Patricia Chilton [eds], *Defence and Dissent in Contemporary France* (Berkenham: Croom Helm, 1984).

Jolyon Howorth and Anand Menon [eds], *The European Union and National Defence Policy* (London: Routledge, 1997).

IRIS, *La Défense de la France dans les Années 90* (Paris: La Documentation Française, 1990).

Russell Johnson, *WEU: Information Report* (Paris: WEU, March 1995).

Russell Johnston, *Resumption of French Nuclear Testing in the Pacific*, WEU Report 1488, WEU (Paris: 7 November 1995).

Steven Philip Kramer, *Does France Still Count?: The French Role in the New Europe*, Washington Papers No 164 (Washington, CSIS/Praeger, 1994).

Pascal Krop, *Les Secrèts de l'Espionage Français* (Paris: JCL, 1996).

Matthias Küntzel, *Bonn und die Bombe: Deutsche Atomwaffenpolitik von Adenauer bis Brandt* (Bonn: Campus Verlag, 1992).

Marie-Hélène Labbé, *La Tentation Nucléaire* (Paris: Payot, 1995).

Pierre Lacoste and Alain-Gilles Minella, *Amiral Lacoste: Un Amiral au Secrèt* (Paris: Flammarian, 1997).

Pierre Lacoste (ed), *Les Renseignement à la Française* (Paris: Economica, 1998).

Jean Lacouture, *De Gaulle: The Ruler 1945–70* (London: Harvill, 1991).

Jean Lacouture, *Mitterrand: Une Histoire de Français* [2 Volumes] (Paris: Seuil, 1998).

Jean-Jacques Langendorf, *Le Bouclier et la Tempête: Aspects Militaire de la Guerre du Golfe* (Paris: Georg, 1995).

Georges Lavroff [ed], *La Politique Africaine du Général de Gaulle* (Paris: Pédone, 1980).

Colin Legum, *Pan Africanism: a Short Political Guide* (London: Pall Mall Press, 1962).

Pierre Lellouche, *Le Nouveau Monde: De l'Ordre de Yalta au Désordre de Nations* (Paris, Editions Grasset, 1992).

Pierre Lellouche, *L'Europe et sa Sécurité*, Rapport d'Information 1294, Assemblée Nationale (Paris: May 1994).

Pierre Lellouche, *Légitime Défense* (Paris: Patrick Banon, 1996).

André Levin [ed], *La France et l'ONU 1945–1995* (Paris: Arléa-Corlet, 1995).

Martene Lignières-Cassou, *Rèussir la Diversification des Industries de Défense*, Report No 911, Assemblée Nationale, (Paris: May 1998).

Roland Louvel, *Quelle Afrique pour Quelle Cooperation?* (Paris: L'Harmattan, 1994).

Xavier Luccioni, *L'Affaire Greenpeace* (Paris: Payot, 1986).

Nic Maclellan and Jean Chesneaux, *After Muroroa: France in the South Pacific* (Sydney: Ocean Press, 1998).

Roy Macridis [ed], *Foreign Policy in World Politics* (Prentice-Hall, 1992).

Jean-Luc Marret, *La France et le Désarmement* (Paris: L'Harmattan, 1997).

André Martel, *Histoire Militaire de la France* [Volume 4] (Paris: PUF, 1994).

Nicholas Martin and Marc Créspin, *L'Armée Parle* (Paris: Fayard, 1983).

Paul Martin-Lalande, *Restructurer l'Industrie de Défense: Un Defi Economique et Social*, Report No 2823, Assemblée Nationale, (Paris: May 1996).

Gilles Martinet, *Le Système Pompidou* (Paris: Editions du Seuil, 1973).

Jean-Luc Mathieu, *Histoire des DOM-TOM* (Paris: Que Sais-Je?/PUF, 1993).

Jean-Luc Mathieu, *La Défense Nationale* (Paris: Que Sais-Je?/PUF, 1996).

Jean-Pierre Maury, *La Construction Européenne, la Sécurité et la Défense* (Paris: PUF, 1996).

Jeffrey McCausland, *The Gulf Conflict: A Military Analysis*, Adephi Paper 282 (London, IISS/Brassey's,1994).

Henri Ménudier, *Le Couple Franco-Allemand en Europe* (Asnières: Institut d'Allemand d'Asnières, 1993).

Serge Michailof, *La France et l'Afrique* (Paris: Karthala, 1993).

The Military Balance 1991–1992 (London: IISS/Brassey's 1992).

The Military Balance 1993–1994 (London: IISS/Brasseys, 1993).

The Military Balance 1994–1995 (Oxford: IISS/Oxford University Press, 1994).

The Military Balance 1996–1997 (Oxford: IISS/Oxford University Press, 1996).

The Military Balance 1998–1999 (Oxford: IISS/Oxford University Press, 1999).

Jules Moch, *La Folie des Hommes* (Paris: Laffont, 1954).

Jules Moch, *En Retard d'une Paix* (Paris: Laffont, 1958).

Jules Moch, *Non à la Force de Frappe* (Paris: Laffont, 1963).

Geneviève Moll, *François Mitterrand: Le Roman de Sa Vie* (Paris: Sand, 1995).

Pierre Montagnon, *Histoire de l'Armée Française* (Paris: Pygmalion/Gerard Watelet, 1997).

Aymeri de Montesquiou, *Les Réactions Internationales à la Reprise des Essais Nucléaires: Beaucoup de Bruit pour Rien?*, Report No 2946, Assemblée Nationale (Paris: 28 June 1996).

Philippe Morillon, *Croire et Oser: Chroniques de Sarajevo* (Paris: Grasset, 1993).

Harald Muller, *Falling into line?: France and the NPT*, PPNN Occasional Papers No 6, (Southampton: University of Southampton, May 1990).

NATO Information Service, *NATO Facts and Figures*, (Brussels: NATO, 1989).

NATO Information and Press Office, *NATO Handbook* (Brussels: NATO 1995) pp. 269–77.

Robert S. Norris *et al*, *Nuclear Weapons Databook, Volume V: British, French and Chinese Nuclear Weapons* (Boulder: Westview Press, 1994).

Guy Numa, *Avenir de Antilles-Guyane* (Paris: L'Harmattan, 1986).

Charles Onana, *Bokassa: Ascension et Chute d'un Empereur* (Paris: Editions Duboirris, 1998).

Arthur Paecht, *Une Nouvelle Donne Pour l'Espace Militaire*, Rapport 1892, Assemblée Nationale (Paris: January 1995).

Diego Ruiz Palmer, *French Strategic Options in the 1990s*, Adelphi Paper 260 (London: IISS, Summer 1991).

Diego Ruiz Palmer, *De Metz à Creil: les Structures de Commandement Françaises de l'Après-Guerre Froide* (Paris: CREST, October 1995).

Henri Paris, *L'Arbalète, la Pierre au Fusil et l'Atome: La France Va-t-elle être Encore en Retard d'une Guerre?* (Paris: Albin Michel, 1997).

Pierre Pascallon [ed], *Quelle Défense Pour La France?* (Paris: IRIS/DUNOD, 1993) p. 189.

Pierre Pascallon [ed], *Les Interventions Extérieures de l'Armée Française* (Brusseles: Bruylant, 1997).

Roland Passevant, *Le Golfe: Tempête Pour la Paix* (Paris: Messidor, 1991).

Pierre Péan, *Affaires Africaines* (Paris: Fayard, 1983).

Pierre Péan, *L'Homme et l'Ombre: Eléments d'Enquête Autour de Jacques Foccart, l'Homme le Plus Mystérieux et le Plus Puissant de la 5e République* (Paris: Fayard, 1990).

Pierre Péan, *Une Jeunesse Française: François Mitterrand 1934–47* (Paris: Fayard, 1994).

Alain Pellet, *Droit d'Ingérence ou Devoir d'Assistance Humanitaire* (Paris: La Documentation Française, 1995).

Pascal Perri, *Comores: les Nouveaux Mercenaries* (Paris: L'Harmattan, 1994).

Robert Pfaltzgraff and Uri Ra'anan [eds], *National Security Policy: The Decision Making Process* (Handon: Archon Books, 1984).

François Platone, *Les Partis Politique en France*, (Paris: Editions Milan, 1995).

Lucien Poirier, *Des Stratégies Nucléaires* (Paris: Hachette, 1977).

Lucien Poirier, *Essais de Stratégie Théoretique* (Paris: Fondation pour les Etudes de Défense Nationale, 1982).

Les Politiques de Défense Franco-Allemandes: Etude Comparée (Paris: FED, 1997).

Douglas Porch, *Histoire des Services Secrèts Françaises* [2 vols] (Paris: Albin Michel, 1997).

Catherine Pourre, *Les Interventions Extérieures de l'Armée Française* [2 vols], Thèse de Doctorat (Paris: Université Paris-Nord, 26 May 1998).

Phillipe le Prestre, *French Security Policy in a Disarming World* (London: Lynne Reinner, 1989).

Dominique Prieur, *Agent Secrèt* (Paris: Fayard, 1995).

Gérard Prunier, *The Rwanda Crisis, 1959–94: History of a Genocide* (London: C. Hurst & Co Press, 1995).

Gérard Prunier, *Rwanda in Zaire: From Genocide to Continental War* (London: C. Hurst & Co Press, 1999).

Quel Avenir Pour la Dissuasion Nucléaire Française?, IHEDN, 123 Session Regional [Grenoble] (Grenoble: September/November 1995).

Paul Quiles and Guy-Michel Chauveau, *L'Industrie Française de Défense: Quel Avenir?*, Report No 203, Assemblée Nationale, (Paris: September 1997).

Paul Quiles [ed], *Enquete sur la Tragédie Rwandaise* [4 Vols], Report No 1271. Assemblée Nationale, (Paris: 1998).

Elfriede Regelsberger *et al.*, *The Foreign Policy of the European Union: From EPC to CFSP and Beyond* (Lynne Reinner, 1997).

La Réunion, IHEDN Report 124 (Paris: January 1996).

David Reyrat, *L'Engagement de la France dans la Guerre du Golfe* (Paris: Jean Picollec, 1992).

François de Rose, *Défendre la Défense* (Paris: René Julliard, 1989).

Yves Salesse, *Mayotte: L'illusion de la France* (Paris: L'Harmattan, 1995).

Alexandre Sanguinetti, *La France et l'Arme Atomique* (Paris: René Juillard, 1964).

Jean-Louis Scaringella, *Les Industries de Défense en Europe* (Paris: Economica, 1998).

Wolfgang F. Schlör, *German Security Policy: an Examination of the Trends in German Security Policy in a New European and Global Context*, Adelphi Paper 277 (London: IISS/Brassey's, 1993).

Maurice Schmitt, *De Diên Biên Phu à Koweit Cité* (Paris: Editions Grasset, 1992).

David Schwartz, *NATO's Nuclear Dilemmas* (Washington: Brookings Institute, 1983).

Michael Scriven and Peter Wagstaff [eds], *War and Society in Twentieth Century France*, (London: Berg Press, 1991).

Richard Shears and Isobelle Gidley, *The Rainbow Warrior Affair* (London: Counterpoint, 1986).

Laura Silber and Allan Little, *The Death of Yugoslavia* (London: Penguin 1995).

Haig Simonian, *The Privileged Partnership: Franco-German Relations in the European Community 1969–1984* (London: Clarendon Press, 1985).

Marie-Claude Smouts, *La France et l'ONU* (Paris: Presses de la Fondation Nationale des Sciences Politique, 1979).

Brigitte Stern, *Guerre du Golfe: Le Dossier d'une Crise Internationale 1990–1992* (Paris: La Documentation Française, 1993).

Brigitte Stern [ed], *La Vision Française des Opérations de Maintien de la Paix* (New York: United Nations University Press, 1997).

Brigitte Stern [ed], *United Nations Peacekeeping Operations: A Guide to French Policies*, (New York: United Nations University Press, 1998).

Anne Stevens, *The Government and Politics of France* (London: Macmillan, 1992).

Strategic Survey 1994–95 (Oxford: IISS/Oxford University Press, May 1995).

Strategic Survey 1998–1999 (Oxford: IISS/Oxford University Press, 1999).

Jane Stromseth, *The Origins of Flexible Response* (London: Macmillan, 1988).

Bruno Tertrais, *The French Nuclear Deterrent After the Cold War*, Report No P8012 (Washington: RAND, 1998).

Bruno Tertrais, *Nuclear Policies in Europe*, Adelphi Paper 327 (Oxford: IISS/Oxford University Press, 1999).

Ramesh Thakur, *The Last Bang before a Total Ban: French Nuclear Testing in the Pacific*, PRC Working Paper 159 (Canberra: September 1995).

Marc Theleri, *Initiation à la Force de Frappe Française 1945–2010* (Paris: Editions Stock, 1997).

Jacques Thobie *et al.*, *Histoire de la France Colonial* [2 Vols] (Paris: Editions du Seuil, 1990).

Jean-Marie Tjibaou, *La Présence Kanak* (Paris: Editions Odile Jacobs, 1990).

Philippe Tronquoy (ed), *La France et sa Défense* (Paris: La Documentation Française, 1997).

Maurice Vaisse, *Alger: Le Putsch* (Brussels: Editions Complexe, 1983).

Georges Valence, *France-Allemagne: le Retour de Bismarck* (Paris: Flammarion, 1990).

Hubert Védrine, *Les Mondes de François Mitterrand* (Paris: Fayard, 1996).

François-Xavier Verschave, *La Francafrique* (Paris: Editions Stock, 1998).

Pieter de Vries and Hon Seur, *Mururoa et Nous: Experiences des Polynésiens au Cours des 30 Années d'Essais Nucléaires dans la Pacifique Sud* (Lyon: CDRPC, 1997).

Paul Webster, *François Mitterrand: l'Autre Histoire* (Paris: Editions du Félin, 1995).

WEU, *La Coopération Transatlantique dans la Domaine de la Défense Antimissile Européenne*, WEU Report 1588 (Paris: November 1997).

Dorothy Shipley White, *Black Africa and De Gaulle* (London: Pennsylvania State University Press, 1979).

Vincent Wright, *The Government and Politics of France* (London: Unwin Hyman/Open University Press, 1989).

David Yost, *NATO Transformed: The Alliance's New Roles in International Security* (Washington: USIP Press, 1998).

David Yost, *The United States and Nuclear Deterrence in Europe*, Adelphi Paper 326 (Oxford: IISS/Oxford, 1999).

John Zametica, *The Yugoslav Conflict*, Adephi Paper 270 (London: IISS/Brassey's, 1992).

Charles Zorgbibe, *La France, l'ONU et le Maintien de la Paix* (Paris: PUF, 1996).

Journal articles

Georges Abi-Saab, 'Le deuxième génération des opérations de maintien de la paix', *Le Trimestre du Monde* (Winter 1994) pp. 87–97.

'L'adaptation du déploiement outre mer', *CID Tribune* (June 1997) pp. 106–9.

'Allocution de M. Valéry Giscard d'Estaing, Président de la République, à l'occasion de sa visite à l'IHEDN', *Défense Nationale* (July 1976) pp. 5–20.

Jacques Andréani, 'Les relations franco-américaines', *Politique Etrangère* (Winter 1995/6) pp. 891–902.

Gilles Andréani, 'Old French problem – or new transatlantic debate?', *RUSI Journal* (February/March 1999) pp. 20–4.

'L'après CEP: quel avenir pour la Polynésie Française?', *Bulletin d'Etudes de la Marine* (July 1996) pp. 48–53.

'L'armée française à Djibouti', *Marchés Tropicaux* (15 December 1995) pp. 2772–3.

Yves Artru, 'La programmation militaire et l'annualité budgétaire: le mariage impossible?', *Défense Nationale* (March 1997) pp. 57–68.

Ronald Asmus *et al.*, 'What will NATO enlargement cost?', *Survival* (Autumn 1996) pp. 5–26.

Heinz Junen Axt, 'Hat genscher Jugoslawien entzweit? mythen und fakten zur außenpolitik des vereiten Deutschland', *Europa Archiv* (Spring 1993) pp. 351–60.

Alyson Bailes, 'European defence and security: the role of NATO, WEU and EU', *Security Dialogue* (Spring 1996) pp. 55–64.

Alyson Bailes, 'Europe's defence challenge: reinventing the Atlantic Alliance', *Foreign Affairs* (January/February 1997) pp. 15–20.

Jean-Marc Balancie, 'La presence française dans l'Océan Indien', *Stratégique* (Spring 1993) pp. 203–33.

Sergio Balanzino, 'A year after Sintra: achieving cooperative security through the EAPC and PfP', *NATO Review* (Autumn 1998) pp. 4–8.

Edouard Balladur, 'La politique de défense: essentielle et permanente', *Défense Nationale* (November 1994) pp. 11–26.

Tony Bank, 'Country survey: France', *Jane's Defence Weekly* (23 June 1990) pp. 1246–65.

Bruno Barrillot, 'French finesse nuclear future', *Bulletin of the Atomic Scientists* (September 1992) pp. 23–6.

Raymond Bassac *et al.*, 'Les interventions extérieures: pourquoi? comment?', *Défense* (September 1996) pp. 31–66.

Dario Bastitella, 'De la democratie en politique extérieure: après guerre froide et domaine reservé', Le Débat (January/February 1996) pp. 117–34.

Jean Benet, 'Aspects majeurs de l'intervention française au Cambodge', *Objectif 21* (Winter 1997) pp. 19–43.

Olivier Bertaux, 'Outre-mer: le service militaire adapté', *Défense Nationale* (August/September 1994) pp. 71–8.

Bernard Bescond, 'Pôle de sécurité en Méditerranée', *Les Cahiers de Chear* (September 1994) pp. 85–98.

Margaret Blunden, 'France after the Cold War: inching towards the Alliance', *Defense Analysis*, (Autumn 1993) pp. 259–70.

Bernard Bombeau, 'La défense sans la loupe de la Cour des Comptes', *Air et Cosmos* (27 June 1997) pp. 46–7.

Pascal Boniface, 'La dissuasion nucléaire dans la relation franco-allemande', *Relations Internationales et Stratégiques* (Summer 1993).

Pascal Boniface, 'Dissuasion et non-prolifération: un équilibre difficile, nécessaire mais rompu', *Politique Etrangère* (Autumn 1995) pp. 707–21.

Pascal Boniface, 'Essais et doctrine nucléaires: la victoire à la Pyrrhus des revisionnistes', *Relations Internationales et Stratégiques* (Winter 1995) pp. 44–9.

Pascal Boniface, 'French nuclear strategy and European deterrence: les rendez-vous manqués', *Contemporary Security Policy*, 17(2) (August 1996) pp. 227–37.

Pascal Boniface, 'L'intervention militaire entre intérets, morale, volonté et réticences', *Relations Internationales et Stratégiques* (Winter 1996) pp. 26–36

Pascal Boniface, 'Le débat française sur l'élargissement de l'Otan', *Relations Internationales at Stratégiques* (Autumn 1997) pp. 33–47.

Pascal Boniface, 'La France, l'autonomie stratégique européenne et l'Otan', *Relations Internationales et Stratégiques* (Winter 1997) pp. 20–8.

François-Xavier Bouchard, 'L'espace militaire; un nouvel atout stratégique', *Les Cahiers de Mars* (Winter 1993) pp. 57–69.

Albert Bourgi, 'La fin de l'épopée coloniale?', *Jeune Afrique* (13–26 August 1997) pp. 3–4.

Yves Boyer and Diego Ruiz Palmer, 'L'Alliance atlantique et la coopération européenne en matière de sécurité: l'âge de maturité', *Politique Etrangère* (Spring 1989) pp. 107–18.

Yves Boyer, 'Les armes nucléaires françaises et l'Europe', *Défense Nationale* (August/September 1996) pp. 47–58.

Yves Boyer, 'This way to the revolution!', *RUSI Journal* (April/May, 1999) pp. 46–51.

Frédéric Bozo, 'La France et l'Otan: vers une nouvelle alliance', *Défense Nationale* (January 1991) pp. 19–34.

Frédéric Bozo and Jérôme Paolini, 'L'Europe entre elle-même et le Golfe', *Politique Etrangère* (Spring 1991) pp. 179–92.

Frédéric Bozo, 'Organisations de sécurité et insécurité en Europe', *Politique Etrangère* (Summer 1993) pp. 447–58.

Frédéric Bozo, 'France and security in the new Europe: between the Gaullist legacy and the search for a new model', Paper presented at the FNSP/Georgetown University Conference on *The New France in the New Europe*, Washington DC (21–22 October 1993).

Frédéric Bozo, 'Le nucléaire entre marginalisation et banalisation', *Politique Etrangère* (Spring 1995) pp. 195–204.

Frédéric Bozo, 'La France et l'Alliance: les limites du rapprochement', *Politique Etrangère* (Winter 1995/6) pp. 865–78.

Frédéric Bozo, 'Dissuasion concertée: le sens de la formule', *Relations Internationales et Stratégiques* (Spring 1996) pp. 93–100.

Giovanni de Briganti, 'France continues to pare down nuclear forces', *Defense News* (14–20 October 1996) p. 40.

Serge Brosselin, 'Vers une système d'alerte satellitaire européenne', *Défense* (September 1995) pp. 30–2.

'Le Budget de la Défense 1989', *Armées d'Aujourd'hui* (February 1989) pp. 28–48.

'Le Budget de la Défense 1990', *Armées d'Aujourd'hui* (February 1990) pp. 28–48.

'Le Budget de la Défense 1991', *Armées d'Aujourd'hui* (February 1991) pp. 38–51.

'Le Budget de la Défense 1992', *Armées d'Aujourd'hui* (February 1992) pp. 39–51.

'Le Budget de la Défense 1993', *Armées d'Aujourd'hui* (February 1993) pp. 38–59.

'Le Budget de la Défense 1994', *Armées d'Aujourd'hui* (February 1994) pp. 36–53.

'Le Budget de la Défense 1995', *Armées d'Aujourd'hui* (February 1995) pp. 36–53.

'Le Budget de la Défense 1996', *Armées d'Aujourd'hui* (February 1996) pp. 38–55.

'Le Budget de la Défense 1997, *Armées d'Aujourd'hui* (February 1997) pp. 36–53.

'Le Budget de la Défense 1998', *Armées d'Aujourd'hui* (February 1998) pp. 34–55.

'Le Budget de la Défense 1999', *Armées d'Aujourd'hui* (February 1999) pp. 33–55.

Hartmut Bühl, 'Le corps Européen', *Les Cahiers de Mars* (Summer 1996) pp. 35–45.

Jean-François Bureau, 'La réforme militaire en France: une mutation identitaire', *Politique Etrangère* (Spring 1997), pp. 69–81.

Nicola Butler, 'Sharing secrets', *Bulletin of the Atomic Scientists* (January/February 1997) pp. 11–12.

François Cailleteau, 'La conscription: les elements du problème', *Défense Nationale* (January 1990) pp. 13–27.

François Cailleteau, 'Quelles menaces?', *Relations Internationales et Stratégiques* (Winter 1993) pp. 90–4.

'Cambodge: mission accomplie', *SIRPA Actualité* (November 1993) pp. 23–35

Jean-Pierre Casamayou, 'Vers l'Europe du transport aérien militaire', *Air et Cosmos*, (10 January 1997) p. 35.

François Chauvancy, 'La doctrine militaire française pour les opérations de maintien de la paix', *Défense Nationale* (June 1995) pp. 144–5.

Jean-Pierre Chevènement, 'Confidence and perseverance', *Military Technology*, (Spring 1990) pp. 30–3.

Jean-Pierre Chevènement, 'Evolution du monde, rôle et politique de défense de la France', *Défense Nationale* (July 1990) pp. 9–28.

'Chevènement: on a baisse la France', *Nouvel Observateur* (24 February 1996) p. 12.

Patricia Chilton, 'Maintien de la paix: les nouveaux concepts stratégiques et les organisations internationales', *Relations Internationales et Stratégiques* (Spring 1995) pp. 15–28.

Jacques Chirac, 'La politique de défense de la France', *Défense Nationale* (August/September 1996) pp. 7–18.

Jacques Chirac, 'Construire une architecture Européenne de sécurité', *Regard Européen* (Summer 1997) pp. 50–5.

Samy Cohen, 'La politique étrangère entre l'Elysée et Matignon', *Politique Etrangère* (Autumn 1989) pp. 487–503.

Simon Cohen, 'Le control parlementaire de la politique de défense', *Revue du Parliament* (Spring 1977) pp. 378–446.

Daniel Colard, 'L'Allemagne unie et le couple franco-allemand', Défense Nationale (May 1991) pp. 99–114.

Daniel Colard, 'Le couple Paris-Londres: un partenariat original mais ambigu', *Défense Nationale* (August 1998) pp. 65–75.

André Collet, 'Le règlementation des armements: de nouvelles voies', *Défense Nationale* (May 1994) pp. 99–107.

Bruno Colson, La culture stratégique de la France', *Stratégique* (Spring 1992) pp. 27–60.

Dominique de Combles de Nayves, 'La nouvelle politique militaire française en Afrique', *Défense Nationale* (August/September 1998) pp. 12–16.

Jean Compagnon, 'La convention sur les armes chimiques remise sur rails', *Défense Nationale* (July 1997) pp. 29–36.

'Composition, organisation et fonction du Comité Stratégique', *Air et Cosmos* (24 November 1995) p. 36.

Dominique Conort, 'Les ressources humaines: les défis de la professionalisation', *Défense Nationale* (July 1996) pp. 91–108.

Paul Cornish, 'European Security: the end of architecture and the new Europe', *International Affairs* (October 1996) pp. 751–70.

Stuart Croft, 'European integration, nuclear deterrence and Franco-British nuclear cooperation', *International Affairs* (October 1996) pp. 771–87.

Ivo Daalder, 'Bosnia after SFOR: options for continued engagement', *Survival* 39(4) (Winter 1997/8) pp. 5–18.

Pierre Dabezies, 'Réflexions sur le Livre blanc', *Défense Nationale* (December 1993) pp. 45–58.

Patrice Dabos, 'Défense nationale: l'effet Saddam', *Politique Internationale* (June 1991) pp. 106–14.

Patrice Dabos, 'Qui fait la politique de défense en France?', *Les Cahiers de Mars* (Spring 1995) pp. 31–42.

Dominique David, 'Culture stratégique et mutations Européennes: l'exemple français', *Les Cahiers de Chear* (Autumn 1992) pp. 27–39.

Dominique David, 'La France et le monde: inventaire après essais', *Etudes* (December 1995) pp. 581–90.

Olivier Debouzy, 'L'avenir de la dissuasion nucléaire française', *Commentaire* (Winter 1994/5) pp. 869–73.

Olivier Debouzy, 'France-OTAN: la fin de l'autre guerre froide', *Commentaire* (Summer 1996) pp. 349–52.

'Défense, sécurité, humanitaire: quelles stratégies pour la France en Méditerranée', *Fondation Méditerranée d'Etudes Stratégiques* (Spring 1996) pp. 41–87.

'Une Défense Nouvelle', *Armées d'Aujourd'hui* [Special Issue] (March 1996) pp. 3–116.

Jacques Delors, 'European Integration and Security', *Survival* (March/April 1991) pp. 99–109.

Jean-Arnault Derens, 'Kosovo: la guerre inévitable?', *Etudes* (June 1998) pp. 739–49.

Jean-Philippe Douin *et al.*, 'Les interventions extérieures: pouquoi? comment?', *Défense* (September 1996) pp. 31–66.

Frédéric Drion, 'France: new defense for a new millenium', *Parameters* (Winter 1996/7) pp. 99–108.

Gérald Dubos, 'A deux pas du Kosovo', *Armées d'Aujourd'hui* (February 1999) pp. 14–15.

Richard Dubos, 'Le rôle des forces prépositionnées', *Les Cahiers de Mars* (Summer 1994) pp. 41–7.

Alain Dumoulin, 'Les forces multinationales', *Défense Nationale* (August 1991) pp. 67–85.

André Dumoulin, 'Vers une dissuasion nucléaire Européenne?', *Le Monde Atlantique* (Spring 1996) pp. 29–33.

Jean-François Durand, 'Les forces multinationales européennes', *Les Cahiers de Mars*, (Summer 1996) pp. 21–3.

Philippe Durteste, 'Le rôle de la marine française en Méditerranée', *Les Cahiers de Mars* (Winter 1996) pp. 25–38.

Mireille Duteil, 'La France va-t-elle perdre l'Afrique?', *Le Point* (July 1997) pp. 38–41.

Marcel Duval, 'Risques et options pour la défense', *Défense Nationale* (May 1992) pp. 59–63.

Marcel Duval, 'La dissuasion française et l'Europe: une approche historique', *Relations Internationales et Stratégiques* (Spring 1996) pp. 81–7.

Marcel Duval, 'Perspectives d'avenir de la dissuasion française', *Défense Nationale* (December 1996) pp. 7–28.

'Enquête sur un génocide', *Armées Aujourd'hui* (March 1999), pp. 22–4.

'Entre faucons et colombes', *Le Point* (22–28 August 1992) pp. 22–4.

'Entretien avec Général Heinrich', *Cols Bleu* (2 September 1995) pp. 4–9.

'Entretien avec Xavier de Villepin', *Relations Internationales et Stratégiques* (Spring 1994) pp. 7–19.

Valéry Giscard d'Estaing, 'Allocution de M. Valéry Giscard d'Estaing, President de la République, a l'occasion de sa visite à l'Institut des Hautes Etudes de Défense Nationale', *Défense Nationale* (July 1976) pp. 5–20.

Jean-Yves Faberon, 'L'accord de Nouméa du 21 Avril 1998', *Regards sur l'Actualité* (May 1998) pp. 19–31.

Maurice Faivre, 'Les renseignement dans la Guerre du Golfe', *Stratégique* (Winter 1991) pp. 373–401.

Maurice Faivre, ' Les renseignement après la Guerre du Golfe, *Stratégique* (Spring 1992) pp. 257–74.

Maurice Faivre, 'Participation française aux missions de paix de l'ONU', *Défense Nationale* (November 1992), pp. 189–91.

Alain Faure-Dufourmantelle, 'Défense de la nation et de défense militaire', *Défense Nationale* (June 1996) pp. 73–86.

'Les FFDj: forces françaises stationnées à Djibouti', *CID Tribune* (June 1996) pp. 212–16.

'Une folie nommée Rafale', *L'Expansion* (30 October–9 November 1995) pp. 58–64.

André Fontaine, 'Diplomatie française: Jacques Chirac et l'ombre du général', *Politique Internationale* (Summer 1995) pp. 81–9.

'Les forces françaises de Djibouti [FFDj]: un emplacement exceptionnel', *Armées d'Aujourd'hui* (April 1996) pp. 24–5.

'Les forces prépositionnées de l'armée de terre', *Armées d'Aujoud'hui* (December 1993/January 1994) pp. 36–65.

Nicole Forgeard, 'Industrie de défense français: regroupement et projection européenne', *Les Cahiers du Chear* (March 1996) pp. 9–25.

Michel Forget, 'Faut-il éliminer Albion?', *Géopolitique* (Autumn 1995) pp. 89–96.

La France et la tragédie rwandaise au histoire devenu plus lisible', *Marchés Tropicaux* (23 January 1998) pp. 156–8.

Alfred Frisch, 'L'Allemagne et la défense européenne', *Documents: Revue des Questions Allemandes* (Spring 1996) pp. 30–6.

Henri Froment-Meurice, 'Quelle politique de défense pour la France?', *Commentaire* (Winter 1994/5) pp. 857–67.

Gérard Fuchs, 'Défense, horizon 2000', *Relations Internationales et Stratégiques* (Winter 1991) pp. 19–33.

Jean de Galard, 'Les actions d'interpositions vont coûts 4 milliard en 1993', *Air et Cosmos* (21 December 1992) p. 29.

Pierre M. Gallois, 'Les changements stratégiques dans le monde', *Défense Nationale* (June 1992) pp. 23–30.

Thierry Garcin, 'L'intervention française au Rwanda', *Le Trimestre du Monde*, (Summer 1996) pp. 63–72.

François Gaulme, 'La France et l'Afrique – de François Mitterrand à Jacques Chirac', *Marchés Tropicaux* (26 May 1995) pp. 1112–14.

James Geary, 'Politics and Massacres', *Time* (24 June 1996) p. 24.

François Géré, 'Quatre généraux et l'apocalypse: Ailleret-Beaufre-Gallois-Poirier, *Stratégique* (Spring 1992) pp. 75–116.

Brenard Gillis, 'La centre d'experimentation du Pacifique', *Les Cahiers de Mars* (Summer 1990) pp. 41–51.

Anne-Marie le Gloannec. 'Europe by other means?', *International Affairs* (January 1997) pp. 83–98.

Olivier Gohin, 'La défense de la France Outre-Mer', *Droit et Défense* (January 1994) pp. 4–13.

Paul-Marie de la Gorce, 'Contexte international et défense', *Défense Nationale* (December 1993) pp. 9–22.

Paul-Marie de la Gorce, 'La deuxième mort de Jacques Foccart', *Jeune Afrique*, (13–16 August 1997) pp. 40–2.

Philip Gordon, 'Does the WEU have a role?', *Washington Quarterly* (Winter 1997) pp. 125–40.

Camille Grand, 'La politique français et le non-prolifération nucléaire', *Défense Nationale* (August/September 1994) pp. 99–112.

Camille Grand, 'La diplomatie nucléaire du président Chirac', Relations Internationales et Stratégiques (Summer 1997) pp. 157–69.

Robert Grant, 'France's new relationship with NATO', *Survival*, 38(1) (Spring 1996) pp. 58–80.

Shaun Gregory, 'France and nuclear weapons testing', *Bulletin of Arms Control*, No 19 (August 1995) pp. 14–25.

Jean-Marie Guehenno, 'L'avenir de l'UEO', *Les Cahiers de Chear* (September 1994) pp. 11–24.

Jean-François Guilhaudis, 'Considérations sur l'élargissement de l'Otan', *Défense Nationale* (November 1995) pp. 57–66.

Philippe Guillot, 'France, peacekeeping and humanitarian intervention', *International Peacekeeping* (Spring 1994) pp. 30–43.

Francis Gutmann, 'Après Madrid, Amsterdam, Luxembourg ... la France, l'Europe et l'Otan', *Défense Nationale* (February 1998) pp. 53–64.

François Heisbourg, 'Défense français: l'impossible status quo', *Politique Internationale*, 36 (Summer 1987) pp. 137–53.

François Heisbourg, 'The British and French Nuclear Forces', *Survival* (July/August 1989) pp. 301–20.

François Heisbourg, 'Réflexions sur la politique de défense de la France', *Politique Etrangère* (Spring 1990) pp. 157–70.

François Heisbourg, 'L'après Guerre du Golfe', *Politique Etrangère* (Summer 1991) pp. 411–64.

François Heisbourg, 'Sécurité: l'Europe livrée à elle-même', *Politique Etrangère* (Spring 1994) pp. 247–60.

François Heisbourg, 'La politique de défense française à l'aube d'un nouveau mandat présidentiel', *Politique Etrangère* (Spring 1995) pp. 73–84.

François Heisbourg, 'La France et sa défense', *Politique Etrangère* (Winter 1995) pp. 983–92.

'L'heure de choix: entretien avec M. Charles Millon, Ministre de la Défense', *Armées d'Aujourd'hui* (October 1995) pp. 10–15.

Mark Hibbs, 'Tomorrow a Eurobomb?', *The Bulletin of Atomic Scientists* (January/February 1996) pp. 16–23.

Stanley Hoffman, 'La France dans le nouvel ordre européen', *Politique Etrangère* (Autumn 1990) pp. 503–12.

Geoffrey Howe, 'Bearing more of the burden: in search of a European foreign and security policy', *The World Today* (January 1996) pp. 23–6.

Jolyon Howorth, 'François Mitterrand and the "domaine réservé" from cohabitation to the Gulf War', *French Politics and Society* (Winter 1992) pp. 43–58.

'L'intervention française en Somalie', *Objectif 21* (Spring 1995) pp. 12–36

'Interview with Jean-Philippe Douin', *Jane's Defence Weekly* (19 June 1996) p. 112.

Jacques Isnard, 'Plan français Armées 2000: un difficile equilibre', *Défense et Armement* (September 1989) pp. 23–5.

Jacques Isnard, 'Programmation militaire française: un pari sur l'avenir', *Défense et Armament International* (September/October 1992) pp. 18–20.

Peter Viggo Jakobsen, 'Myth-making and Germany's unilateral recognition of Croatia and Slovenia', *European Security* (Autumn 1995) pp. 400–16.

François Jolivald, 'Regards sur la politique africaine de François Mitterrand', *Marchés Tropicaux* (4 July 1997) pp. 1474–5.

Robert Joseph, 'Proliferation, counter-proliferation and NATO', *Survival*, (Spring 1996) pp. 111–30.

Alain Joxe, 'Etats des lieux de la RAM [Révolution dans les Affairs Militaires]' *L'Armement* (March 1996) pp. 137–42.

Alain Juppé, 'La dissuasion nucléaire dans le nouveau contexte international', *Politique Etrangère* (Autumn 1995) pp. 743–51.

Alain Juppé, 'Politique de défense et dissuasion nucléaire', *Défense Nationale* (November 1995) pp. 7–18.

Alain Juppé, 'Quel horizon pour la politique étrangère de la France?', *Politique Etrangère* (Spring 1996) pp. 245–59.

Alain Juppé, 'Une défense inscrite dans une perspective européenne et internationale', *Défense Nationale* (November 1996) pp. 3–10.

John Keiger, 'France and international relations in the post-Cold War era: some lessons of the past', *Modern and Contemporary France*, (Autumn 1995) pp. 263–74.

Michel Klen, 'L'enfer somalien', *Défense Nationale* (February 1993) pp. 135–43.

Edward Kolodzeij, 'British-French nuclearisation and European denuclearisation: challenge for US policy', *Atlantic Community Quarterly* (Fall 1988) pp. 305–11.

Henri Labrousse, 'La stratégie française dans l'Océan Indien', *Stratégique* (Winter 1992) pp. 227–56.

Jean Labrousse, 'Guyane Française et le Centre Spatial Guyanaise', *Défense* (December 1997) pp. 17–19.

Franck Laffaile, 'Défense nationale et cohabitation', *Revue Droit et Défense* (Autumn 1997) pp. 32–8.

'Lancement du missile stratégique M51', *Air et Cosmos* (3 May 1996) p. 34.

Patrice Langereux, 'Spot a bien servi les alliés dans la tempête du désert', *Air et Cosmos* (22 April 1991) pp. 36–7.

Pierre Langereux, 'Le centre satellitaire de l'UEO commencera a fonctionner en mars', *Air et Cosmos* (7 February 1993) p. 42.

Pierre Langereux *et al.*, 'Un nouveau modèle de défense à l'horizon 2015', *Air et Comsos* (1 March 1996) p. 24.

Jacques Lanxade, 'Quelle armée pour demain?' *Politique Internationale* (Summer 1991) pp. 117–26.

Jacques Lanxade, 'La doctrine militaire français face aux bouleversements stratégi ques en Europe', *Relations Internationales et Stratégiques* (Winter 1991) pp. 34–45.

Jacques Lanxade, 'Orientations par la conception, la préparation, la plannification, le commandement et l'emploi des forces françaises dans l'opérations militaires fondées sur un résolution du Conseil de Sécurité des Nations Unis', *Objectif 21* (Spring 1995) pp. 4–9.

Jacques Lanxade, 'L'opération Turquoise', *Défense Nationale* (February 1995) pp. 7–15.

Jacques Lanxade, 'La défense française dans le nouveau cadre géostratégique', *Défense Nationale* (August/September 1995) pp. 7–22.

Jacques Lanxade, 'Stepping into the breach; France's global role', *International Defence Review* (Winter 1995) pp. 25–7.

Christian Lardier, 'Première visite au centre de Torrejon', *Air et Cosmos* (31 January 1997) p. 40.

Yves-Marie Laulan, 'L'armée française entre l'humanitaire et l'opérationnel', *Défense Nationale* (August 1996) pp. 75–81.

Jean-François Lazerges, 'Armées 2000', *Défense Nationale* (May 1991) pp. 25–40.

Pierre Lellouche, 'France in search of security', *Foreign Affairs* (Spring 1993) pp. 122–31.

Pierre Lellouche, 'Essais nucléaires: leçons d'un psychodrame', *Politique Etrangère* (Winter 1995/6) pp. 95–111.

Jean-Marie Lemoine, 'Les forces terrestres dans les DOM-TOMs', *Les Cahiers de Mars* (Summer 1990) pp. 20–5.

François Léotard, 'Une nouvelle culture de la défense' *Défense Nationale* (July 1993) pp. 9–20.

J.A.C. Lewis, 'Country briefing: France', *Jane's Defence Weekly* (21 October 1995) pp. 23–31.

J.A.C. Lewis, 'All change for the future: how the big shake-out will shape up', *Jane's Defence Weekly* (13 March 1996) pp. 19–20.

J.A.C. Lewis, 'Fitter, leaner, forces for a multi-polar world', *Jane's Defence Weekly* (11 June 1997) pp. 68–83.

Philippe Leymarie, 'En Afrique: la fin des ultimes "chasses gardées"', *Le Monde Diplomatique* (December 1996) pp. 4–5.

Michel Loridon, 'L'Armée française an Cambodge', *Les Cahiers de Mars* (Autumn 1992) pp. 78–84.

Robin Luckham, 'French militarism in Africa', *Review of African Political Economy* (May 1982) pp. 55–84.

Bernard Lugan and François-Xavier Rocchi, 'Comores', *L'Afrique Reelle*, (Summer 1997) pp. 34–52.

Eric de la Maisonneuve, 'Le soldat et la politique', *Défense Nationale* (June 1996), pp. 61–72.

Pascal Maraval, 'Le projet CREDO', *Défense Nationale* (August/September 1996) pp. 175–7.

Roland Marchal, 'Somalie: autopsie d'une intervention', *Politique Internationale* (Spring 1993) pp. 121–208.

Hanns W. Maull, 'Germany in the Yugoslav crisis', *Survival* (Winter 1995/6), pp. 99–130.

François Maurin, 'L'originalité français et le commandement', *Défense Nationale* (July 1989) pp. 45–57.

François Maurin, 'Avenir de la défense militaire française', *Défense Nationale* (December 1993) pp. 23–34.

Marcel Maymil, 'Avenir de la presence française à Djibouti', *Bulletin d'études de la marine* (July 1996) pp. 43–7.

Michael Mazarr, 'Virtual nuclear arsenals', *Survival* (Autumn 1995) pp. 7–26.

Anand Menon, Anthony Forster and William Wallace, 'A common European defence', *Survival* (Autumn 1992) pp. 98–118.

Anand Menon, 'From independence to co-operation: France, NATO and European security', *International Affairs* (Spring 1995) pp. 19–34.

Guy Méry, 'Une armée pourquoi faire et comment?', *Défense Nationale* (June 1976) pp. 11–24.

Stephanie Mesnier, 'Le rôle du Quai d'Orsay de mai 1986 à mai 1988', *Revue Administrative* (December 1990) pp. 489–98.

Holger H. Mey and Andrew Denison, 'View From Germany: France's nuclear tests and Germany's nuclear interests', *Comparative Strategy*, No 15 (Autumn 1996) pp. 169–71.

Charles Millon, 'La France et la rénovation de l'Alliance Atlantique', *Revue de l'OTAN* (May 1996) pp. 13–16.

Charles Millon, 'Towards a European defence pillar', *Defence Review* (Annual 1996) pp. 18–19.

Charles Millon, 'Vers une défense nouvelle', *Défense Nationale* (July 1996) pp. 13–20.

Charles Millon, Défense: rompre avec les routines et les préjugés!', *Défense Nationale* (February 1997) pp. 7–20.

Christian Millotat and Jean-Claude Philippot, 'Le jumelage Franco-Allemand pour la sécurité de l'Europe', *Défense Nationale* (October 1990) pp. 67–80.

Carlos Miranda, 'La position de l'Espagne sur les questions nucléaires', *Relations Internationales et Stratégiques* (Spring 1996) pp. 121–5.

Dominique Moisi and Pierre Lellouche, 'French policy in Africa: the lonely battle against destabilisation', *International Security* (Spring 1979) pp. 108–33.

Dominique Moisi, 'De Mitterrand à Chirac', *Politique Etrangère* (Autumn 1995) pp. 849–56.

Dominique Moisi, 'Chirac of France', *Foreign Affairs* (November/December 1995) pp. 8–13.

Dominque Moisi, 'The trouble with France', *Foreign Affairs* (May/June 1998) pp. 94–104.

Claude Monier, 'Orientation de l'alliance atlantique vers le partenariat pour la paix', *Défense Nationale* (March 1995) pp. 166–9.

Henri Motte, 'Saint-Pierre et Miquelon: l'espoir d'être au le neant', *La Revue Maritime* (June 1995) pp. 33–47.

Harald Müller, 'L'Europe, l'Allemagne et le debat sur le désarmement nucléaire', *Relations Internationales et Stratégiques* (Summer 1998) pp. 112–34.

Klaus Naumann, 'NATO's new military structure', *NATO Review* (Spring 1998) pp. 10–14.

Robin Niblett, 'The European disunion: competing visions of integration', *Washington Quarterly* (Winter 1997) pp. 94–107.

'Nouveau depart pour l'industrie de défense', *Air et Cosmos* (1 March 1996) pp. 14–19.

Michael O'Hanlon, 'Transforming NATO: the role of European forces', *Survival* (Autumn 1997) pp. 5–15.

David Omand, 'Nuclear deterrence in a changing world: the view from the UK perspective', *RUSI Journal* (June 1996) pp. 15–22.

Arséne Orain, 'The defence posture of France: an overview', *Military Technology*, 5 (1990) pp. 34–7.

'L'Outre-mer', *L'Administration* [Special Issue], (January 1996) pp. 2–144.

Arthur Paecht, 'La défense: le temps des choix', *Défense Nationale* (February 1996) pp. 7–16.

Arthur Paecht, 'L'abandon de service militaire obligatoire: une solution sans alternative', *Défense* (September 1996) pp. 11–14.

Jérôme Paolini, 'La France, l'Europe et la bombe', *Commentaire*, No 54 (Summer 1991) pp. 247–55.

Henri Paris, 'Les armées de l'an 2000', *Défense Nationale* (November 1989) pp. 31–42.

Henri Paris, 'Armée de metier où de conscription?', *Défense Nationale* (May 1993) pp. 89–96.

Henri Paris, 'Domaine réservé: la conspiration du silence en matière de défense', *Revue Politique et Parlementaire* 976, (March/April 1995) pp. 45–9.

'Le partenariat pour la paix', *Défense Nationale* (March 1998) pp. 168–70.

Pierre Pascallon, 'L'evolution de nos forces classiques d'intervention', *Revue Politique et Parlementaire* (September/October 1993) pp. 47–54.

Pierre le Peillet, 'Les treize missions militaires de l'ONU', *Défense Nationale* (April 1993) pp. 165–9.

Jean-Pierre Philippe, 'Helios 1A: France's [and Europe's] first spy satellite', *Military Technology* (October 1995) pp. 40–5.

Daniel Pichard, 'Le désarmement', *Les Cahiers du Chear* (December 1994) pp. 89–103.

Andrew J. Pierre and Dmitri Trenin, 'Developing NATO-Russian relations', *Survival* (Spring 1997) pp. 5–18;

Sylvie Pierre-Brossolette, 'Chirac sur deux fronts', *L'Express* (15 June 1995) pp. 8–9.

Frédéric Praeter, 'La France et la crise du Golfe', *Politique Etrangère*, (Summer 1991) pp. 441–54.

Gérard Prunier, 'L'inconcevable aveuglement de l'ONU', *Le Monde Diplomatique* (November 1993) p. 7.

'Querelle sur le nucléaire', *L'Express* (7–13 October 1993) pp. 52–3.

Paul Quiles, 'L'OTAN et la défense Européenne', *Relations Internationales et Stratégiques* (Autumn 1996) pp. 13–17.

Daniel Reydellet, 'Histoire des missiles nucléaires françaises', *L'Armement* (December 1996/January 1997) pp. 132–5.

Gabriel Robin, 'A quoi sert l'Otan?', *Politique Etrangère* (Spring 1995) pp. 171–80.

Michel Rocard, 'Les orientations de la politique de défense de la France', *Défense Nationale* (November 1989) pp. 13–30.

Claude Roche, 'Une pays nucléaire comme la France doit-il se doter d'une défense active contre les missiles balistiques?', *Les Cahiers de la Fondation* (September 1998) pp. 13–18.

John Ruggie, 'Consolidating the European pillar: the key to NATO's future', *The Washington Quarterly* (Winter 1997) pp. 109–23.

Hugo Sada, 'Réexamen de la politique militaire française en Afrique', *Défense Nationale* (June 1997) pp. 183–5.

Yves Salkin, 'Regards sur les petites Antilles', *Défense Nationale* (January 1997) pp. 135–44.

Alexandre Sanguinetti, 'Considérations sur la réforme des armées; *Etudes* (July 1996) pp. 27–33.

Jacques Santer, 'The European Union's security and defence policy', *NATO Review* (November 1995) pp. 3–9.

Alex Sauder, 'Les changements de la politique de défense française et la coopération franco-allemande', *Politique Etrangère* (Autumn 1996) pp. 583–98.

Jean Saulnier, 'Missions et engagements des forces françaises', *Défense Nationale* (July 1992) pp. 9–16.

Philippe Séguin, 'La défense de la France' *Défense Nationale* (April 1994) pp. 7–20.

Stanley R. Sloan, 'French defense policy: Gaullism meets the post Cold War World', *Arms Control Today* (April 1997) pp. 3–8.

'Spécial Caraïbe: Les tensions dans la "Mediterranée américaine"', *Revue Politique et Parliamentaire* (July 1986) pp. 10–66.

Henri Starck, 'Dissonances Franco-Allemandes sur fond de guerre Serbo-Croat', *Politique Etrangère* (Summer 1992), pp. 339–47.

Henri Starck, 'France-Allemagne: entente et mésentente', *Politique Etrangère* (Winter 1993/4), pp. 989–1001.

Barbara Starr, 'USA warns of three-tier NATO technology rift', *Jane's Defence Weekly* (1 October 1997), p. 15.

Michael Sutton, 'Chirac's foreign policy: continuity with adjustment', *The World Today* (July 1995) pp. 135–8.

Michel Tatu, 'Après Muroroa', *Politique Internationale* (Autumn 1995) pp. 143–60.

Bruno Tertrais, 'Le printemps des relations franco-britanniques', *Relations Internationales et Stratégiques* (Spring 1995) pp. 7–14.

Bruno Tertrais, 'Faut-il croire à la révolution dans les affaires militaires?', *Politique Etrangère* (Autumn 1998) pp. 611–30.

Thierry Tardy, 'La France et l'ONU: 50 ans de relations contrastées', *Regards sur l'Actualité* (November 1995) pp. 3–23.

'Touring the Inflexible', *Revue Aerospatiale* [English language version], No 44, (November 1987) p. 27.

Jacques Touvenin, 'La condition militaire aujourd'hui', *Défense Nationale* (February 1992) pp. 13–28.

Adrian Treacher, 'New tactics, same objectives: France's relationship with NATO', *Contemporary Security Policy* (August 1998) pp. 91–110.

'Trois ans pour demanteler le Plateau d'Albion', *Air et Cosmos* (3 May 1996) p. 33.

Francis Tusa, 'Who will rule the roost in France?', *Armed Forces Journal International* (May 1995) p. 40.

Richard Ullman, 'The covert French connection', *Foreign Policy* (Summer 1989) pp. 3–33.

Maurice Vaisse, 'La politique de défense de la France 1945–1995', *Les Cahiers de Chear* (March 1996) pp. 88–99.

Hubert Védrine and Jean Musitelli, 'Les changements des années 1989–1990 et l'Europe de la prochaine décennie', *Politique Etrangère* (Spring 1991) pp. 165–78.

Hubert Védrine, 'Non, la France n'a pas à rougir', *Le Nouvel Observateur* (17–23 February 1994) pp. 42–5.

Alain Vigarie, 'Méditerranée: mer aux multiples dangers', *La Revue Maritime* (Autumn 1995) pp. 23–51.

Xavier de Villepin, 'Les hommes sont-ils plus sage?: reflexions sur la revision de la loi de programmation militaire', *Défense Nationale* (August/September 1989) pp. 13–22.

Claude Wauthier, 'La politique africaine de la France 1988–93', *Relations Internationales et Stratégiques* (Spring 1993) pp. 198–205.

Claude Wauthier, 'La politique africaine de Jacques Chirac', *Relations Internationales et Stratégiques*, (Spring 1997) pp. 121–8.

Klaus-Uwe Wolff, 'La force navale franco-allemande', *Les Cahiers de Mars*, (Summer 1996) pp. 46–9.

Pia Christina Wood, 'France and the post Cold War order: the case of Yugoslavia', *European Security* (Spring 1994) pp. 537–43.

David Yost, 'France and western European defence identity', *Survival* (July/August 1991) pp. 327–51.

David Yost, 'France in the new Europe', *Foreign Affairs* (Winter 1990/1) pp. 107–28.

David Yost, 'Europe and nuclear deterrence', *Survival* (Autumn 1993) pp. 97–120.

David Yost, 'La France, les Etats-Unis et la révolution militaro-technique', *L'Armement* (May/June 1994) pp. 135–41.

David Yost, 'Nuclear debates in France', *Survival* (Winter 1994/5) pp. 113–39.

David Yost, 'France's nuclear dilemmas', *Foreign Affairs*, (January/February 1996) pp. 108–18.

Jean-Claude Zarka, 'Le "domaine réservé" à l'épreuve de la seconde cohabitation', *Revue Politique et Parlementaire* (1994) pp. 40–4.

Charles Zorgbibe, 'La Méditerranée: nouvelle ligne de front', *Revue Politique et Parliamentaire* (November/December 1995) pp. 67–72.

Charles Zorgbibe, 'Condamner l'action humanitaire?', *Revue Politique et Parlementaire* (July 1996) pp. 62–6.

Charles Zorgbibe, 'La France et le maintien de la paix: propositions', *Les Cahiers du Cedsi* (October 1998) pp. 57–64.

In addition articles were taken from the following newspapers: *L'Echos, Figaro, Financial Times, The Independent, International Herald Tribune, Le Monde, Libération,* and *The Times.*

Index